State-of-the-Art Sensors Technology in Spain 2015

Volume 1

Special Issue Editor

Gonzalo Pajares Martinsanz

Special Issue Editor
Gonzalo Pajares Martinsanz
Department Software Engineering and Artificial Intelligence
Faculty of Informatics
University Complutense of Madrid
Spain

Editorial Office
MDPI AG
St. Alban-Anlage 66
Basel, Switzerland

This edition is a reprint of the Special Issue published online in the open access journal *Sensors* (ISSN 1424-8220) from 2015–2016 (available at: http://www.mdpi.com/journal/sensors/special_issues/state-of-the-art-spain-2015).

For citation purposes, cite each article independently as indicated on the article page online and as indicated below:

Author 1; Author 2; Author 3 etc. Article title. *Journal Name.* **Year**. Article number/page range.

Vol 1 ISBN 978-3-03842-370-6 (Pbk)   Vol 1-2 ISBN 978-3-03842-312-6 (Pbk)
Vol 1 ISBN 978-3-03842-371-3 (PDF)   Vol 1-2 ISBN 978-3-03842-313-3 (PDF)

Table of Contents

About the Guest Editor

Gonzalo Pajares received his Ph.D. degree in Physics from the Distance University, Spain, in 1995, for a thesis on stereovision. Since 1988 he has worked at Indra in critical real-time software development. He has also worked at Indra Space and INTA in advanced image processing for remote sensing. He joined the University Complutense of Madrid in 1995 on the Faculty of Informatics (Computer Science) at the Department of Software Engineering and Artificial intelligence. His current research interests include computer and machine visual perception, artificial intelligence, decision-making, robotics and simulation and has written many publications, including several books, on these topics. He is the co-director of the ISCAR Research Group. He is an Associated Editor for the indexed online journal *Remote Sensing* and serves as a member of the Editorial Board in the following journals: *Sensors, EURASIP Journal of Image and Video Processing, Pattern Analysis and Applications*. He is also the Editor-in-Chief of the *Journal of Imaging*.

Preface to "State-of-the-Art Sensors Technology in Spain 2015"

Since 2009, three Special Issues have been published on sensors and technologies in Spain, where researchers have presented their successful progress. Thirty-one high quality papers demonstrating significant achievements have been collected and reproduced in this book.

They are self-contained works addressing different sensor-based technologies, procedures and applications in several areas, including measurement devices, wireless sensor networks, robotics, imaging, optical systems or electrical/electronic devices among others.

Readers will find an excellent source of resources for the development of research, teaching or industrial activity.

Although the book is focused on sensors and technologies in Spain, it describes worldwide developments and references on the covered topics. Some works have been or come from international collaborations.

Our society demands new technologies for data acquisition, processing and transmission for immediate actuation or knowledge, and with important impact on one's welfare when required.

The international, scientific and industrial communities worldwide will also be an indirect beneficiary of these works. Indeed, the book provides insights and solutions for the varied problems covered. Also, it lays the foundation for future advances toward new challenges and progress in many areas. In this regard, new sensors will contribute to the solution of existing problems, and, where the need arises for the development of new technologies or procedures, this book paves the way.

We are grateful to all the people involved in the preparation of this book. Without the invaluable contributions of the authors together with the excellent help of reviewers, this book would not have reached fruition. More than 120 authors have contributed to this book.

Thanks also to the *Sensors* journal editorial team for their invaluable support and encouragement.

Gonzalo Pajares Martinsanz
Guest Editor

Article

A Validation of the Spectral Power Clustering Technique (SPCT) by Using a Rogowski Coil in Partial Discharge Measurements

Jorge Alfredo Ardila-Rey [1,*], Ricardo Albarracín [2], Fernando Álvarez [3] and Aldo Barrueto [1]

[1] Departamento de Ingeniería Eléctrica, Universidad Técnica Federico Santa María, Av. Vicuña Mackenna 3939, Santiago de Chile 8940000, Chile; aldo.barrueto@usm.cl

[2] Generation and Distribution Network Area. Dept. Electrical Engineering. Innovation, Technology and R&D, Boslan S.A. Consulting and Engineering, Calle de la Isla Sicilia 1, Madrid 28034, Spain; rasbarracin@gmail.com

[3] Departamento de Ingeniería Eléctrica, Universidad Politécnica de Madrid, Ronda de Valencia 3, Madrid 28012, Spain; fernando.alvarez@upm.es

* Correspondence: jorge.ardila@usm.cl; Tel.: +56-22-303-7231; Fax: +56-22-303-6600.

Academic Editor: Gonzalo Pajares Martinsanz
Received: 4 September 2015; Accepted: 6 October 2015; Published: 13 October 2015

Abstract: Both in industrial as in controlled environments, such as high-voltage laboratories, pulses from multiple sources, including partial discharges (PD) and electrical noise can be superimposed. These circumstances can modify and alter the results of PD measurements and, what is more, they can lead to misinterpretation. The spectral power clustering technique (SPCT) allows separating PD sources and electrical noise through the two-dimensional representation (power ratio map or PR map) of the relative spectral power in two intervals, high and low frequency, calculated for each pulse captured with broadband sensors. This method allows to clearly distinguishing each of the effects of noise and PD, making it easy discrimination of all sources. In this paper, the separation ability of the SPCT clustering technique when using a Rogowski coil for PD measurements is evaluated. Different parameters were studied in order to establish which of them could help for improving the manual selection of the separation intervals, thus enabling a better separation of clusters. The signal processing can be performed during the measurements or in a further analysis.

Keywords: partial discharges (PD); Rogowski coil; wideband PD measurements; clustering techniques; condition monitoring; electrical insulation condition; on-line PD measurements; pattern recognition; signal processing

1. Introduction

The electrical generation, transmission and, even distribution infrastructures require large financial investments, so their long-term profitability must be optimized. In this context, there has been a growing interest on the one hand, to reduce maintenance cycles applied to electrical machinery and power cables when they are very aged and secondly, to adequately plan their replacement when its operation becomes unreliable [1].

For these reasons, it is assumed that electrical equipment must be replaced every certain period of time, close to 30–40 years [2,3], and that the maintenance cycles must be fixed in advance (preventive maintenance). However, the progress made in basic electrical insulation research and the increase in the availability of historical failure data allows choosing new maintenance strategies. Through these strategies, it is possible to know the operation condition of the electrical assets by performing in service (on-line) measurements in high-voltage installations. This procedure extends the lifespan of the equipment, as well as their periods of scheduled maintenance [4]. Thus, the lack of investment, that is often required in the replacement of equipment [3], could be compensated with the implementation of a proper Condition-Based Maintenance (CBM) program [5]. For these reasons, PD measurement has become a major diagnostic method used in the maintenance of electrical installations, in order to establish the degradation of the insulation systems, since its lifetime is determined by the degree of degradation present [6].

In measurements made on-site or even in controlled environments, such as high-voltage laboratories, the pulses from multiple PD sources and electrical noise may be overlapped, thus creating complex phase-resolved PD (PRPD) patterns. In some cases, the noise signals can have magnitudes greater than the PD pulses, so the raising the trigger level of the acquisition systems is not a valid PD separation technique. This problem has increased due to the growing use of electronic power converters in electrical systems (variable frequency drives, power supplies switching, rectifiers, inverters, converters, *etc.*). Consequently, source separation has become a fundamental requirement in obtaining an effective diagnostic, such that avoid erroneous assessments in the equipment or system insulation.

Many modern measuring instruments are equipped with pulse classification tools that are based on characterization of the waveforms of the acquired pulses, inasmuch as noise and PD pulses generated by different sources present different shapes.

The classification procedures require broadband sensors capable of detecting ranges up to tens of MHz [7,8]. Commonly, inductive sensors are used for PD measurements. These sensors are capable of measuring according to the standard detection circuits. The most widely used are the high-frequency current transformers (HFCT), inductive loop sensors (ILS) and the Rogowski coils (RC) [9–14]. Recent studies have shown that the SPCT applied to the pulses obtained with HFCT and ILS sensors measuring in different test objects, have been successfully characterized and its effectiveness to separate different PD sources and electrical noise has been proven, even when these sources are simultaneously active [15].

In this paper, the ability of clustering by applying the SPCT to PD pulses and electrical noise acquired with a RC for various test objects in two different environments is evaluated. The aim of the paper is to show the benefits of SPTC technique, even when sensors with a poor transfer impedance are used. To this end, the RC was used, since it has an air-core of non-magnetic material, which provides linearity and low self-inductance [13]. This type of core allows designing sensors with lower weights, cheap and more flexible, allowing more applicability and easy-to-use.

Additionally, the behaviour of the different frequency bands is studied when multiple sources are present during the acquisition. This is done, in order to establish some important indicators, that allow the operator of the classification tool (based on SPCT), to evaluate whether the selected frequency ranges are the most appropriate, when it comes to separate the different sources that may be present during the measurement.

2. Rogowski Coil

The RC, as well as the different inductive sensors commonly used for PD detection (HFCT or ILS) [9,15], operates on the basic principle of the Faraday's Law and can be applied to measure PD [16]. Accordingly, the air-core coil is placed around the conductors through which the current pulses associated to PD and electrical noise can be propagated. This variable current produces a magnetic field, which links the secondary of the coil and induces a voltage directly proportional to the rate of change of the current in the conductor and the mutual inductance between the coil and the conductor. The RC designed and used in this paper, is based on a toroidal transformer with an air-core of transversal rectangular section, made of 12 identical turns as on the geometry indicated in Figure 1a. This configuration is modelled with the equivalent circuit shown in Figure 1b, with the induced voltage represented by the source voltage V_{coil} and the electrical effects of the winding represented by R, L and C parameters, which correspond with the resistance, self-inductance of the winding cable and capacitance between the coil turns and the return cable, respectively. The geometric and electrical parameters for this design are indicated in Table 1.

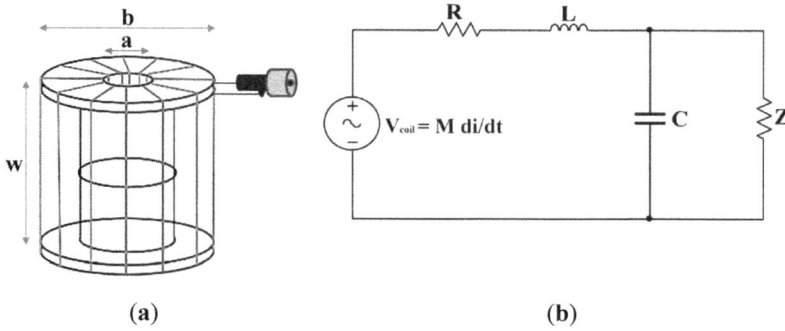

(a) **(b)**

Figure 1. (a) Rogowski Coil; and (b) its electric equivalent circuit.

Table 1. Parameters of the Rogowski Coil.

a (cm)	b (cm)	w (cm)	M (nH)	R (Ω)	L (nH)	Z (Ω)	C (pF)
0.5	3	2	86	0.038	1032	50	18.1

Considering the equivalent circuit and the electric characteristics of this sensor, its transfer function is defined by Equation (1):

$$\frac{V_{out}(s)}{I(s)} = \frac{4.30 \cdot 10^{-6} s}{9.38 \cdot 10^{-16} s^2 + 1.032 \cdot 10^{-6} s + 50.038} \tag{1}$$

It is important to indicate that Z is usually 50 Ω and represents the input impedance of the measuring instrument, where the sensor is connected. More details about this type of sensor and the calculation of the electric parameters can be found in [12,13]. Finally, the frequency response for this sensor, calculated from Equation (1), is shown in Figure 2. The results indicate that the coil has a derivative behaviour up to 9 MHz approximately, and then the output signal turns into to a voltage proportional to the current. Moreover, the sensitivity is around 12 dB. The frequency analysis for this sensor is presented up to 60 MHz, since the observation of the average spectra for the signals measured in the experiments presented in the following sections, led to the conclusion that the power above 60 MHz was very low.

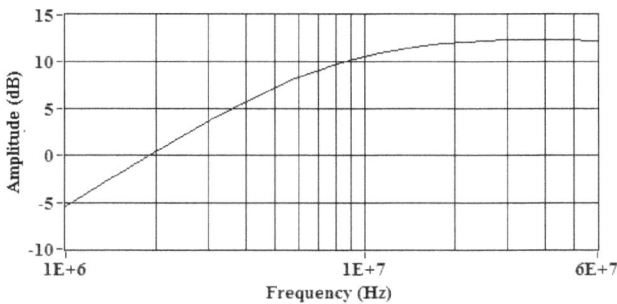

Figure 2. Frequency response of the Rogowski coil.

3. Experimental Setup and PD Sources Separation

3.1. Experimental Setup

Due to the importance of PD phenomenon to estimate the insulation lifetime, a procedure to measure PD including the circuits implemented for their detection is detailed in the standard IEC 60,270 [6]. Although the measured PD signals are different to the original signals originated in the PD sources, due to the attenuation and dispersion effects until the sensor captures them, much information of the pulse shape to distinguish their source type (internal, surface or corona) can be obtained [17–19].

Accordingly, in this paper all data analysed have been collected experimentally with an indirect detection circuit based on the standard IEC 60,270. The circuit consists of a 750 VA transformer that provides high-voltage to several test objects, where PD are created. A capacitive divider with a high-voltage capacitor (1 nF), connected in series with a measuring impedance, provides a path for the high-frequency currents generated by the PD pulses, see Figure 3. Pulses flowing through the capacitive path are measured using the Rogowski sensor presented in Section 2. The measuring impedance gives the synchronization signal from the grid frequency to the measuring instrument, so PD pulses can be plotted in conventional PRPD patterns.

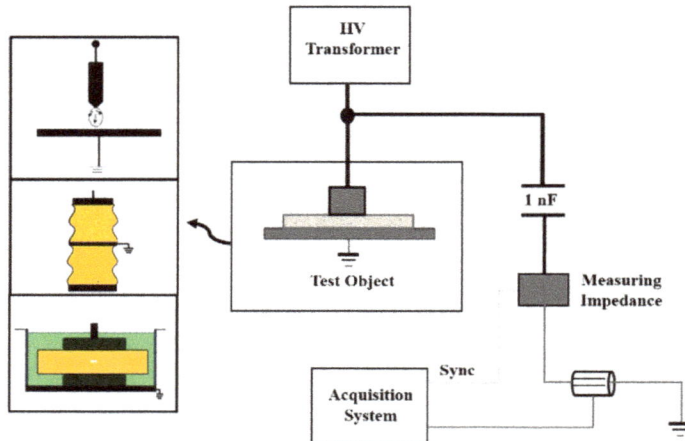

Figure 3. Experimental setup for PD measurements.

A NI-PXI-5124 digitizer was programmed to get the information from each experiment. The technical characteristics of this digitizer are 200 MS/s of sampling rate, 12 bits of vertical resolution and 150 MHz of bandwidth. The channel 0 was used for the 50 Hz reference voltage signal measurement and the channel 1 was used to get the waveform of the high-frequency pulses. This acquisition system acquires data each 20 ms (network cycle). These data are divided in time windows of 1 μs or 4 μs, depending on the duration of the PD pulses. Only the data that have peaks higher than the trigger level in the time windows are considered; the rest are discarded. Each signal measured is represented by 200 or 800 samples corresponding to the time window widths of 1 μs or 4 μs and is stored in vectors, in order to calculate contents in frequency up to 100 MHz. As shown in Figure 3, some basic PD sources such as corona effect, a surface defect and an internal defect were created through some simple test objects (see [10] for more details):

- Corona effect: point-plane experimental test object. A 0.5 mm thick needle was placed above a metallic ground plane. The distance between the needle and the plane was adjusted to 1 cm.

- Surface defect: Contaminated ceramic bushing. A 15 kV ceramic bushing was contaminated by spraying a solution of salt in water to create ionization paths along the surface. In order to avoid unstable PD activity, the measurements are carried out once moisture has been disappeared.
- Internal defect: Insulating sheets immersed in mineral oil. This setup was designed to generate internal discharges and consists of eleven insulating sheets of NOMEX paper (polyimide 0.35 mm thick film). The central paper sheet was pierced with a needle (1 mm in diameter) to create an air void inside this dielectric.

As it will be indicated later in the experimental results, the experimental setup was implemented identically in two different high-voltage laboratories, one that is completely shielded and another that is unshielded. This was done in order to characterize the ability of the PR maps to separate PD sources in two different environments measuring with the RC sensor. In the first laboratory, controlled, the noise signals present a low magnitude and in the second one, less controlled, the noise signals present similar characteristics to those found in industrial environments: high-levels of amplitude and high-spectral variability.

3.2. Spectral Power Clustering Technique (SPCT)

Separation and identification of PD sources are stages that must be approached sequentially, due to separation is a prerequisite fundamental and obligatory for a successful and accurate identification.

When PD measurements are carried out with inductive sensors, such as the RC, the waveform of the carrying currents sensed as a result of PD activity cannot be universally identified with a particular type of PD source (corona, surface or internal), due to the stochastic behaviour of PD phenomena and due to the distortion caused in the pulse transmission from the source to the measuring point and in the coupling system itself. However, PRPD patterns allow to successful identifying PD sources [19]. Therefore, a generic solution widely used in most PD measuring instruments, is based on the analysis of the entire PRPD pattern containing all sources measured and on the separation of this pattern into sub-PRPD patterns, each corresponding to a specific source. The separation is accomplished by assuming that each PD source exhibits similar waveforms, while the signals produced by different sources are different. Following this premise, this paper attempts to prove that the SPCT allows separating different PD and noise sources, mapping for each of the measured signals the value of the relative spectral power calculated for two intervals: PRL (power ratio for low-frequencies) and PRH (power ratio for high-frequencies).

In this approach, the pulses are analysed in the frequency domain therefore, the fast Fourier transform is applied to each detected pulse, obtaining its spectral magnitude distribution $s(f)$ [10]. Then, the spectral power of each pulse is calculated in two frequency intervals, $[f_{1L}, f_{2L}]$ and $[f_{1H}, f_{2H}]$. Since the total spectral power or amplitude of the signals may influence the pulse characterization, these spectral powers are divided into the overall spectral power calculated up to the maximum analysed frequency f_t. The obtained quantities are defined as power ratios (%), one for the higher frequency interval, PRH, and another for the lower frequency interval, PRL, as shown in Equations (2) and (3). These two parameters are represented in a two dimensional map, where each pulse source showed a different cloud of points (clusters) with different positions, (see Figure 4).

For all measurements, the frequency analysis was made up to 100 MHz, however, the observation of the average spectra for all the experiments led to the conclusion that the power above 60 MHz was very low, so this last value was used as f_t.

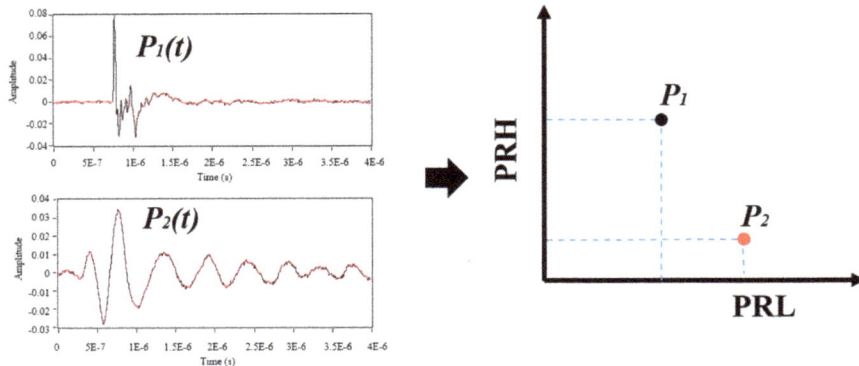

Figure 4. Example of PR map for two pulse sources (PD and noise).

Thus, the power ratio for low-frequencies (PRL) and the power ratio for high-frequencies (PRH) are calculated as follows:

$$PRL = \frac{\sum_{f_{1L}}^{f_{2L}} |s(f)|^2}{\sum_0^{f_t} |s(f)|^2} \cdot 100 \tag{2}$$

$$PRH = \frac{\sum_{f_{1H}}^{f_{2H}} |s(f)|^2}{\sum_0^{f_t} |s(f)|^2} \cdot 100 \tag{3}$$

For all measurements presented in this paper, the frequencies for the PRL and PRH calculation were set to: $f_{1L} = 10$ MHz, $f_{2L} = f_{1H} = 30$ MHz, $f_{2H} = 50$ MHz, $f_T = 60$ MHz. The interval, [0, 10] MHz was not taken into account in the analysis due to the derivative behaviour of the sensor in this frequency band. In addition, these intervals are the same as those used in [15] for the initial analysis of clusters.

4. Experimental Results

In the first part of the experimental results, the measurements were carry out using the indirect circuit described in Section 3.1, but housed in the shielded high-voltage laboratory. The sources were initially characterized individually and were measured in the following way:

1. The noise signals present in the laboratory were registered, by performing measurements with a low trigger level and by applying a low-voltage to the test object.
2. Then, the voltage and the trigger levels were increased until a stable PD activity noise-free was found for each of the PD sources.

To obtain statistically significant results and guarantee the reliability of the phenomenon observed, the number of pulses acquired by the measuring instrument for each of these measurements must be high and was set in 1500. Then, the clusters associated with PD and noise were characterized again, but when the different type of PD sources were simultaneously emitting. In this case, the measurements were performed for a high-voltage level, but with a reduced trigger level, in order to enable the acquisition of PD and noise simultaneously. For this last measurement, over 3000 pulses were acquired, since it was present more than one type of pulse sources.

For each experiment, the PR maps and the average spectral power of the pulses are described. Additionally, the dispersion obtained in PRL and PRH parameters was compared for each of the clusters (PD or noise), in order to evaluate which type of source has greater dispersion when a RC sensor is used. This information is of great interest because, as described in [15], the location and shape adopted by clusters in the PR maps depends, not only on the frequency intervals, on the equivalent capacitance of the test object, and on the pulse nature, but also on the type of sensor used. Therefore, the more homogeneous are the clusters associated with a particular source the better is its characterization when different sources are detected, especially if there is a clear separation between them.

In the second part of the experimental results, the measurements were made using the same indirect measurement circuit but in this case, this was housed in the unshielded laboratory. Again, the clusters associated to PD and electrical noise were characterized in these tests, when different types of sources were present simultaneously.

4.1. Experimental Measurements Performed in the Shielded Laboratory

The measurements described below, were made, in order to obtain a controlled environment to minimize the influence of electrical noise, and to facilitate the characterization of the dispersion of the clusters of pulses for each type of PD source. However, as it will be indicated in Section 4.1.3, it is not possible to completely minimize the influence of noise during the acquisitions, especially when the measurements are performed with a RC sensor, whose signal-to-noise ratio (SNR) is low compared to other types of inductive sensors [20–22].

4.1.1. Noise Characterization

In order to characterize the noise sources in the three test objects described in the experimental setup, a low-voltage (800 V) level was applied to each test object. Additionally, the trigger level in the acquisition system was set at low level (0.4 mV). This procedure ensures that the pulses obtained with the RC correspond to sources of electrical noise and not to PD sources, since the voltage level is very low to start PD activity. It can be confirmed, for the data acquired in the case of the point-plane experimental test object (see Figure 5), that the PRPD pattern obtained is the typical pattern of the electrical noise (uncorrelated in phase). In this case, the maximum noise levels found were close to 1 mV.

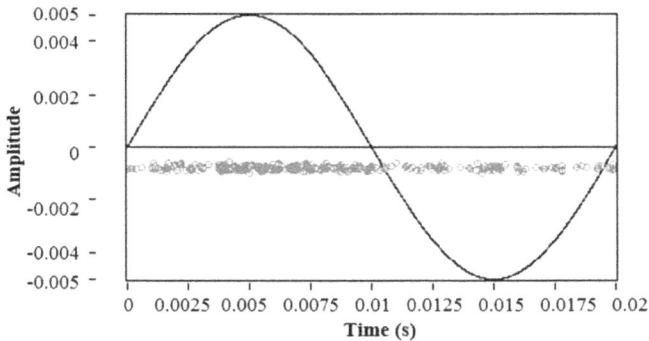

Figure 5. PRPD for noise acquisition in the point-plane experimental test object.

Figure 6a–c show the PR maps for the signals associated with electrical noise that were obtained for each test object. In all measurements, the noise was clearly characterized as a cloud of points in the lower right part of the map. This position is coherent with Figure 6d corresponding to the average spectral power to each of the signals, where the spectral power content in the interval [10, 30] MHz (PRL), is higher than the obtained in the interval [30, 50] MHz (PRH). The high spectral power obtained for the interval PRH in each of the measurements of noise is due to the presence of two peaks of power around 12 MHz and 18 MHz. These characteristics are typical of conventional noisy environment, whose behaviour is narrow-band. In all noise measurements that were made for each test objects the average spectral power has the same behaviour in PRL and PRH, see Figure 6d.

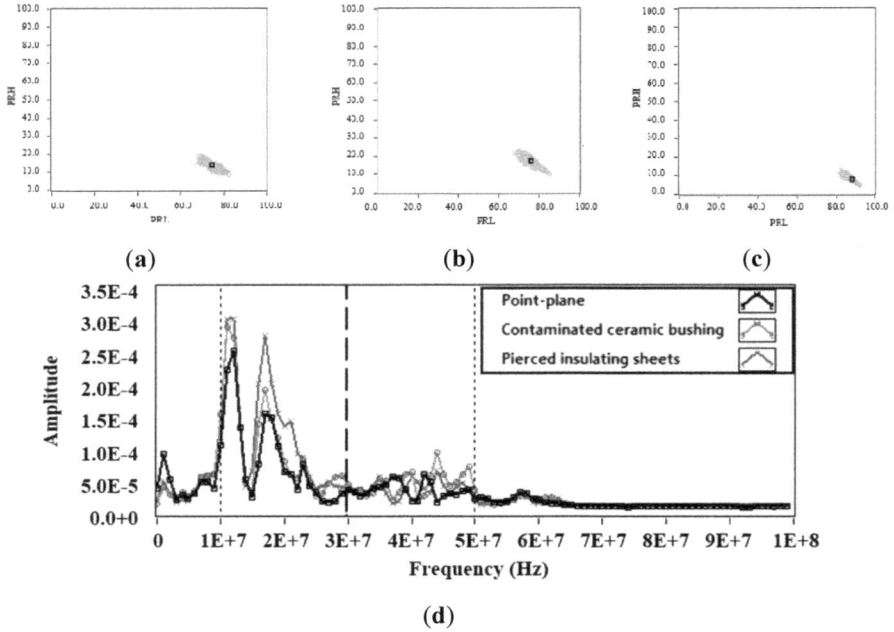

Figure 6. Noise acquisition 800 V. (**a**) Power ratio map for noise signals in the point-plane experimental test object; (**b**) Power ratio map for noise signals in a contaminated ceramic bushing; (**c**) Power ratio map for noise signals in pierced insulating sheets; (**d**) Average spectral power of the signals obtained with the three test objects.

In order to evaluate the statistical dispersion of the clusters obtained in each experiment, the standard deviation for PRL and PRH was calculated. The standard deviation was obtained from the centroid for each cluster according to the following mathematical expression:

$$\sigma_{PRL} = \sqrt{\sum_{i=1}^{n} \frac{(X_i - CX)^2}{n}} \tag{4}$$

$$\sigma_{PRH} = \sqrt{\sum_{i=1}^{n} \frac{(Y_i - CY)^2}{n}} \tag{5}$$

where n is the number of points in each clusters, X_i and Y_i is the position (PRL-PRH) of each point i in the PR map and (CX, CY) is the centroid of the cluster, which is obtained according to described in [15]. Table 2 summarizes the results of the dispersion obtained for each cluster associated with electrical noise.

Table 2. Results obtained of the standard deviation for each cluster associated with electrical noise.

Indicator	Point-Plane Experimental Specimen	Contaminated Ceramic Bushing	Pierced Insulating Sheets
σ_{PRL}	3.15	3.10	3.02
σ_{PRH}	2.23	2.52	2.35
$\sigma_{PRL}/\sigma_{PRH}$	1.41	1.23	1.28

Analysing the results in Table 2, the dispersion in PRL and PRH for each of the clusters has very similar values. When comparing the dispersion between PRL and PRH for each cluster, it is observed that the dispersion in PRL is higher than that obtained by PRH, this last is most notable in the case of the point-plane experimental test object, where the $\sigma_{PRL}/\sigma_{PRH}$ ratio is higher than for other test objects (1.41). This ratio will allow us to identify during the measurements in which axis of the PR map, is more dispersed one cluster, *i.e.* if $\sigma_{PRL}/\sigma_{PRH} > 1$; PRL has a greater dispersion, otherwise if $\sigma_{PRL}/\sigma_{PRH} < 1$ this means that PRH has the greater dispersion.

From the point of view of source separation, an ideal cluster is the one that has a ratio $\sigma_{PRL}/\sigma_{PRH} \approx 1$ and a high homogeneity (low dispersion in PRL and PRH). Therefore, if the clusters located on the classification map are obtained with these characteristics (*i.e.*, $\sigma_{PRL}/\sigma_{PRH} \approx 1$, low σ_{PRL} and low σ_{PRH}), it will be obtained a very homogeneous clouds of points that facilitates the application of any method of clusters identification (K-means, K-medians, Gaussian, *etc.*), after the application of the SPCT, so a better separation and identification of points associated with each cluster is achieved when multiple sources are present.

It is important, for the operator of the classification tool, to consider this information once each of the sources in the classification map have been characterized, as this can help to verify if the separation intervals manually selected should be modified slightly or completely changed in order to enhance the clusters separation. This will facilitate, in a later stage, the sources identification process, that can be performed through visual inspections or applying automatic identification algorithms. Furthermore, it must be emphasized that this information is only useful once the intervals of separation have been selected, because these indicators alone cannot estimate the frequency bands where the separation intervals should be located. They only allow assessing homogeneity, dispersion and shape of the clusters for each of the previously demarcated intervals.

However, in Section 4.1.3, an additional graphic indicator is presented. This is based on the variability of the spectral power of the captured signals, which does allow identifying areas of interest where the user must locate the separation intervals, in order to obtain an initial characterization which may, or may not, be improved by modifying slightly the position of the separation intervals or evaluating the dispersion in PRL (σ_{PRL}), PRH (σ_{PRH}) and its ratio ($\sigma_{PRL}/\sigma_{PRH}$), for each case, up to a better characterization of each source.

4.1.2. Partial Discharge Source Characterization

In order to find stable PD activity and to avoid the acquisition of noise signals, the following measurements were made for high-voltages applied and high-trigger levels (1.2 mV). Figure 7a–d represent the PR maps and the average spectral power of the signals measured by applying 5 kV to the point-plane experimental test object, 8.3 kV to the ceramic bushing and 9 kV to the pierced insulating sheets respectively.

Figure 7d shows that the spectral power components detected by the RC in the intervals [10, 30] MHz and [30, 50] MHz are higher for the pulses associated to internal PD (PD in the pierced insulating sheets), this can be confirmed in Figure 7c, where the position of the cluster with respect to the PRL and PRH axis is higher than that obtained with the clusters of corona and surface PD. Figure 7d also shows, that for the pulses associated to corona and surface PD, the spectral power detected is very similar in the interval [10, 30] MHz, while in the interval [30, 50] MHz the spectral power is slightly higher for surface PD, which is consistent with the position of the clusters in both classification maps, see Figure 7a (corona PD) and b (surface PD).

(a) (b) (c)

(d)

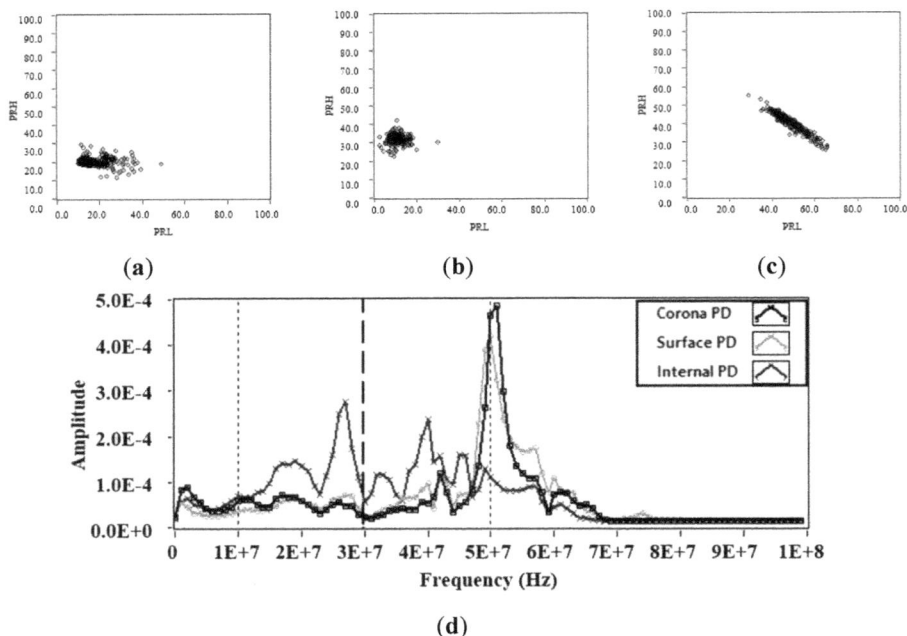

Figure 7. Partial discharges (**a**) Power ratio map for PD in point-plane experimental test object; (**b**) Power ratio map for PD in contaminated ceramic bushing; (**c**) Power ratio map for PD in pierced insulating sheets; (**d**) Average spectral power of the PD signals obtained with the three test objects.

Regarding the dispersion of the clusters associated to PD (see Table 3), the results indicate that the internal PD clusters present the higher spectral power dispersion, both in PRL and PRH (5.41 and 4.40, respectively). On the contrary, the dispersion was considerably low for the surface PD clusters, thus together with the fact that the $\sigma_{PRL}/\sigma_{PRH}$ ratio is close to 1, makes the cluster for this type of source be very homogeneous and with low dispersion. The latter is particularly important because, as will be shown in the next section, a high homogeneity in clusters facilitates the separation task and their subsequent display of the PRPD pattern if there are present several sources acting simultaneously.

In these measurements, it is noted that the dispersion in PRL continues to be higher than that obtained in PRH. Again, the $\sigma_{PRL}/\sigma_{PRH}$ ratio is much greater for the PD obtained for the point-plane experimental test object (3.18).

Table 3. Results obtained of the standard deviation for each cluster associated with PD.

Indicator	Corona PD	Surface PD	Internal PD
σ_{PRL}	5.12	2.11	5.51
σ_{PRH}	1.61	1.63	4.40
$\sigma_{PRL}/\sigma_{PRH}$	3.18	1.29	1.25

4.1.3. PD and Noise Characterization

In this section, measurements were performed for each of the test objects with the same voltage level used in the previous section, but with a reduced trigger level (0.4 mV). This was made to enable the acquisition of PD and noise simultaneously. Figure 8 (left), shows the PR map with the clusters associated to PD (black cluster) and noise (grey cluster). Figure 8 (right) shows the average power spectrum densities, for each of PD and noise pulses that are represented on the PR map and that are acquired for each test object.

Considering that for this experiment, there are two sources simultaneously acting, the average power spectrum density is presented in order to see if in the selected intervals ($f_{1L} = 10$ MHz, $f_{2L} = f_{1H} = 30$ MHz, $f_{2H} = 50$ MHz, $f_T = 60$ MHz) are included the bands where greater variability of spectral power is presented. If these are included, a clear separation of sources (PD and noise) could be achieved in the PR maps, since for these bands, the captured pulses have less similarity. Otherwise, if the bands with less variability of spectral power are selected, the clusters could be overlapped and the separation of sources could result more difficult. The average of the spectrums is plotted in central thick line; the shaded area corresponds to the area at one standard deviation of the mean that was obtained for the pulses in each measurement.

For each of the experiments, the *K-means* algorithm [23] has been used to identify the clusters and its centroids after applying the SPCT to the pulses measured.

As expected, the clusters associated to PD and noise tend to take similar positions as those observed in Figure 6 (only noise) and Figure 7 (only PD). However, the position of the clusters no longer matches with the average spectral power obtained (central thick line), since the spectral content dependent on the spectral power of both types of sources (PD and noise). Therefore, the spectral components will be affected by the two sources acting simultaneously during the acquisition. On the other hand, as was described above, the standard deviation (shaded area) helps to indicate the frequency bands where there is greater statistical variability. With this information, the user can select or modify the frequency bands, in order to improve the separation of sources. Accordingly, for this experiment it is observed that both PRL and PRH include some frequency bands where the standard deviation of the frequency spectra was high. This allows the identification in the three cases, the presence of the two clusters, one associated with PD and other with electrical noise.

Note that, in these intervals, also the frequency bands where the standard deviation is minimal were included; therefore, this separation can be improved if only the bands with the higher standard deviation are selected. When evaluating the dispersion values in PRL and PRH, which are summarized in Table 4, it is found that the dispersion in each of the clusters associated with PD was increased. This occurs due to the low trigger level during the acquisition, so the noise pulses can be added to the PD pulses, generating signals with combined spectral power components that cause an increase of the dispersion in the clouds of points. For example, in the case of internal PD and noise in Figure 8c, that can be considered the most extreme case for having the largest dispersion in PRL and PRH, if it is represented one of the points of the PR map that is located between the two cluster (see Figure 9), clearly it is observed that the spectral power content of this pulse is formed by components of both sources, see Figures 6d and 7d. Due to this, the pulses tend to be located in an intermediate region of the two clusters (critical zone), hindering the separation process and the subsequent identification of the sources.

Table 4. Results obtained for the standard deviation and its increase in each cluster associated with PD and electrical noise.

Indicator	Corona PD and Noise			Surface PD and Noise			Internal PD and Noise		
	PD	Increase (%)	Noise	PD	Increase (%)	Noise	PD	Increase (%)	Noise
σ_{PRL}	6.69	**30.66**	3.10	6.02	**185.30**	4.28	6.71	**21.77**	2.42
σ_{PRH}	1.97	**22.36**	2.27	2.71	**66.25**	2.68	5.19	**17.95**	1.88
$\sigma_{PRL}/\sigma_{PRH}$	3.39	**6.60**	1.36	2.22	**72.02**	1.59	1.29	**32.00**	1.28

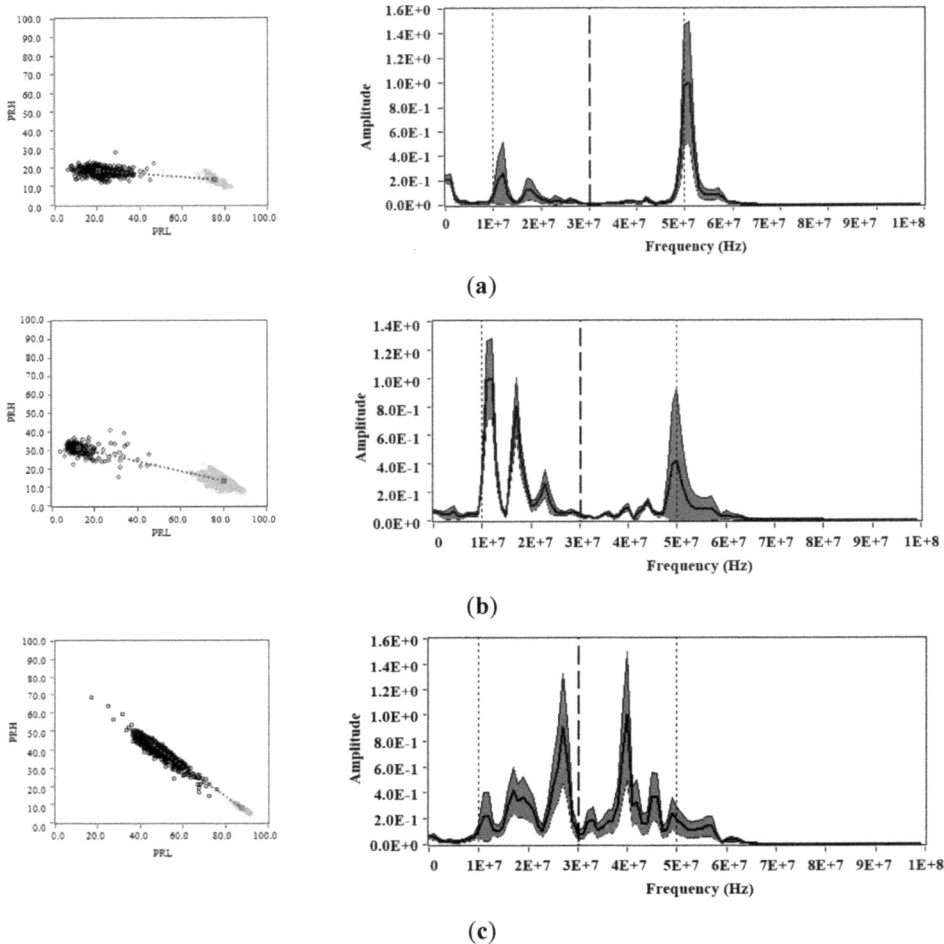

Figure 8. Power ratio maps for PD (black cluster) and noise (grey cluster), **left**: Average power spectrum densities, **right**: Obtained for (**a**) corona PD and noise; (**b**) surface PD and noise and; (**c**) internal PD and noise.

However, despite this increase in σ_{PRL} and σ_{PRH} for each of the clusters associated with PD due to the presence of an additional source of electrical noise, it has been possible to show that the RC allows to characterize adequately in different areas of the map both types of sources by applying the SPCT.

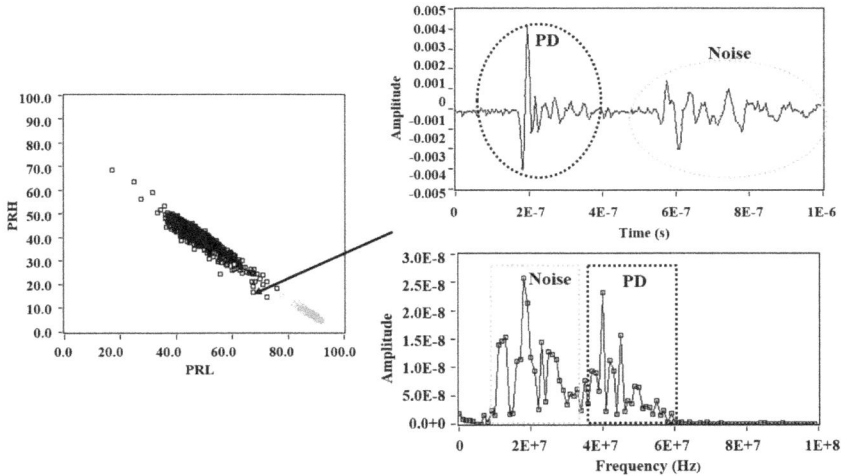

Figure 9. Example of a pulse formed by components of PD and noise.

4.2. Experimental Measurements Performed in the Unshielded Laboratory

In order to evaluate the performance of the RC in a less controlled environment, where the noise present has completely different characteristics in time and frequency to those found in the previous experiments, new measurements were carried out in a second high-voltage laboratory. In this second emplacement, there is not any type of shielding that can minimize the presence of external noise sources generated. Additionally, the laboratory is in an area of industrial activity (surrounded by industrial facilities), where the noise level can vary depending on the external activity during the measurement process.

On the other hand, the experimental setup used in this section was prepared to simultaneously generate three different sources: one associated with corona another to internal PD and the last one associated with electrical noise. Thus, the pulse sources were measured with the RC working in a very similar environment to that found in on-site measurements, where most of times it is necessary to detect and separate simultaneous PD and noise sources in order to identify the insulation defects involved.

For this purpose, the tests objects used for internal and corona PD were modified to obtain a stable PD activity for the same voltage level on both test objects (5.2 kV). In this case, the separation from the needle to the ground in the point-plane configuration was 1.5 cm. For the internal defect, the insulation system was composed by eleven insulating sheets of NOMEX where the five central sheets were pierced. Then, an air cylinder with 5×0.35 mm in height inside the solid material is obtained. For this experiment, the trigger level was set at 1.3 mV (low), because the maximum noise levels found were close to 2.1 mV, this level of noise is greater than that in the laboratory shielded (1 mV).

Finally, both test objects were electrically connected in parallel and subjected to 5.2 kV, measuring thousands of PD and noise pulses waveforms. The resulting PR map for this experiment, maintaining the same intervals of separation as previously (f_{1L} = 10 MHz, $f_{2L} = f_{1H}$ = 30 MHz, f_{2H} = 50 MHz, f_T = 60 MHz) is presented in Figure 10. For this case, three different clouds of points can be easily selected (since they are clearly separated) to identify the PD source type through the PRPD patterns.

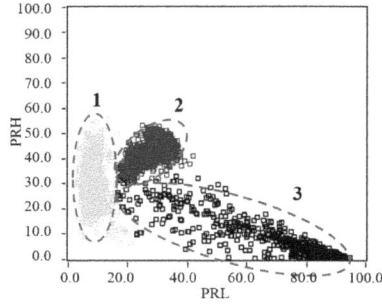

Figure 10. PR map obtained for noise (cluster 1), corona PD (cluster 2) and internal PD (cluster 3).

The position of the cluster associated with the new source of electrical noise, indicates that the spectral power in the range [10, 30] MHz is lower than that obtained in the previous experiments, in which the source of noise had a high spectral power for the same interval. As for the values of dispersion in PRL and PRH shown in Table 5, it is seen that the dispersion in PRL is 2.62, lower than previously values previously obtained in Table 2 for the noise in all the test objects. Contrary to this, the dispersion in PRH is increased almost 292% for the point-plane experimental test object, 247% for the contaminated ceramic bushing and 272% for the pierced insulating sheets; which is consistent with the form taken by this cluster on the PR map and (see Figure 10). In addition, the $\sigma_{PRL}/\sigma_{PRH}$ ratio for this new source happened to be well below 1 (0.30).

Table 5. Results obtained of the standard deviation for each cluster associated with PD and noise.

Indicator	Noise	Corona PD	Internal PD
σ_{PRL}	2.62	4.08	11.51
σ_{PRH}	8.75	3.68	4.45
$\sigma_{PRL}/\sigma_{PRH}$	0.30	1.10	2.58

For the cluster associated with corona PD pulses, it was also observed a variation in the shape and the position on the PR map, which differs greatly from previous experiments. The values shown in Table 5 indicate that the dispersion in PRL and PRH suffered significant changes (σ_{PRH} increases and σ_{PRL} decreases). For this cluster, the new relation $\sigma_{PRL}/\sigma_{PRH}$ was 1.10. A value close to unit means that the cluster takes a more "symmetric" shape. As mentioned throughout this paper, when it has this kind of geometries or forms in the clusters it facilitates the identification process when the operator have to select the cluster to be represented its respective PRPD pattern, improving the process of identifying the type of source.

Finally, the cluster associated with internal PD also presents great changes, both in position and in shape, according to its PR map characterization. The most notable change, in terms of dispersion, is observed for PRL, which it is increased by almost 108% over the value of PRL obtained in previous experiments (see Table 3), this is easily seen in the PR map in Figure 10, where the cluster occupies a large map space due to its lack of homogeneity in this axis. In this case, the relationship $\sigma_{PRL}/\sigma_{PRH}$ (2.58) indicates an increase of almost 106% compared to the values previously obtained when the size of the vacuole was lower.

These results confirm those described in [10,15], where is disclosed that any change of the equivalent capacity in the measuring circuit can vary the shapes and positions of the clusters on the classification map. These variations can be due to the process of manufacturing the test objects or by the fact of perform measurements in an environment where noise levels are different in nature and magnitude than those found in more controlled environments. Therefore, it is very important to properly select the separation intervals depending on the scenario to obtain a clear characterization of all sources present during the measurements.

The complete PRPD pattern, for the three PD sources acting simultaneously, is shown in Figure 11 (Up), where it is clear that PRPD interpretation seems to be quite complex even for an expert in the field. However, if the PRPD pattern for each cluster is represented individually, it can be clearly identify each of the sources present during the acquisition. As it can be seen in Figure 11 (Down), the PRPD associated to the cluster 1 represents the captured electrical noise during the acquisition (uncorrelated pulses in phase). The PRPD of the cluster 2 corresponds to the typical PRPD for corona discharges, where highly stable PD magnitudes are observed for the negative maxima of the applied voltage. Finally, when selecting the cluster 3 from the PR map provides a "clean" PRPD representation typical from internal PD, where the high-magnitude discharges occur in the phase positions where the voltage slope is maximum.

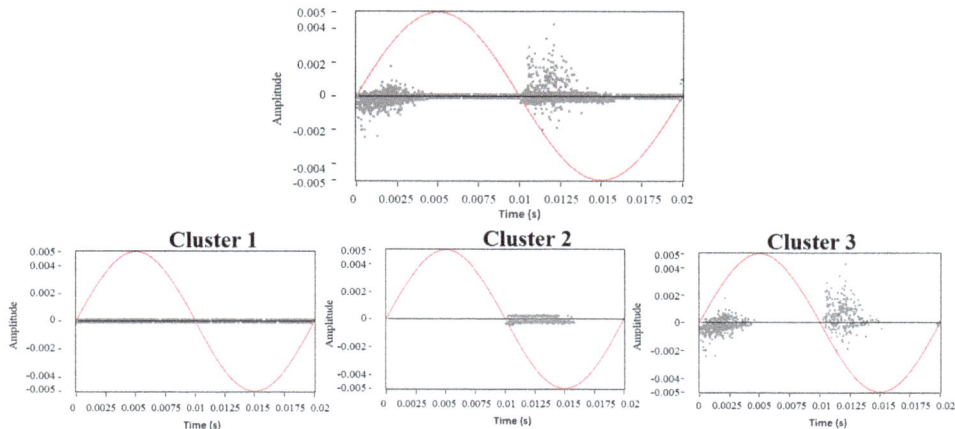

Figure 11. PRPD pattern for noise and PD simultaneously, **Up**: PRPD patterns for **Down**: cluster 1 (noise), cluster 2 (corona PD), and cluster 3 (internal PD).

Accordingly, when assessing the average power spectrum density of the pulses obtained in this experiment, it is observed that the selected intervals for PRL [10, 30] MHz and PRH [30, 50] MHz match bands where greater variability of spectral power occurs, that it is shown in the enlarged view in Figure 12. This justifies the separation obtained for each source on the PR map shown (see Figure 10).

On the contrary, if the PRL and PRH had other bands, where the variability of spectral power was lower, the separation of the sources would be impossible. This can be demonstrated if the intervals [20, 40] MHz for PRL and [40, 60] MHz for PRH are used, for example. As shown in the PR map in Figure 13 (Left), using these new intervals of separation (including bands where the variability of spectral power is low), only two different clusters can be identify, which they are also very close each other. Therefore, two types of sources are superimposed in a single cluster, this can be checked representing the PRPD patterns for each cluster, see Figure 13 (Right). In the PR map of the Figure 13, the cluster 2 corresponds to the electrical noise, while the cluster 1 corresponds with the two types of the two remaining sources (corona and Internal PD), which are clearly overlapping.

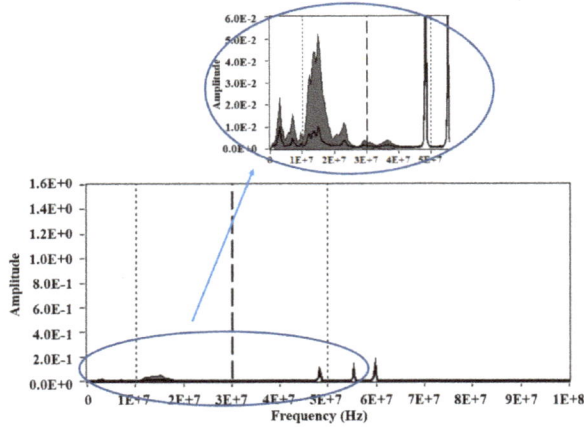

Figure 12. Average power spectrum density for PD and noise simultaneously measured.

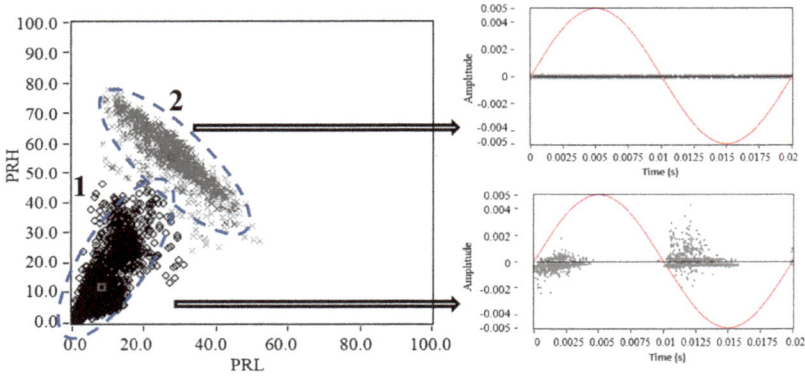

Figure 13. PR map for simultaneous PD (corona and internal) and noise activity. Frequency intervals for the power ratios calculations: f_{1L} = 20 MHz, f_{2L} = 40 MHz, f_{1H} = 40 MHz, f_{2H} = 60 MHz.

Likewise, if other intervals, in which the frequency bands present greater variability of spectral power are selected, the sources separation will become more effective. For example, if f_{1L} = 2 MHz, f_{2L} = 25 MHZ, f_{1H} = 15 MHz, f_{2H} = 38 MHz (f_T = 60 MHz), intervals are used, a better separation between clusters is achieved (see Figure 14), compared with the separation obtained with the intervals used previously, see Figures 10 and 13.

In this new separation map, the cluster 1 is associated with electrical noise pulses, cluster 2 with corona PD and cluster 3 with internal PD. Analysing the values of dispersion for PRL and PRH that are presented in Table 6, except σ_{PRL} for corona PD cluster and σ_{PRH} for internal PD, a clear decrease in the dispersion for each of the clusters using these new separation intervals it is shown. Additionally, a marked improvement in the $\sigma_{PRL}/\sigma_{PRH}$ relation to the case of clusters associated with internal PD and electrical noise was achieved, since values close to 1 were obtained compared to the values shown in Table 5. For the cluster associated with corona PD, this relationship is increased, but not very significantly (1.35) compared to the previous value obtained in Table 5 (1.10).

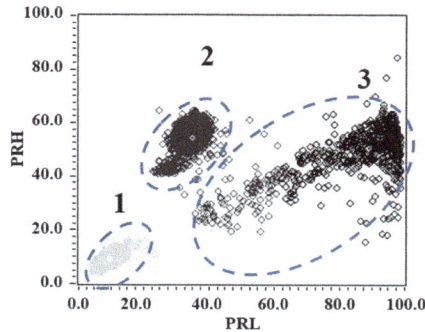

Figure 14. PR map for simultaneous PD and noise activity. Frequency intervals for the power ratios calculations: $f_{1L} = 2$ MHz, $f_{2L} = 25$ MHz, $f_{1H} = 15$ MHz, $f_{2H} = 38$ MHz.

Table 6. Results obtained of the standard deviation for each cluster associated with PD and noise, using $f_{1L} = 2$ MHz, $f_{2L} = 25$ MHZ, $f_{1H} = 15$ MHz, $f_{2H} = 38$ MHz, $f_T = 60$ MHz intervals.

Indicator	Noise	Corona PD	Internal PD
σ_{PRL}	2.35	4.66	6.55
σ_{PRH}	2.56	3.45	9.72
$\sigma_{PRL}/\sigma_{PRH}$	0.91	1.35	0.67

5. Conclusions

In this paper, the clustering capacity and the dispersion of the clusters obtained by the application of the SPCT to the pulses measured with a Rogowski coil have been studied. Results indicate that with this simple and inexpensive sensor, without magnetic core, the separation of different types of pulse sources using PR maps can be made adequately, even when there are several PD sources acting simultaneously.

Furthermore, this paper proposes using four different indicators, in order to find the separation intervals that allow a better separation of the PD sources and electrical noise present during the measurements. Three of these indicators (σ_{PRL}, σ_{PRH} and $\sigma_{PRL}/\sigma_{PRH}$) assist the operator of the classification tool to identify the intervals that enable to obtain those clouds of points more homogeneous. This allows an easier selection of the clusters on the PR maps for the further representation of the respective PRPD patterns. The fourth proposed indicator is based on a graphic tool that represents the average power spectrum density of the measured signals. This indicator allows selecting the bands of interest where the variability of the pulses is high and makes more feasible the separation between different groups of pulses. Analysing these indicators the manual selection of the PRL and PRH intervals can be easily improved, especially in those cases where the clusters obtained appear overlapped and/or there are suspicions of the presence of multiple sources during the measurements. The authors propose the use of these indicators, even when any other type of UHF sensor (ILS, HFCT, *etc.*) is employed.

Acknowledgments: This work was supported by Basal Project FB0008. The tests were carried out in the High-Voltage Research and Tests Laboratory at Universidad Carlos III de Madrid (LINEALT) and in the High-Voltage Research and Test Laboratory at Universidad Técnica Federico Santa Maria (LIDAT) de Santiago de Chile.

Author Contributions: The presented research was developed by the authors with the contribution of each one in a specific task. Jorge Ardila defined the test so as to be performed and carried out the experimental measurements. Besides, Jorge Ardila and Ricardo Albarracín were in charge of the programming and the instrumentation tools. Aldo Barrueto, designed the Rogowski coil sensor and studied its frequency response. Fernando Álvarez did the statistical study. Finally, all the authors collaborated in the analysis of results.

Conflicts of Interest: The authors declare no conflict of interest.

References

1. Willis, H.; Schrieber, R. *Aging Power Delivery Infrastructures, Second Edition*; Power Engineering (Willis), Taylor & Francis: Boca Raton, FL, USA, 2013.
2. Ravi, S.; Ward, J. *A Novel Approach for Prioritizing Maintenance of Underground Cables. Technical report, Power Systems Engineering Research Center (PSERC)*; Arizona State University: Tempe, AZ, USA, 2006.
3. Zhang, X.; Gockenbach, E.; Wasserberg, V.; Borsi, H. Estimation of the Lifetime of the Electrical Components in Distribution Networks. *IEEE Trans. Power Deliv.* **2007**, *22*, 515–522. [CrossRef]
4. James, R.; Su, Q. *Condition Assessment of High Voltage Insulation in Power System Equipment*; The Institution of Engineering and Technology: London, UK, 2008.
5. Jardine, A.K.; Lin, D.; Banjevic, D. A review on machinery diagnostics and prognostics implementing condition-based maintenance. *Mech. Syst. Signal Process.* **2006**, *20*, 1483–1510. [CrossRef]
6. IEC-60270. High-Voltage Test Techniques—Partial Discharge Measurements. Available online: https://webstore.iec.ch/publication/1247 (accessed on 10 October 2015).
7. Sahoo, N.; Salama, M.; Bartnikas, R. Trends in partial discharge pattern classification: A survey. *IEEE Trans. Dielectr. Electr. Insul.* **2005**, *12*, 248–264. [CrossRef]
8. Cavallini, A.; Montanari, G.; Puletti, F.; Contin, A. A new methodology for the identification of PD in electrical apparatus: properties and applications. *IEEE Trans. Dielectr. Electr. Insul* **2005**, *12*, 203–215. [CrossRef]
9. Cavallini, A.; Montanari, G.; Contin, A.; Pulletti, F. A new approach to the diagnosis of solid insulation systems based on PD signal inference. *IEEE Electr. Insul. Mag.* **2003**, *19*, 23–30. [CrossRef]
10. Ardila-Rey, J.; Martinez-Tarifa, J.; Robles, G.; Rojas-Moreno, M. Partial discharge and noise separation by means of spectral-power clustering techniques. *IEEE Trans. Dielectr. Electr. Insul.* **2013**, *20*, 1436–1443. [CrossRef]
11. Rojas-Moreno, M.V.; Robles, G.; Tellini, B.; Zappacosta, C.; Martinez-Tarifa, J.M.; Sanz-Feito, J. Study of an inductive sensor for measuring high frequency current pulses. *IEEE Trans. Instrum. Meas.* **2011**, *60*, 1893–1900. [CrossRef]
12. Argüeso, M.; Robles, G.; Sanz, J. Implementation of a Rogowski coil for the measurement of partial discharges. *Rev. Sci. Instrum.* **2005**, *6*, 065107. [CrossRef]
13. Robles, G.; Martinez, J.; Sanz, J.; Tellini, B.; Zappacosta, C.; Rojas, M. Designing and Tuning an Air-Cored Current Transformer for Partial Discharges Pulses Measurements. In Proceedings of the Instrumentation and Measurement Technology Conference, Victoria, BC, USA, 12–15 May 2008; pp. 2021–2025.
14. Álvarez, F.; Garnacho, F.; Ortego, J.; Sánchez-Urán, M. Application of HFCT and UHF sensors in on-line partial discharge measurements for insulation diagnosis of high voltage equipment. *Sensors* **2015**, *15*, 7360–7387. [CrossRef] [PubMed]
15. Ardila-Rey, J.A.; Rojas-Moreno, M.V.; Martínez-Tarifa, J.M.; Robles, G. Inductive sensor performance in partial discharges and noise separation by means of spectral power ratios. *Sensors* **2014**, *14*, 3408–3427. [CrossRef] [PubMed]
16. Samimi, M.; Mahari, A.; Farahnakian, M.; Mohseni, H. The rogowski coil principles and applications: A review. *IEEE Sens. J.* **2015**, *15*, 651–658. [CrossRef]
17. Okubo, H.; Hayakawa, N. A novel technique for partial discharge and breakdown investigation based on current pulse waveform analysis. *IEEE Trans. Dielectr. Electr. Insul.* **2005**, *12*, 736–744. [CrossRef]
18. Cavallini, A.; Montanari, G.; Tozzi, M.; Chen, X. Diagnostic of HVDC systems using partial discharges. *IEEE Trans. Dielectr. Electr. Insul.* **2011**, *18*, 275–284. [CrossRef]
19. IEC-TS-60034-27-2. Rotating Electrical Machines—Part 27-2: On-Line Partial Discharge Measurements on the Stator Winding Insulation of Rotating Electrical machines. Available online: https://webstore.iec.ch/publication/131 (accessed on 10 October 2015).
20. Zhang, Z.; Xiao, D.; Li, Y. Rogowski air coil sensor technique for on-line partial discharge measurement of power cables. *IET Sci. Meas. Technol.* **2009**, *3*, 187–196. [CrossRef]

21. Hewson, C.; Ray, W. The Effect of Electrostatic Screening of Rogowski Coils Designed for Wide-Bandwidth Current Measurement in Power Electronic Applications. In Proceedings of the 2004 IEEE 35th Annual Power Electronics Specialists Conference, Aachen, Germany, 20–25 June 2004; pp. 1143–1148.

22. Hashmi, G.M. Partial Discharge Detection for Condition Monitoring of Covered-Conductor Overhead Distribution Networks Using Rogowski Coil. Ph.D. Thesis, Helsinki University of Technology, Helsinki, Finland, 22 August 2008.

23. Hastie, T.; Tibshirani, R.; Friedman, J. *The Elements of Statistical Learning: Data Mining, Inference, and Prediction; Springer Series in Statistics*; Springer: New York, NY, USA, 2013.

Article

Analysis of Uncertainty in a Middle-Cost Device for 3D Measurements in BIM Perspective

Alonso Sánchez [1,*], José-Manuel Naranjo [2], Antonio Jiménez [3] and Alfonso González [1]

[1] University Centre of Mérida, University of Extremadura, 06800 Mérida, Spain; agg@unex.es
[2] Polytechnic School, University of Extremadura, 10003 Cáceres, Spain; jnaranjo@unex.es
[3] Development Area, Provincial Council of Badajoz, 06071 Badajoz, Spain; ajimenez.atm@dip-badajoz.es
*   Correspondence: schezrio@unex.es; Tel.: +34-924-289-300; Fax: +34-924-301-212

Academic Editor: Gonzalo Pajares Martinsanz
Received: 19 April 2016; Accepted: 19 September 2016; Published: 22 September 2016

Abstract: Medium-cost devices equipped with sensors are being developed to get 3D measurements. Some allow for generating geometric models and point clouds. Nevertheless, the accuracy of these measurements should be evaluated, taking into account the requirements of the Building Information Model (BIM). This paper analyzes the uncertainty in outdoor/indoor three-dimensional coordinate measures and point clouds (using Spherical Accuracy Standard (SAS) methods) for Eyes Map, a medium-cost tablet manufactured by e-Capture Research & Development Company, Mérida, Spain. To achieve it, in outdoor tests, by means of this device, the coordinates of targets were measured from 1 to 6 m and cloud points were obtained. Subsequently, these were compared to the coordinates of the same targets measured by a Total Station. The Euclidean average distance error was 0.005–0.027 m for measurements by Photogrammetry and 0.013–0.021 m for the point clouds. All of them satisfy the tolerance for point cloud acquisition (0.051 m) according to the BIM Guide for 3D Imaging (General Services Administration); similar results are obtained in the indoor tests, with values of 0.022 m. In this paper, we establish the optimal distances for the observations in both, Photogrammetry and 3D Photomodeling modes (outdoor) and point out some working conditions to avoid in indoor environments. Finally, the authors discuss some recommendations for improving the performance and working methods of the device.

Keywords: photogrammetry; point clouds; uncertainty; constrained least squares adjustment; middle-cost device

1. Introduction

The three-dimensional modeling of an object begins with the required data acquisition process for the reconstruction of its geometry and ends with the formation of a virtual 3D model that can be viewed interactively on a computer [1]. The information provided by the display of these models makes its application possible for different uses [2], such as the inspection of elements, navigation, the identification of objects and animation, making them particularly useful in applications such as artificial intelligence [3], criminology [4], forestry applications [5,6], the study of natural disasters [7,8], the analysis of structural deformation [9,10], geomorphology [11,12] or cultural heritage conservation [13,14].

In particular, the generation of point clouds and 3D models has important applications, especially in Building Information Modeling (BIM). This digital representation of the physical and functional characteristics of the buildings serves as an information repository for the processes of design and construction, encouraging the use of 3D visualizations [15]. In the future, devices could include different types of sensors to capture all kind of information for BIM applications. In addition, important technological advances in automated data acquisition has led to the production of more specific models tailored to Historic Building Information Modeling (HBIM) for the preservation of historical or artistic heritage [16,17].

In recent years, different techniques have been developed to acquire data [18]. On the one hand, there are active measurement techniques, carrying out modeling based on scans (range-based modeling), which uses instruments equipped with sensors that emit a light with a structure defined and known by another sensor that has to capture it [19]. On the other hand, there are passive measurement techniques, with modeling based on images (image-based modeling), which use optical or optical-electronic capture systems for extracting geometric information in the construction of 3D models [19]. The former uses different types of laser scanners, while the latter employs photogrammetric or simple conventional cameras. In each case, specific software for data processing is used.

One of the most important geometric aspects is the verification of the accuracy and reliability of measurements with which data are acquired and the resulting 3D models are obtained, since, according to the tolerances and maximum permissible errors required for the use of certain models, for example in BIM working environment, the final accuracy and reliability obtained with a specific device will determine its suitability for certain works [20]. Many studies have carried out such analysis for active measurement techniques [21–23] as in the case of passive measurement techniques [4,24,25]. These are deduced in the first case for objects of medium format, with the use of handheld laser scanners, where an accuracy up to 0.1 mm can be achieved [26]; in the second case, using techniques of automated digital photogrammetry, precision is of the order of about 5 mm [27], but with the advantage of a smaller economic cost.

There are instruments equipped with a low-cost sensor on the market: the David laser scanner [28], Microsoft Kinetic v1 and v2 sensors, and RGB-D cameras. These cameras are easy to manage and they are being used for applications that require a precision of about 5 mm to a measured distance of 2 m [29]. There are also middle-cost devices, based on structured light technology, such as the DPI-8 Handheld Scanner (from DotProduct LLC, Boston, MA, USA) and the FARO Freestyle^{3D} Scanner (FARO, Lake Mary, FL, USA).

Nowadays, there are new projects that are trying to enter the market using instruments based on a smartphone or a tablet including a range imaging camera and special vision sensor, which are user-friendly, affordable and offer accuracy for a wide range of applications. These include Google's Tango project from 2014, the Structure Sensor from Occipital from 2015 and EyesMap (EM) carried by e-Capture Research and Development from 2014.

Nonetheless, one of the main problems encountered when performing 3D modeling is to determine the accuracies obtained with these devices, especially when taking into account the rate of information uptake and the intended product. Normally, the two products we are trying to obtain are geometric models and 3D point clouds. The first is used to describe the shape of an object, by means of an analytical, mathematical and abstract model. The second produces very dense and elaborate coordinate data points for the surfaces of a physical object [30,31]. For this reason, one objective of this paper is to perform an analysis of the accuracy of the EM, in two modes of data capture: (1) Photogrammetry to get 3D point coordinates; and (2) Photomodeling to get 3D point cloud and the color of the object observed.

This accuracy was evaluated by comparison with the EM measurements and the data acquired by a Total Station. On the other hand, operator error was estimated by comparison with the coordinates of symmetrical target centers measured by EM and by a Scanstation. Additionally, to investigate the feasibility of coordinates, measurements and point cloud acquisition from a BIM perspective, further evaluation was performed in reference to the guidelines of the GSA for BIM Guide for 3D Imaging [32].

2. Materials and Methods

This study was conducted with an EM tablet from e-capture Research & Development Company. It has dimensions of $303 \times 194 \times 56$ mm^3, a weight of 1.9 kg and a screen of 11.6 inches. The device has a processor Intel Core i7, 16 gigabytes of RAM and runs on the Windows 8 operating system.

It has an Inertial Measurement Unit and a GNSS system, which allow for measuring speed, orientation and gravitational forces, as well as the positioning of the instrument in real time. To capture the three-dimensional information, on the back of the tablet (Figure 1), there is a depth sensor and two cameras with a focal length of 2.8 mm and a 13 megapixel resolution, that form a base line of 230 mm, with a field of view up to 67°.

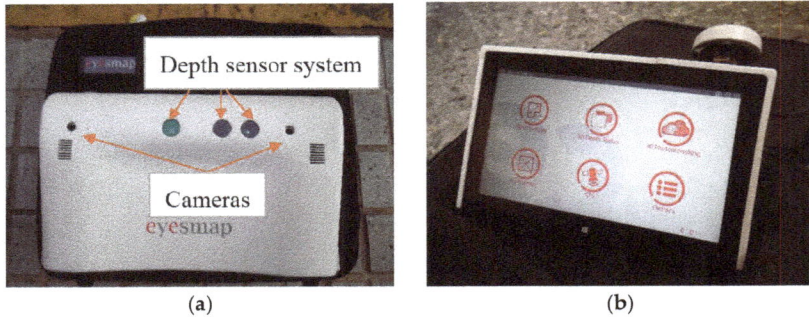

(a) (b)

Figure 1. EyesMap (EM): (**a**) back; and (**b**) front.

The beta version of EM costs around €9500. The basic principle of operation is based on photogrammetry techniques, which reconstruct a scene in real time. The precision indicated by the manufacturer (Table 1) for both measurement modes are:

Table 1. EyesMap (EM) precision specified by the manufacturer.

Range	Accuracy STD [1]	Accuracy STD Optimized Scale
3 m	3 mm	2.6 mm
15 m	15 mm	11 mm
30 m	30 mm	23 mm

[1] Standard deviation (STD).

Precisely in order to achieve the precisions expressed in the previous table, the recommendations of the manufacturer for the Photogrammetry measurement are: (1) take at least 2 pictures; (2) 80% overlap/2 pictures; and (3) capture in parallel or convergent. In the case of measurement by 3D Photomodeling, the same recommendations apply, but take at least five pictures instead of two. EM uses a computer vision approach based on general method of Photogrammetry [33].

In this sense, obtaining coordinates (X_P, Y_P, Z_P), is computed by Digital Image Correlation (DIC). In this way, 3D cloud points are achieved to a very high density from the surface of the studied object, moreover, storing color information (RGB). The calculation process of the coordinates of the points that compose the cloud, from a pair of oriented pictures is carried out by the method of triangulation [34].

The continuous evolution of algorithms that perform DIC has been reaching very high levels of precision and automation. Currently, the most effective are Structure from Motion (SFM) and the algorithms of Reconstruction in 3D in high density named Digital Multi-View 3D Reconstruction (DMVR) which produce 3D models of high precision and photorealistic quality from a collection of disorganized pictures of a scene or object, taken from different points of view [35].

2.1. EM Workflow

The processes of calibration and orientation of cameras are implemented in the EM software. The orientation of pictures can be done in three ways: (1) automatic orientation, matching homologous points that the system finds in both pictures; (2) manual orientation, in which the user chooses at least 9 points in common in both pictures; and (3) automatic orientation by means of targets, which require the existence of at least 9 asymmetrical targets in common. The latter one offers major precision and requires a major processing time. The information obtained can also be viewed in real dimension by means of the target named the Stereo target. EM offers the following options: Photogrammetry, 3D Photomodeling, 3D Modeling with Depth Sensor and Orthophoto. Photogrammetry allows for measuring coordinates, distances and areas between points, as well as exporting its coordinates in different formats (*.txt and *.dxf) so other computer aided design programs can be used. 3D Photomodeling and 3D Modeling with Depth Sensor allow 3D point clouds with XYZ and color information (PLY and RGB formats respectively), from an object. However, modeling with the support of the depth sensor is restricted for indoor work, offering less precise results than the 3D Photomodeling. The last gives an orthophotograph of the work area.

In Photogrammetry (Figure 2a), pictures can be either captured or loaded. Secondly, pictures have to be managed and the desired pictures selected. Thirdly, we can choose: (1) automatic orientation; (2) manual orientation; or (3) automatic target orientation, in order to achieve the relative orientation of the pictures. In this regard, an automatic scale is made by means of automatic target orientation and the Stereo target is used. After this, the following measurements can be obtained: (1) coordinates of points; (2) distances; or (3) areas. Finally, the geometric model is obtained.

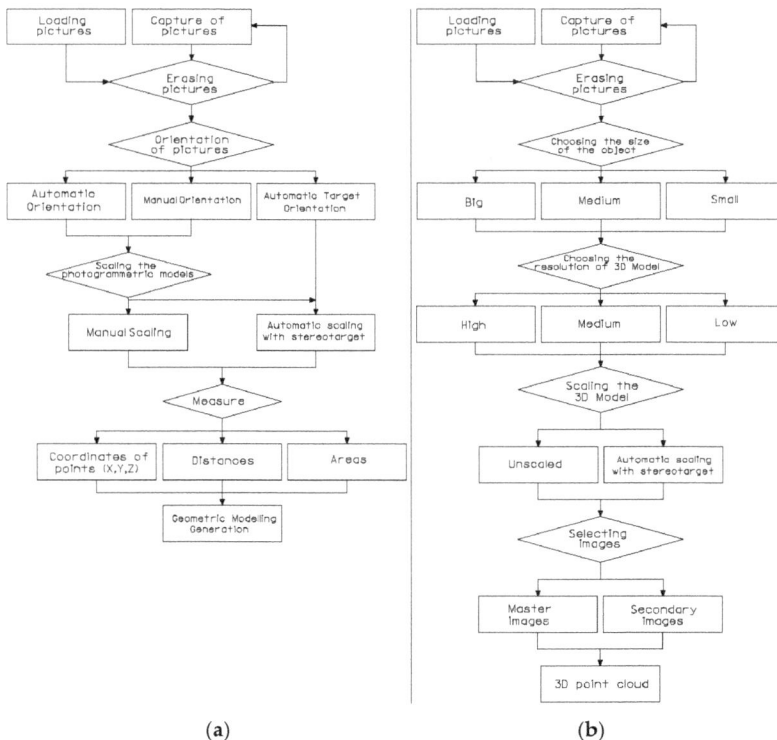

(a)   (b)

Figure 2. Workflow of EM measurement: (**a**) Photogrammetry; and (**b**) 3D Photomodeling.

In 3D Photomodeling (Figure 2b), pictures are managed in the same way as Photogrammetry. Secondly, the object to be measured according to its size is selected: small if dimensions are less than one meter, medium-sized if the dimensions are below 10 m and large for all other dimensions.

Consequently, high, medium or low resolution must be selected. The final model will be scaled or unscaled by means of the Stereo target. After this, the master picture can be selected.

In each of these four options, different working procedures are followed, depending on capture methodology, shooting planning, and the size and characteristics of the object to measure. Figure 2 shows the two options that were used in this study.

2.2. Data Acquisition Systems

This work is going to determine the instrumental errors of EM for two of the measurement options available: (1) Photogrammetry; and (2) 3D Photomodeling. To achieve it, we have resorted to two other, more precise, measurement instruments [28,31,35]. The Geomax Zoom 80 (GeoMax AG, Widnau, Switzerland) high precision Total Station, with a standard deviation of 2" (0.6 mgon) for the angular measures and 2 mm ± 2 ppm for the distance measurements (Figure 3a), and the Scanstation Leica P30 (Leica Geosystems AG, Heerbrug, Switzerland), with a standard deviation in the 3D position of 3 mm (Figure 3b).

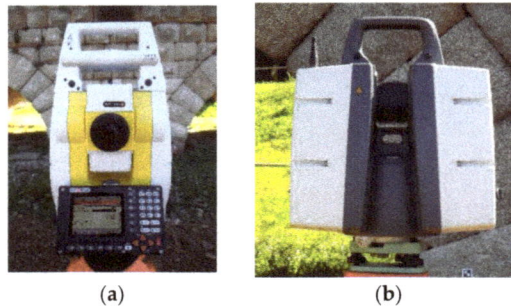

(a)   (b)

Figure 3. Used equipment: (a) Geomax Zoom 80 high precision Total Station; and (b) Leica ScanStation P30.

Regarding Photogrammetry, the coordinates of the center of the symmetrical targets (Figure 4) were measured by EM, on a canvas 1, 2, 3, 4, 5 and 6 m away. Subsequently, these measurements and the measurements obtained by means of the high precision Total Station were compared.

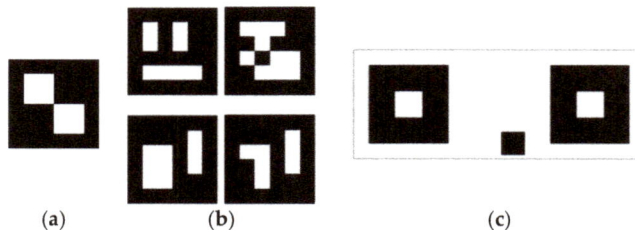

(a)   (b)   (c)

Figure 4. Targets provided by EM: (a) symmetric target; (b) asymmetric targets; and (c) stereo target.

Symmetrical targets were used with asymmetric targets and the Stereo target. The asymmetric targets served for the automatic orientation of the stereoscopic pairs, because this is the most accurate way according to the manufacturer. The Stereo target was also used to scale the obtained measurements.

Regarding the measurement by 3D Photomodeling, high-resolution point clouds were achieved by EM from 1–6 m to the canvas. Subsequently, the centers of symmetrical targets were measured from the point clouds by means of CloudCompareV2 and they were compared with the coordinates obtained by the high precision Total Station. In any case, no point of the clouds obtained by EM coincides exactly with the center of a target and it is necessary to locate and measure the closest point to this center (not the real center) using CloudCompareV2. On the other hand, only the coordinates of the targets that could be correctly identified were measured.

2.3. Data Processing

The coordinates measured by EM (x, y, z); and those obtained by the Total Station and the Scan station (X, Y, Z) are geo referenced on different coordinate systems. To be able to compare them, the coordinates obtained by EM were transformed to the coordinate system provided by the Total Station. The transformation that was used was the so-called Helmert or 7 parameters. The three steps of this transformation are: (1) three rotation angles (Ω, Φ, K); (2) three translations (Tx, Ty, Tz); and (3) a change of scale (λ), which except for the last step were calculated using the EM coordinates system. Both systems of coordinates were parallel. Through the translations, both systems would have the same origin of coordinates. Finally, the scale factors of both systems of coordinates have the same measurement units. Nonetheless, the application of the scale factor may alter the measurements [36], which was not applied for this reason.

$$
\begin{bmatrix} X \\ Y \\ Z \end{bmatrix} = \begin{bmatrix} a_{11} & a_{12} & a_{13} \\ a_{21} & a_{22} & a_{23} \\ a_{31} & a_{32} & a_{33} \end{bmatrix} \begin{bmatrix} x \\ y \\ z \end{bmatrix} + \begin{bmatrix} T_X \\ T_Y \\ T_Z \end{bmatrix} \tag{1}
$$

where:

$$
\begin{array}{lll}
a_{11} = \cos\Phi\cos K & a_{12} = -\cos\Phi\sin K & a_{13} = \sin\Phi \\
a_{21} = \cos\Omega\sin K + \sin\Omega\sin\Phi\cos K & a_{22} = \cos\Omega\cos K - \sin\Omega\sin\Phi\sin K & a_{23} = -\sin\Omega\cos\Phi \\
a_{31} = \sin\Omega\sin K & a_{32} = \sin\Omega\cos K + \cos\Omega\sin\Phi\sin K & a_{33} = \cos\Omega\cos\Phi
\end{array} \tag{2}
$$

The equations were linearized for a point P by means of the development in Taylor series to the first term:

$$
X_P = (X_P)_0 + \left(\frac{\partial X_P}{\partial\Omega}\right)_0 d\Omega + \left(\frac{\partial X_P}{\partial\Phi}\right)_0 d\Phi + \left(\frac{\partial X_P}{\partial K}\right)_0 dK + \left(\frac{\partial X_P}{\partial T_X}\right)_0 + dT_X + \left(\frac{\partial X_P}{\partial T_Y}\right)_0 dT_Y + \left(\frac{\partial X_P}{\partial T_Z}\right)_0 dT_Z \tag{3}
$$

$$
Y_P = (Y_P)_0 + \left(\frac{\partial Y_P}{\partial\Omega}\right)_0 d\Omega + \left(\frac{\partial Y_P}{\partial\Phi}\right)_0 d\Phi + \left(\frac{\partial Y_P}{\partial K}\right)_0 dK + \left(\frac{\partial Y_P}{\partial T_X}\right)_0 + dT_X + \left(\frac{\partial Y_P}{\partial T_Y}\right)_0 dT_Y + \left(\frac{\partial Y_P}{\partial T_Z}\right)_0 dT_Z \tag{4}
$$

$$
Z_P = (Z_P)_0 + \left(\frac{\partial Z_P}{\partial\Omega}\right)_0 d\Omega + \left(\frac{\partial Z_P}{\partial\Phi}\right)_0 d\Phi + \left(\frac{\partial Z_P}{\partial K}\right)_0 dK + \left(\frac{\partial Z_P}{\partial T_X}\right)_0 + dT_X + \left(\frac{\partial Z_P}{\partial T_Y}\right)_0 dT_Y + \left(\frac{\partial Z_P}{\partial T_Z}\right)_0 dT_Z \tag{5}
$$

On the basis of the linearized equations of general expression and knowing the coordinates in both systems of at least two points, the following equations were formed:

$$
r_{11}^n d\Omega + r_{12}^n d\Phi + r_{13}^n dK + r_{14}^n dT_X + r_{15}^n dT_Y + r_{16}^n dT_Z = X_n - (X_n)_0 \tag{6}
$$

$$
r_{21}^n d\Omega + r_{22}^n d\Phi + r_{23}^n dK + r_{24}^n dT_X + r_{25}^n dT_Y + r_{26}^n dT_Z = Y_n - (Y_n)_0 \tag{7}
$$

$$
r_{31}^n d\Omega + r_{32}^n d\Phi + r_{33}^n dK + r_{34}^n dT_X + r_{35}^n dT_Y + r_{36}^n dT_Z = Z_n - (Z_n)_0 \tag{8}
$$

Expressing the system of equations in matrix form:

$$
\begin{bmatrix} r_{11}^n & r_{12}^n & r_{13}^n & r_{14}^n & r_{15}^n & r_{16}^n \\ r_{21}^n & r_{22}^n & r_{23}^n & r_{24}^n & r_{25}^n & r_{26}^n \\ r_{31}^n & r_{32}^n & r_{33}^n & r_{34}^n & r_{35}^n & r_{36}^n \end{bmatrix} \begin{bmatrix} d\Omega \\ d\Phi \\ dK \\ dT_X \\ dT_Y \\ dT_Z \end{bmatrix} - \begin{bmatrix} X_n - (X_n)_0 \\ Y_n - (Y_n)_0 \\ Z_n - (Z_n)_0 \end{bmatrix} = \begin{bmatrix} V_{X_n} \\ V_{Y_n} \\ V_{Z_n} \end{bmatrix} \tag{9}
$$

Applying the adjustment by least squares, the system of equations is solved and 6 transformation parameters were obtained (Ω, Φ, K, Tx, Ty, and Tz). Nevertheless, half of the coordinates of the center of the symmetrical measured targets were used. These were called Transformation Points. Subsequently, with the transformation parameters obtained, the other half of the coordinates of the center of symmetrical targets measured were transformed from the system of coordinates of EM to the system of coordinates of Total Station. The resulting Validation Points have two sets of coordinates in the coordinate system established by the Total Station: (1) coordinates transformed to the Total Station coordinate system; from the measured performed by EM; and (2) coordinates of reference directly measured by Total Station.

2.4. Uncertainty Assessment

The measurements were made at the Roman Bridge in Merida (Spain), on a canvas of approximately 6×5 m^2 (Figures 5 and 6). This bridge, being of granite, presents an optimal texture for automatic correlation of images. EM was evaluated according to how correctly it measured elements placed at different depth levels.

Figure 5. Used targets.

(a)   (b)

Figure 6. Data capture using EM: (**a**) front view; and (**b**) back view.

The metric quality of measurements obtained by EM was evaluated using the method proposed by Hong et al. [31]. The three-dimensional coordinate measurements and point clouds obtained by EM (Figure 7) were compared to a set of Total Station point measurements used as reference points. In the mapping accuracy assessment, comparisons were based on identifiable target centers. These targets were distributed across the canvas. The accuracy assessment was based on the well-distributed and clearly identifiable point targets. A number of reference points were measured for each test side. In addition, using as a reference the tolerances established in the guidelines of the GSA for BIM Guide for 3D Imaging [32], the viability and acceptability of this measurement device for BIM generation was determined. According to [32], tolerance is the dimensional deviation allowed as error from the true value in the specified coordinate frame. The true value is a measurement obtained by other means.

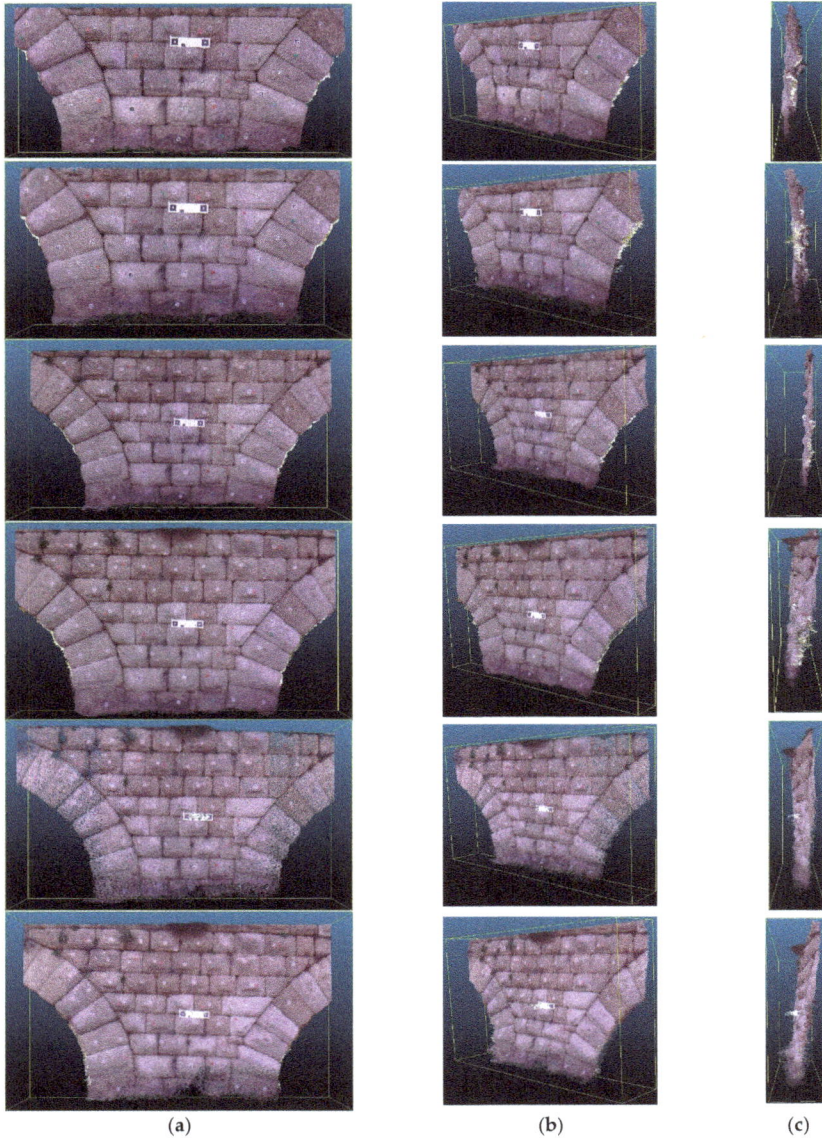

Figure 7. 3D point clouds obtained by EM for 1, 2, 3, 4, 5 and 6 m, from up to down: (**a**) front view; (**b**) middle-side view; and (**c**) right side view.

Firstly, the precision of the measurements made by EM by Photogrammetry and 3D Photomodeling are evaluated through the Euclidean average distance error (δ_{avg}).

$$\delta_{avg} = \frac{1}{n}\sum_{i=1}^{n}|Ra_i - T - b_i| \tag{10}$$

where a_i corresponding to the measurement is carried out by EM for the i-th check point in one case by Photogrammetry and in the other case by 3D Photomodeling and b_i is the measurement made for the same point by Total Station. In addition, the rotation and translation parameters of the 3D Helmert transformation are R and T, respectively. Note that scale was not considered in this transformation [37].

Secondly, the corresponding average error is calculated, together with the error vectors in the x, y and z directions. The Root Mean Square Error (RMSE) is then calculated to assess the quality of the points captured by EM and measured by means of photogrammetry and 3D Photomodeling.

$$\text{RMSE} = \sqrt{\frac{1}{n}\sum_{i=1}^{n}\left(a_i^t - b_i\right)^2} \tag{11}$$

where a_i^t shows the point transformed to the coordinates of the Total Station.

Thirdly, the quality of the points measured by EM is also assessed by calculating the Spherical Accuracy Standard (SAS). The SAS, which represents the spherical radius of a 90% probability sphere [38], is defined as

$$\text{SAS} = 2.5 \times 0.3333 \times \left(\text{RMSE}_x + \text{RMSE}_y + \text{RMSE}_Z\right) \tag{12}$$

This represents a positional accuracy of the coordinates obtained by Photogrammetry and the point cloud obtained by 3D Photomodeling with a 90% confidence level. The calculation of errors was repeated for the measurements carried out by EM from 1–6 m to the object measured by Photogrammetry.

3. Results

Results were tabulated (Tables 2 and 3) and shown in graphs. Different points were chosen in each case (3D Photomodeling and Photogrammetry modes) depending on the correct identification of target centers by the operator.

Table 2. Accuracy assessment result for photogrammetry data measured from 3 m to measured object (unit: mm).

Point ID	Error Vector X	Error Vector Y	Error Vector Z	Error
7	3	0	0	3
9	−10	0	−3	10
11	−9	0	−2	9
13	15	2	4	16
16	−6	−4	−2	7
18	−14	3	−5	15
20	−7	3	−2	8
22	−13	2	−3	13
25	−15	1	−1	15
27	12	−3	−4	13
29	11	2	3	12
31	12	1	3	12
34	12	1	3	12
Average error				11
RMSE	11	2	3	12
SAS				13

Table 3. Accuracy assessment result for 3D Photomodeling data measured from 3 m to measured object (unit: mm).

Point ID	Error Vector X	Error Vector Y	Error Vector Z	Error
3	12	−1	5	13
7	31	−2	−5	31
9	18	−2	5	19
11	−2	5	6	8
16	−4	12	8	15
18	16	0	0	16
20	16	0	0	16
22	0	8	2	8
25	0	5	1	5
27	19	2	−3	20
29	−5	5	0	7
32	−69	5	6	69
34	−11	7	−9	16
36	−5	7	1	8
Average error				18
RMSE	23	5	5	24
SAS				27

Similarly, there are estimates for observation distances from 1 to 6 m, obtaining the general results shown in Figure 8.

The value of Average Error, RMSE, SAS and STD (Figure 8) varied depending on the distance of separation between the object to be measured and the position from which we perform data capture by means of EM.

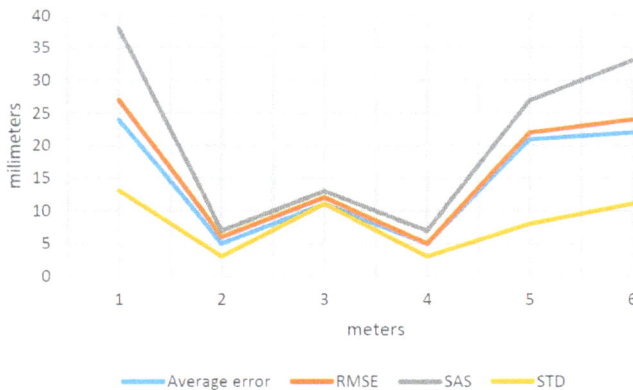

Figure 8. Average error, RMSE, SAS and STD for 1–6 m.

Nonetheless, as shown in Figure 8, error does not increase progressively as the separation distance increases. In fact, the optimum separation distances are 2, 3 and 4 m and not 1 m, as could be supposed. At 1, 5 and 6 m, the errors increase considerably.

The results obtained demonstrate that geometric models from between 2 and 4 m of distance to the object measured, satisfy the requirements of the GSA for BIM Guide for 3D Imaging [32] (Section 2.3. types of deliverables from 3D data) for projects in urban design, architectural design, room space measurement, historic documentation, renovation (level 2) and above ceiling condition capture. Subsequently, the quality of the point clouds obtained by EM by Photomodeling was evaluated. The point clouds were obtained from 1–6 m to the measured object. However, it was not possible to obtain errors for 5 and 6 m, since the low density of the mesh does not allow for correctly identifying the centers of the symmetrical targets. As a result, it was impossible to measure the coordinates of these targets.

As before, the measurements carried out by EM and measurements made by 3D Photomodeling are evaluated through the Euclidean average distance error (δ_{avg}) (see Equation (10)).

Similarly, there are estimates for distances of observation from 1 to 4 m, producing the general results that are shown in Figure 9.

The value of Average Error, RMSE, SAS and STD (Figure 9) varies depending on the distance of separation between the object to be measured and the position from which we perform data capture by EM.

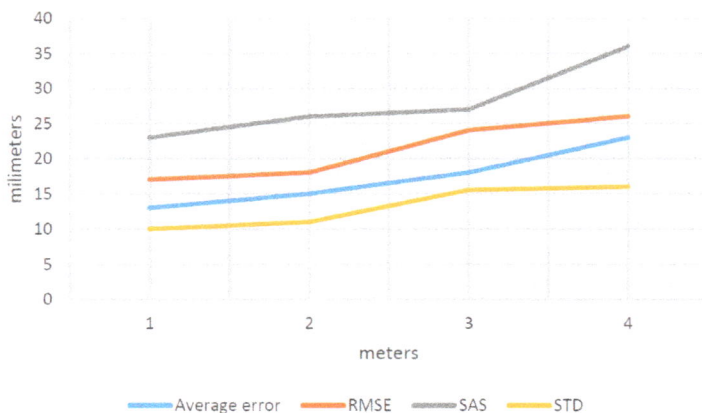

Figure 9. Average error, RMSE, SAS and STD for 1, 2, 3 and 4 m.

As shown in Figure 9, error increases in proportion to the increase in separation distance from the object being measured. Therefore, the most appropriate distance for taking measurements is 1 m.

These errors show that point clouds between 1 and 4 m of distance from the measured object satisfy the requirements of the GSA for BIM Guide for 3D Imaging [32] for level 1 projects, urban design and historic documentation.

Nonetheless, errors for measurements obtained by both Photogrammetry and 3D Photomodeling are influenced by the operator error. This error is produced by the visual acuity of the operator, when the operator identifies a dot that appears in each picture. The identification of these types of point is done for different purposes, such as the adjustment of photographic pairs and the generation of geometric models. The estimate of this error allows evaluation of their influence on the previously obtained errors. To estimate error, the centers of symmetrical targets were identified at the 3D point clouds achieved by the Scanstation (only considering targets with a point measured close to their centers), the coordinates are measured and these are compared with the coordinates measured by Total Station, since these data are correct and associated errors are known.

The differences between coordinates are used to calculate the error for the vectors x, y, z. Error for each target is measured. In this case, we use the average distance of separation from the measured object, 3 m, in order to determine the standard deviation for the point cloud (Table 4). In this manner, the standard deviation of the measurements for the targets STD_T is equal to 11 mm.

Table 4. Estimation of the standard deviation of the measurements for the targets. Distance: 3 m (unit: mm).

Point ID	Error Vector X	Error Vector Y	Error Vector Z	Error
4	13	17	−4	25
5	45	−3	−16	48
7	16	−4	−11	19
8	6	−5	−3	9
11	12	2	−35	38
12	25	4	−5	26
13	41	6	−2	41
18	25	17	−1	30
20	39	20	5	44
21	16	26	−6	31
24	−26	25	2	36
26	19	24	5	31
28	9	47	2	48
29	6	37	7	38
30	9	39	2	40
31	5	38	1	38
STD_T				11

In addition, STD_T is related to: (1) the standard deviation for the Scanner Station $STD_{SC} = \pm3$ mm in X, Y, Z coordinates; (2) the standard deviation for the Total Station $STD_{ST} = 2$ mm ± 2 ppm and 2″ (0.6 mgon) also supplied by the manufacturer; and (3) the standard deviation of the operator STD_{OP} when the operator measure the targets:

$$STD_T = \sqrt{STD_{SC}^2 + STD_{ST}^2 + STD_{OP}^2} \tag{13}$$

The estimation of the error committed by the operator in the identification of the targets, in this case, is equal to 10 mm (Table 5). Likewise, if we take into account the standard deviation for measurements by Photogrammetry STD_{PH} and Photomodelling STD_{CP} (Figures 8 and 9) and STD_{OP} estimated previously, it was observed (Table 5) that there is a huge influence for this error on the measures carried out.

Table 5. Relation between errors of measurements by Photogrammetry, 3D Photomodeling and the estimated error of the operator (unit: mm).

Distance (Meters)	STD$_{PH}$	STD$_{CP}$	STD$_{OP}$	$\dfrac{STD_{OP}}{STD_{PH}}$	$\dfrac{STD_{OP}}{STD_{CP}}$
3	11	16	10	91%	62%

In this respect, the estimated error of the operator is roughly 91% of the total error measured by Photogrammetry and 62% when we measure with 3D Photomodeling. Note that the color of targets should be in black and white, since when we carried out tests with red and green targets, the error estimate for the operator was even higher.

4. 3D Modeling of Indoor Environments with EM

The instrument under study (EM) allows obtaining 3D models inside buildings and structures (indoor). For this, the manufacturer recommends using the option of working with the depth sensor system of the instrument (Figure 1), with which we can create a complete and scaled point cloud of an indoor environment in real time, with an object to device distance less than 4 m.

In order to perform the analysis of the accuracy of the EM in this mode of data capture, we have designed two experiments: the first, which was conducted inside a historic building that presents an optimal texture for the automatic correlation of images (Figure 10), and the second, which has the purpose of checking the operation of the equipment in unfavorable working conditions, has been carried out in premises where we have placed a metal aluminum structure, with a smooth, bright and white surface placed in front of a smooth blue wall and separated at a distance of 0.5 m (Figure 11).

(a)   (b)

Figure 10. Walls built with granite ashlars (**a**,**b**). Interior of Santa María Church, Guareña (Spain).

Figure 11. Aluminum structure used in the second experiment placed in front of the blue wall.

Table 6 and the models of Figure 12 show the results obtained in the first experiment.

Table 6. First experiment. Results obtained with 17 control points (unit: mm).

	Error Vector X	Error Vector Y	Error Vector Z	Error
Average error				22
RMSE	16	11	11	22
SAS (90% probability)				32

However, the tests conducted in the second experiment have not been successful because the resulting models are not acceptable (Figure 13) with these working conditions.

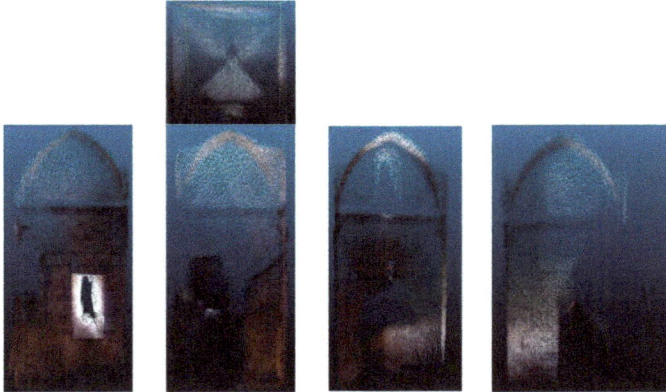

Figure 12. Point clouds obtained in the first experiment: different sections and ceiling.

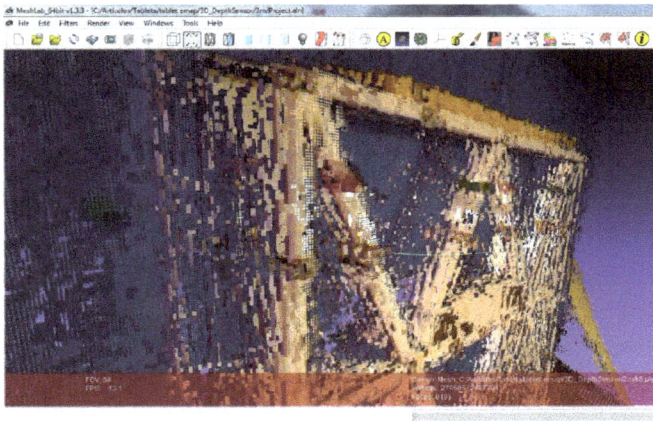

Figure 13. Erroneous point cloud obtained in the second experiment.

5. Conclusions

The tests show that precisions indicated by the EM manufacturer are greater than those obtained. Likewise, errors could not be quantified for measurements exceeding four meters from the object to be measured, as it was impossible to identify the center of symmetrical targets.

Errors vary in the distance of separation when capturing data by means of EM, a key factor in the precision of measurements. Error obtained following GSA requirements for the BIM Guide for 3D Imaging [32] shows that measurements by Photogrammetry are suitable for urban design projects, room space measurement, historical documentation, renovation and above ceiling condition. The measurements obtained by 3D Photomodeling (outdoor) and 3D Modeling with Depth Sensor (indoor) are conducive to level 1 projects for urban design and historical documentation.

Nonetheless, to reduce this error, an algorithm within the software for automatic recognition of the center of symmetrical targets or singular homologous points that serves to take some measurements is proposed. In this way, the estimated error produced by the operator would be minimized.

In addition, an error report that comments on the adjustment of photogrammetric models is recommended prior to obtaining the coordinates by Photogrammetry or the cloud points using 3D

Photomodeling. Thus, the user would know whether the dimension of error in the photogrammetric adjustment is acceptable when performing a particular task.

Furthermore, it would be convenient for EM to report on what parameter values were used for internal, relative and absolute orientation for each picture once the adjustment has been made. In this sense, EM should also enter the precise value of these parameters. Thus, a user can resume a working session without having to start the entire process of adjusting each picture. Users could even work with historical pictures where orientation parameters were known.

Finally, the convenient portability of EM and its calculation of error make it complementary to the Scanstation, particularly with measurements difficult to obtain by the latter device.

Acknowledgments: The authors would like to acknowledge the delegation of Leica Geosystems in Seville for its provision of technical equipment. The anonymous reviewers are kindly acknowledged for their contribution to the improvement of the paper with their valuable comments and suggestions.

Author Contributions: Alonso Sánchez Ríos conceived and designed the methodologies; Alonso Sánchez Ríos, José Manuel Naranjo Gómez, Antonio Jiménez Serrano and Alfonso González González performed the experiments; Alonso Sánchez Ríos and José Manuel Naranjo Gómez analyzed the data; and all aforementioned authored this paper.

Conflicts of Interest: The authors declare no conflict of interest.

References

1.   Remondino, F.; El-Hakim, S. Image-based 3D modelling: A review. *Photogramm. Rec.* **2006**, *21*, 269–291. [CrossRef]

2.   Remondino, F.; El-Hakim, S.; Μπαλτσαβιάς, E.; Picard, M.; Γραμματικόπουλος, Λ. Image-Based 3D Modeling of the Erechteion, Acropolis of Athens. Available online: http://hypatia.teiath.gr/xmlui/handle/11400/11019?locale-attribute=en (accessed on 20 September 2016).

3.   Patil, H.; Kothari, A.; Bhurchandi, K. 3-D face recognition: Features, databases, algorithms and challenges. *Artif. Intell. Rev.* **2015**, *44*, 393–441. [CrossRef]

4.   Zancajo-Blazquez, S.; Gonzalez-Aguilera, D.; Gonzalez-Jorge, H.; Hernandez-Lopez, D. An automatic image-based modelling method applied to forensic infography. *PLoS ONE* **2015**, *10*, e0118719. [CrossRef] [PubMed]

5.   Liang, X.; Wang, Y.; Jaakkola, A.; Kukko, A.; Kaartinen, H.; Hyyppä, J.; Honkavaara, E.; Liu, J. Forest data collection using terrestrial image-based point clouds from a handheld camera compared to terrestrial and personal laser scanning. *IEEE Trans. Geosci. Remote Sens.* **2015**, *53*, 5117–5132. [CrossRef]

6.   Behmann, J.; Mahlein, A.K.; Paulus, S.; Kuhlmann, H.; Oerke, E.C.; Plümer, L. Calibration of hyperspectral close-range pushbroom cameras for plant phenotyping. *ISPRS J. Photogramm. Remote Sens.* **2015**, *106*, 172–182. [CrossRef]

7.   Abellán, A.; Oppikofer, T.; Jaboyedoff, M.; Rosser, N.J.; Lim, M.; Lato, M.J. Terrestrial laser scanning of rock slope instabilities. *Earth Surf. Process. Landf.* **2014**, *39*, 80–97. [CrossRef]

8.   Ghuffar, S.; Székely, B.; Roncat, A.; Pfeifer, N. Landslide displacement monitoring using 3D range flow on airborne and terrestrial lidar data. *Remote Sens.* **2013**, *5*, 2720–2745. [CrossRef]

9.   Akavci, S.S.; Tanrikulu, A.H. Static and free vibration analysis of functionally graded plates based on a new quasi-3D and 2D shear deformation theories. *Compos. Part B Eng.* **2015**, *83*, 203–215. [CrossRef]

10. Cabaleiro, M.; Riveiro, B.; Arias, P.; Caamaño, J.C. Algorithm for beam deformation modeling from lidar data. *Meas. J. Int. Meas. Confed.* **2015**, *76*, 20–31. [CrossRef]

11. Lotsari, E.; Wang, Y.; Kaartinen, H.; Jaakkola, A.; Kukko, A.; Vaaja, M.; Hyyppä, H.; Hyyppä, J.; Alho, P. Gravel transport by ice in a subarctic river from accurate laser scanning. *Geomorphology* **2015**, *246*, 113–122. [CrossRef]

12. Harpold, A.A.; Marshall, J.A.; Lyon, S.W.; Barnhart, T.B.; Fisher, B.A.; Donovan, M.; Brubaker, K.M.; Crosby, C.J.; Glenn, N.F.; Glennie, C.L.; et al. Laser vision: Lidar as a transformative tool to advance critical zone science. *Hydrol. Earth Syst. Sci.* **2015**, *19*, 2881–2897. [CrossRef]

13. Cacciari, I.; Nieri, P.; Siano, S. 3D digital microscopy for characterizing punchworks on medieval panel paintings. *J. Comput. Cult. Herit.* **2015**, *7*. [CrossRef]

14. Jaklič, A.; Erič, M.; Mihajlović, I.; Stopinšek, Ž.; Solina, F. Volumetric models from 3D point clouds: The case study of sarcophagi cargo from a 2nd/3rd century ad roman shipwreck near sutivan on island brač, Croatia. *J. Archaeol. Sci.* **2015**, *62*, 143–152. [CrossRef]

15. Johansson, M.; Roupé, M.; Bosch-Sijtsema, P. Real-time visualization of building information models (bim). *Autom. Constr.* **2015**, *54*, 69–82. [CrossRef]

16. Brilakis, I.; Fathi, H.; Rashidi, A. Progressive 3D reconstruction of infrastructure with videogrammetry. *Autom. Constr.* **2011**, *20*, 884–895. [CrossRef]

17. Murphy, M.; McGovern, E.; Pavia, S. Historic building information modelling–adding intelligence to laser and image based surveys of european classical architecture. *ISPRS J. Photogramm. Remote Sens.* **2013**, *76*, 89–102. [CrossRef]

18. Forlani, G.; Roncella, R.; Nardinocchi, C. Where is photogrammetry heading to? State of the art and trends. *Rend. Lincei* **2015**, *26*, 85–96. [CrossRef]

19. Guidi, G.; Russo, M.; Beraldin, J.-A. *Acquisizione 3D e Modellazione Poligonale*; McGraw-Hill: New York, NY, USA, 2010.

20. Remondino, F.; Spera, M.G.; Nocerino, E.; Menna, F.; Nex, F. State of the art in high density image matching. *Photogramm. Rec.* **2014**, *29*, 144–166. [CrossRef]

21. Muralikrishnan, B.; Ferrucci, M.; Sawyer, D.; Gerner, G.; Lee, V.; Blackburn, C.; Phillips, S.; Petrov, P.; Yakovlev, Y.; Astrelin, A.; et al. Volumetric performance evaluation of a laser scanner based on geometric error model. *Precis. Eng.* **2015**, *40*, 139–150. [CrossRef]

22. Pejić, M.; Ogrizović, V.; Božić, B.; Milovanović, B.; Marošan, S. A simplified procedure of metrological testing of the terrestrial laser scanners. *Measurement* **2014**, *53*, 260–269. [CrossRef]

23. Polo, M.-E.; Felicísimo, Á.M. Analysis of uncertainty and repeatability of a low-cost 3D laser scanner. *Sensors* **2012**, *12*, 9046–9054. [CrossRef] [PubMed]

24. He, F.; Habib, A.; Al-Rawabdehb, A. Planar constraints for an improved uav-image-based dense point cloud generation. *Int. Arch. Photogramm. Remote Sens. Spat. Inform. Sci.* **2015**, *40*, 269–274. [CrossRef]

25. Percoco, G.; Salmerón, A.J.S. Photogrammetric measurement of 3D freeform millimetre-sized objects with micro features: An experimental validation of the close-range camera calibration model for narrow angles of view. *Meas. Sci. Technol.* **2015**, *26*, 095203. [CrossRef]

26. Meetings: Laser Technik Journal 4/2015. Available online: http://onlinelibrary.wiley.com/doi/10.1002/latj.201590046/abstract (accessed on 20 September 2016).

27. Martínez Espejo Zaragoza, I. *Precisiones Sobre el Levantamiento 3D Integrado con Herramientas Avanzadas, Aplicado al Conocimiento y la Conservación del Patrimonio Arquitectónico*; Universitat Politècnica de València: Valencia, Spain, 2014.

28. Dupuis, J.; Paulus, S.; Behmann, J.; Plümer, L.; Kuhlmann, H. A multi-resolution approach for an automated fusion of different low-cost 3D sensors. *Sensors* **2014**, *14*, 7563–7579. [CrossRef] [PubMed]

29. Lachat, E.; Macher, H.; Landes, T.; Grussenmeyer, P. Assessment and calibration of a RGB-D camera (kinect v2 sensor) towards a potential use for close-range 3D modeling. *Remote Sens.* **2015**, *7*, 13070–13097. [CrossRef]

30. Anil, E.B.; Akinci, B.; Huber, D. Representation requirements of as-is building information models generated from laser scanned point cloud data. In Proceedings of the International Symposium on Automation and Robotics in Construction (ISARC), Seoul, Korea, 29 June–2 July 2011.

31. Hong, S.; Jung, J.; Kim, S.; Cho, H.; Lee, J.; Heo, J. Semi-automated approach to indoor mapping for 3D as-built building information modeling. *Comput. Environ. Urban Syst.* **2015**, *51*, 34–46. [CrossRef]

32. General Services Administration. *BIM Guide for 3D Imaging*; Version 1.0; U.S. General Services Administration: Washington, DC, USA, 2009; p. 53.

33. Luhmann, T.; Robson, S.; Kyle, S.; Boehm, J. *Close-Range Photogrammetry and 3D Imaging*; De Gruyter: Vienna, Austria, 2013; p. 702.

34. Hartley, R.I.; Sturm, P. Triangulation. *Comput. Vis. Image Underst.* **1997**, *68*, 146–157. [CrossRef]

35. Koutsoudis, A.; Vidmar, B.; Ioannakis, G.; Arnaoutoglou, F.; Pavlidis, G.; Chamzas, C. Multi-image 3D reconstruction data evaluation. *J. Cult. Herit.* **2014**, *15*, 73–79. [CrossRef]

36. Gordon, S.; Lichti, D.; Stewart, M.; Franke, J. Modelling point clouds for precise structural deformation measurement. *Int. Arch. Photogramm. Remote Sens.* **2004**, *35*, B5.

37. Reit,. B. The 7-parameter transformation to a horizontal geodetic datum. *Surv. Rev.* **1998**, *34*, 400–404. [CrossRef]

38. Greenwalt, C.R.; Shultz, M.E. *Principles of Error Theory and Cartographic Applications*; ACIC Technical Report No. 96; Aeronautical Chart and Information Center, U.S. Air Force: St. Louis, MO, USA, 1968; pp. 46–49.

Article

On the Use of Monopole Antennas for Determining the Effect of the Enclosure of a Power Transformer Tank in Partial Discharges Electromagnetic Propagation

Ricardo Albarracín [1,*], Jorge Alfredo Ardila-Rey [2] and Abdullahi Abubakar Mas'ud [3]

[1] Generation and Distribution Network Area. Department of Electrical Engineering. Leader in Innovation, Technology and R&D, Boslan Engineering and Consulting S.A., Madrid 28034, Spain; rasbarracin@gmail.com

[2] Department of Electrical Engineering, Universidad Técnica Federico Santa María, Santiago de Chile 8940000, Chile; jorge.ardila@usm.cl

[3] Department of Electrical and Electronic Engineering Technology, Jubail Industrial College, Road No. 6, 8244 Al Huwailat, Al Jubail 35718, Saudi Arabia; abdullahi.masud@gmail.com

* Correspondence: rasbarracin@gmail.com; Tel.; +34-653-204-691; Fax: +56-22-303-6600

Academic Editor: Gonzalo Pajares Martinsanz
Received: 8 December 2015; Accepted: 21 January 2016; Published: 25 January 2016

Abstract: A well-defined condition-monitoring for power transformers is key to implementing a correct condition-based maintenance (CBM). In this regard, partial discharges (PD) measurement and its analysis allows to carry out on-line maintenance following the standards IEC-60270 and IEC-60076. However, new PD measurements techniques, such as acoustics or electromagnetic (EM) acquisitions using ultra-high-frequency (UHF) sensors are being taken into account, IEC-62478. PD measurements with antennas and the effect of their EM propagation in power transformer tanks is an open research topic that is considered in this paper. In this sense, an empty tank model is studied as a rectangular cavity and their resonances are calculated and compared with their measurement with a network analyser. Besides, two low cost improved monopole antennas deployed inside and outside of the tank model capture background noise and PD pulses in three different test objects (Nomex, twisted pair and insulator). The average spectrum of them are compared and can be found that mainly, the antenna frequency response, the frequency content distribution depending on the PD source and the enclosure resonances modes are the main factors to be considered in PD acquisitions with these sensors. Finally, with this set-up, it is possible to measure PD activity inside the tank from outside.

Keywords: power transformer tank; VHF and UHF sensors; monopole antennas; partial discharges; condition monitoring

1. Introduction

Measurement of partial discharges using very-high frequency (VHF), 30–300 MHz, and ultra-high frequency, 300–3000 MHz, sensors has become an important tool when monitoring and assessing the state of the insulation system on gas-insulated substations (GIS) and power transformers [1,2]. The current pulses that occur as a consequence of PD activity within this electrical equipment can generate the presence of electromagnetic emissions with bandwidths ranging from MHz to several GHz [3]. All information contained in these frequency bands can be captured and stored by using wideband sensors (antennas) which also does not require galvanic contact with any terminal of the equipment under test [4–7].

Besides, one of the main drawbacks when using externally such sensors in electrical equipment with self-shielding, according with its constructive nature, is that it may modify and/or mitigate the

EM generated by PD activity inside the enclosure [8]. This happens in the case of power transformers, where part of the insulation system susceptible to the presence of PD is within this physical structure. In this regard, some studies have been addressed in order to establish the effect of shielding in the attenuation suffered by EM waves when pulses from inside are acquired from outside the transformer tank [8,9]. However, these papers have not taken into account the design of the sensor, and the resonance frequencies of the transformer tank, which vary according to the geometry and size of the structure [10], and the activity of different PD sources inside and outside the power tank.

In this paper, resonant wire antennas are justifiably selected to study the frequency behaviour of a transformer tank model when measuring PD, both from outside and inside the enclosure. With this aim, sequentially, test objects generating internal PD and surface PD inside the tank (vacuole in Nomex and twisted pair) and surface PD outside the tank (insulator) are acquired from antennas deployed inside and outside the enclosure to study their emissions in the UHF range. With post-processing and analysis of the measurements it will be possible to identify the main effects of a rectangular enclosure in PD propagation through a shielding such as a tank of a power transformer, which it contributes to a better characterization of the type of sources in this type of electrical equipment.

2. Selection of Antennas

To select accurately an antenna to measure EM emissions due to PD activity, it is necessary to know the main parameters that define its behaviour. Some of the most important are: *Directivity*, *Gain*, Bandwidth and *Scattering parameters* or *S*-parameters. Being $S_{11} = \frac{(Z_a - Z_0)}{(Z_a + Z_0)}$ the reflection coefficient, that indicates how much power is reflected from the sensor (frequency characteristic), where Z_a is the load input impedance of the antenna and Z_0 is the 50 Ω impedance from the acquisition system. And the transmission parameter S_{21}, to obtain the frequency response of an emitter-transmitter radiofrequency (RF) system [11,12].

As it is known, disc couplers are the most common RF sensors for PD measurements in GIS and they are being slightly used in research models of power transformers [13]. However, the absence in the High-Voltage (HV) laboratory of a real power transformer including these sensors in which carry out the measurements, makes as requirement the use of a model of a transformer tank. A disc sensor has much sensitive response than a monopole antenna. However, the aim of the paper in this regard is to find an economical, alternative and easy to be implemented design antenna model to test with it. For this reason, the authors use low cost antennas such as monopoles due to their easy manufacturing. These kinds of antennas have been used in research for PD measurements such as in transformers [14,15], as well as, in GIS substation models [16] and inverters feeding electrical motors [17].

According to the antenna theory and design [12,18], the length of the antenna defines its main frequency resonance. Besides, a monopole antenna is half of a dipole antenna, thus, the definition of its first resonance is depicted as the Equation (1) and its first resonance is at $\lambda/4$.

$$f_r = \frac{c}{l4} \tag{1}$$

where $\lambda = c/f_r$ and l is the length of the antenna. Being c the speed of light, 3×10^8 ms^{-1}, and f_r the resonance frequency of the antenna in Hz.

Besides, the resonances for a monopole, given from the first one ($\lambda/4$) occur for $\lambda/2$, $3\lambda/4$ and λ, according to [12].

With the aim of selecting the best configuration of a monopole, in this paper four lengths are analysed and their theoretical results are shown in Table 1. The requirements for the antenna selection were a suitable frequency response with the smallest possible length to avoid electrical hazards.

Sensors **2016**, *16*, 37–54

Table 1. Theoretical resonance frequencies for monopole antennas.

Monopole antenna	$f_r(\lambda/4)$ **(MHz)**	$f_r(\lambda/2)$ **(MHz)**	$f_r(3\lambda/4)$ **(MHz)**	$f_r(\lambda)$ **(MHz)**
1 cm	7500	15000	22500	30000
5 cm	1500	3000	4500	6000
10 cm	750	1500	2250	3000
16.5 cm	454	910	1364	1820

As described in experimental setup (Section 5), the measurement equipment used can reach up to 2.5 GHz, so the resonances that can be measured are gray as marked in Table 1. The frequency resonances for the 1 cm monopole cannot be measured, so its used is discarded. In addition, the oscilloscope only can represent the first resonance of the 5 cm monopole, thus this antenna is also rejected. With respect to the 16.5 cm monopole, although it has their frequency resonances into the analysed bandwidth, its high length can produce electric hazard due to contact with the feeding of the setup or with energized electrode of the test object. Moreover, this drawback can be increased in real power transformers with the possibility of electrical contact with their windings, so this sensor is also discarded. Thus, the length of the monopole elected is 10 cm, even more, it has its three first resonances below than 2.5 GHz band, so it could measure energy from PD with less risk of electric hazard than the 16.5 cm monopole.

An Agilent E8364B network analyser (NA) is used to measure the frequency response (S_{11} parameter) for the monopole antenna. Likewise, this kind of resonant antennas could be improved by introducing a ground plane at its terminal so that it works as dipole antenna. Thus, to evaluate the performance of the sensor, the measurements with the NA was carried out with and without a ground plane. The main advantage of a dipole antenna is to have the double gain and directivity than a monopole antenna with the same length. Thus, for a $\lambda/4$ dipole its gain is $G_{\frac{\lambda}{4}} \simeq e_{cd} 3.286$ dB and its directivity $D_{\frac{\lambda}{4}} = 5.167$ dB, only applicable for their resonance frequencies at $\lambda/4$, $\lambda/2$, $3\lambda/4$ and λ. Where e_{cd} is the radiation efficiency of the antenna, being $e_{cd} \simeq 1$ at the frequency resonances of monopoles and dipoles sensors [12,18].

In Table 2, it is shown the resonance frequency values for the 10 cm antenna with and without a ground plane together with the relative error (RE) between the theoretical value and the experimental measure. Note that a $\lambda/4$ monopole or dipole antenna has the same theoretical frequency resonances, so the comparison between the results of calculations for the monopole (dipole) can be carry out with those obtained from measurements on dipole. Besides, it can be observed that the match between the resonances calculated theoretically, Table 1, and the experimental results using a ground plane is better than those obtained without the ground plane, Table 2. In addition, when a monopole or dipole type antenna is used, sometimes, the resonant frequencies are shifted with respect to the theoretical ones [11].

In this case, the dipole antenna has a better match of the frequency resonances with the theoretical calculation made in Table 2, with an error between $[-15,+15]\%$. These results are contrasted with those obtained in [11], where the frequency response of a 10 cm monopole antenna with and without ground plane is quite similar, however the power spectrum is greater for the dipole antenna than that obtained at their analogous resonant frequencies (approximately at 750 MHz and 1500 MHz) for the monopole.

Table 2. Experimental values of the resonance frequencies of the 10 cm antenna working as dipole or monopole antenna for $\lambda/4$, $\lambda/2$, $3\lambda/4$ and λ.

		$f_r(\lambda/4)$		$f_r(\lambda/2)$		$f_r(3\lambda/4)$		$f_r(\lambda)$	
Antenna		Monopole	Dipole	Monopole	Dipole	Monopole	Dipole	Monopole	Dipole
10 cm	MHz	530	640	1060	1730	1385	2150	2137	2600
	%	29	15	29	-15	38	4	29	13

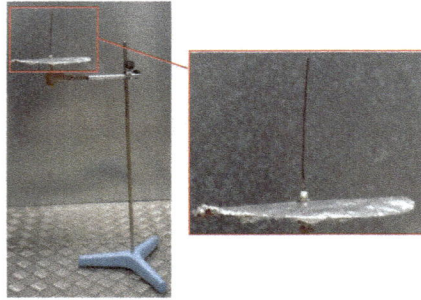

Figure 1. 10 cm monopole with ground plane.

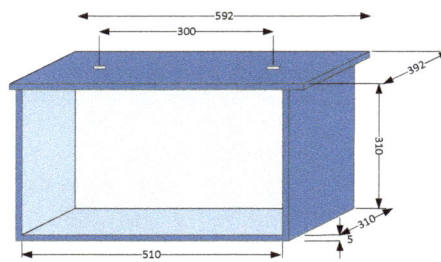

Figure 2. Dimensions of the tank, in mm.

In Table 2, RE below than 15% are gray marked. Negative values mean that the theoretical value is below than the experimental one, while a positive number is interpreted as the theoretical frequency has a value above the measure. The 10 cm monopole with ground plane has a RE below than 15% for all the resonance frequencies measured, while a RE above 29% is obtained for the same antenna without ground plane. Thus, a 10 cm monopole with ground plane dipole antenna sustained with a tripod, as in Figure 1, is selected for the measurements.

3. Power Transformer Tank

In order to have a representative model of a real enclosure for an oil-immersed power transformer, [19], a $30 \times 30 \times 50$ cm tank made of steel plates of 5 mm thickness are constructed to carry out PD measurements in RF. The material and thickness are selected according to those used in a power transformer 15 kV/420 V of 25 kVA deployed in distribution networks.

For a first approximation, this scale construction is suitable to reproduce the same electromagnetic effects that can be found in an empty transformer tank. The upper cover has two entrances separated 30 cm, which allow one, the passage of the high-voltage electrode in the tank to feed the indoor test object and the other input for an antenna deployment. These holes are 3 cm in diameter and are protected by a Teflon hollow cylinder serves to prevent electrical contact between the housing, grounded, and the test object that is placed inside the tank. Finally, a cork layer is used as a mechanical fit between the tank and the cover, just as in distribution transformer tanks. The Teflon hollow cylinder and the layer between metal and cork acts as dielectric windows for signals in RF. The model is represented in Figure 2 with the front wall erased for a better understanding.

In next subsections, it is shown the behaviour of the power transformer tank implemented considered as rectangular resonant cavity, their theoretical frequency resonances and their measurements when PD activity it is taken into account.

The Tank as a Rectangular Resonant Cavity

A transformer tank is a rectangular structure that has an electromagnetic behaviour as a resonant cavity [12]. So, when there is PD activity inside the cavity, multiple reflections create stationary waves between its conductive metal walls which generate resonance frequencies. These stationary waves are the transverse magnetic (TM) and transverse electric (TE). TM has no components of magnetic fields in the direction of propagation $H_z = 0$, but being nonzero its electric field component, E_z. While, TE wave, has no components of the electric field in the direction of propagation $E_z = 0$, but with nonzero components of magnetic field, H_z.

To designate the distribution of a stationary wave TM and TE for axis x, y, z of a resonant cavity, subindex mnp, are used. For the calculation of these modes, the following equation is used [10,20]:

$$(f_{rc})_{mnp} = \frac{1}{2\pi\sqrt{\mu\epsilon}}\sqrt{\left(\frac{m\pi}{a}\right)^2 + \left(\frac{n\pi}{b}\right)^2 + \left(\frac{p\pi}{c}\right)^2} \tag{2}$$

where f_{rc} is the resonance frequency of the cavity, $\mu = 4\pi \times 10^{-7}$ Hm/A the permeability of vacuum, $\epsilon = 8.85 \times 10^{-12}$ F/m the permittivity of vacuum and a, b, c the dimensions of the tank in m.

Equation (2) is used to calculate the frequencies in which the rectangular cavity resonates and to compared these frequencies with their measurement in next subsection.

Assuming that the maximum length of a transformer tank corresponds to the z-axis propagation, width with the x-axis and height with the y-axis, then it is possible to calculate the results of the resonance frequencies $(f_{rc})_{mnp}$, according to Equation (2), for transverse electric modes, TE$_{mnp}$, and transverse magnetic, TM$_{mnp}$. This equation is applied to the tank geometry studied for calculating their resonant frequencies. The first resonance frequency is obtained at 583 MHz.

To have a measure of these frequencies to be compare with its theoretical calculations, the transmission parameter S_{21} is measure to obtain the frequency response of the tank. For the acquisitions, the Agilent E8364B network analyser and two antennas are used, thus, one injects energy in the tank by sweeping the frequency of the NA between 500 MHz and 2500 MHz and the other acts as a receiver. Figure 3 shows the positions of the antennas deployed inside the tank through the cavities on the top cover.

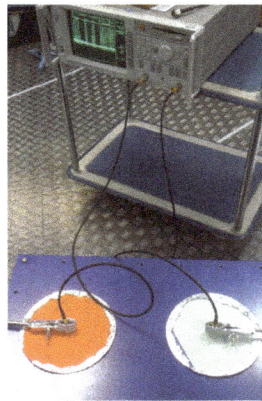

Figure 3. Two 10 cm monopole antennas with ground plane and an Agilent E8364BD network analyser for the measurement of the S_{21} parameter of the tank.

In Figure 4 the S_{21} parameter measured with two 10 cm monopoles with ground plane together with the network analyser is shown.

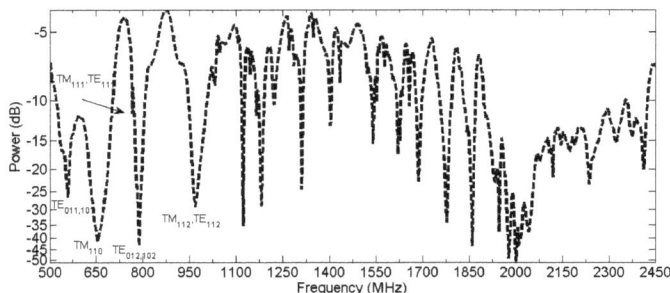

Figure 4. S_{21} parameter (frequency response of the tank) measured with two 10 cm dipoles.

The calculated frequency resonances, their modes, the resonances measured $(f_{rc})_{mnp}$ (MHz), their power (dB) and the ER between frequencies calculated and measured are shown in Table 3 up to 2000 MHz. When there is no mode or no measurement of a resonance frequency it is represented by a dash (−) in Table 3.

Table 3. Frequency resonances for a $30 \times 30 \times 50$ cm cavity in 500–2000 MHz.

$(f_{rc})_{mnp}$ (MHz) Calculated	TM_{mnp}	TE_{mnp}	$(f_{rc})_{mnp}$ (MHz) Measured	Power (dB)	RE (%)	$(f_{rc})_{mnp}$ (MHz) Calculated	TM_{mnp}	TE_{mnp}	$(f_{rc})_{mnp}$ (MHz) Measured	Power(dB)	RE (%)
583	–	011, 101	560	−26	3.89	1535	222	222	1540	−15	−0.31
707	110	–	655	−41	7.30	1561	–	024, 204	1552, 1560	−10, −6	0.57, 0.06
768	111	111	765	−11	0.34	1580	130, 310	015, 105	1578, 1590	−7, −6	0.13, −0.63
780	–	012, 102	787	−43	−0.83	1608	131, 311	131, 311	1602	−8	0.39
927	112	112	967	−29	−4.35	1614	032, 302	–	1620	−17	−0.34
1029	–	013, 103	1032	−8	−0.31	1639	124, 214	124, 214	1628, 1635	−15, −10	0.67, 0.24
1043	–	021, 201	1048	−6	−0.45	1657	115	115	1658	−11	−0.05
1117	120, 210	–	1105, 1122	−6, −35	1.10, −0.42	1675	225	225	1665	−6	0.61
1144	113	113	1145	−8	−0.11	1690	132, 312	132, 312	1688	−23	0.12
1157	121, 211	121, 211	1165	−12	−0.71	1748	–	033, 303	1778	−34	−1.71
1165	–	022, 202	1180, 1222	−29, −10	−1.25, −4.86	1802	230, 320	025, 205	–	–	–
1268	122, 212	122, 212	1270	−6	−0.16	1818	133, 313	133, 313	–	–	–
1299	–	014, 104	1285, 1310	−7, −24	1.09, −0.84	1826	231, 321	231, 321	1842	−14	−0.86
1344	–	023, 203	1348	−5	−0.27	1853	224	224	1842	−14	0.62
1392	114	114	1402	−13	−0.73	1867	–	016, 106	1860	−43	0.37
1413	220	–	–	–	–	1870	125, 215	125, 215	–	–	–
1434	123, 213	123, 213	1435	−8	−0.05	1899	232, 322	232, 322	1902	−12	−0.17
1445	221	221	–	–	–	1920	–	034, 304	–	–	–
1529	–	031, 301	1522	−10	0.43	1933	116	116	1932	−23	−0.03

Experimentally, almost all modes calculated theoretically are obtained, except at high-frequencies, where resonances are not reproduced.

The dominant resonance mode corresponds to TE_{011} and TE_{101}. The first resonance for the dominant transverse mode corresponds to TM_{110}. In this cavity, these modes are reproduced at 583 MHz and 707 MHz, TE and TM, respectively. When they are measured with the network analyser, the first TE mode is obtained at 560 MHz with a RE of 3.89%, compared with the theoretical values, and −26 dB, where the negative value indicates emitted power inside the tank. Besides, the first TM mode has its resonance frequency at 655 MHz with a RE of 7.30% and power of −41 dB, this high energy value received, is because the antenna has its first resonance at 640 MHz as it is mentioned above.

As it is shown in Figure 4, experimentally, not all modes are excited in the resonant cavity for frequencies above 2300 MHz. Assuming that the network analyser emits the same power at all frequencies, then the receiving or transmitting antenna, are not able to excite the frequencies within the tank. However, for lower frequencies, the tank is capable of resonating in the most theoretically calculated modes in Table 3.

When the tank resonance modes are measured, there are mainly two factors that affect the result. First, the modes defined by the structure. To obtain these resonances, a frequency sweep is done with the network analyser up to 2500 MHz. The second is the frequency response of the antenna used, that must be matched for all the bandwidth required. However, the hole size to accommodate the antenna on the tank and its geometry do not allow to deploy an antenna that meets these requirements. By restriction of size, and to have a cheaper antenna than a disc-coupler to do the measures, a monopole antenna is used because can also be deployed inside the empty model tank through the holes and can acquire energy up to 2250 MHz for the 10 cm in length antenna.

In the experimental measurements with the transformer tank model, almost all their own theoretical frequencies of a rectangular resonant cavity of the same dimensions are obtained, as shown in Table 3 [10]. To study only the effect that the tank has on PD propagation, the tank model unfilled of oil and without placing a magnetic core and windings therein it is used for measuring discharge with the antennas. Inside, it is expected to measure the direct wave of the discharge and the pulses reflected in the shield walls that excite their resonant frequencies. Outside, it is intended to receive power content from frequencies with the higher energy, mitigated by the enclosure, which goes through the holes in the top and the joints between the cover and the walls. These components depend on the frequency response of the test object emitting and the resonance frequencies that allows the cavity.

4. Test Objects

To study the effect of the transformer tank in the RF propagation of the PD, three types of test objects are used and measurements are carried out sequentially to ensure repeatability. First, Nomex paper is located inside the tank ensuring that the activity from internal PD occurs inside the tank. The second test object generates surface PD in a twisted pair deployed into the tank. Finally, PD surface are measured on an insulator located on the tank cover.

- Internal PD are the most harmful for the insulation system and their detection is essential to carry out proper maintenance based on the condition of the electrical equipment [21,22]. Internal discharges occur in defects, holes, into the insulating systems with low dielectric strength. The accomplishment of vacuoles in a solid insulation can be a difficult task, for example in epoxy, due to requires raising its temperature and inject air bubbles through a needle. The main drawback is that it is not possible to control the geometry of the hole and, also the needle always leaves a return path in the material. As a solution to these problems, the idea of segmenting the solid insulation in several layers is used, drilling by a needle the intermediate sheet to generate a cylindrical imperfection in the insulation. Each of these layers must be cut it with the same shape to be fitted inside the oil vessel in which it is housed, and they are glued so that when performed the hole it is aligned. Finally, when the vacuole is done, the needle leaves a burr to be removed from the material.

  The test object for internal PD is housed in a glass vessel with a steel bases, that can be connected to ground. This vessel contains mineral dielectric oil Nytro Taurus, used in high-power transformers. Inside the vessel are 11 insulating layers and above them an electrode made of steel which connects a cable to apply voltage to the sample. Insulating layers are laminated flexible Nomex, F-20.08 Triplex electrical insulation manufacturer *Royal-Diamond*, Figure 5a. This material is a sheet coated on both sides with polyester fibres. In its practical application, is inserted in windings of transformers, motors and generators subjected to high-mechanical, dielectric and thermal requirements. Each of those 11 layers has a thickness of 0.35 mm, 3 core layers are perforated with a needle to create a small cylindrical cavity diameter of 1 mm and a length of 1.05 mm. In addition, the assembly is placed in a plastic bag and the air is removed with a vacuum machine. Thus, it is to ensure that this vacuole has the dielectric constant in vacuum that is lower than that of the insulating material. Therefore, internal PD can occur in the cavity at voltage levels relatively small.

- Surface PD appear between two dielectrics, usually between the insulation system and air. Pollution or moisture, e.g., in insulators, may accelerate the process fetterless of this type of discharge and can even lead to a short-circuit between the overhead line and the tower connected to ground. As another example, when PD are produced in the insulated phase feeding a power transformer, they can cause a fault between the input line and the tank, leading to equipment out of service.

A twisted pair of copper wire type Pulse Shield SD and manufactured by Rea Magnet Wire Company is used Figure 5b. Its enamel consists of an insulating layer of resin polyamide-imide modified and an over layer of modified polyester (THEIC) trishydroxyethyl isocyanurate. The set has a thickness of 1 mm and can withstand temperatures up to 220 °C with a lifetime of 20,000 h and voltage levels until failure 5.7–11 kV.

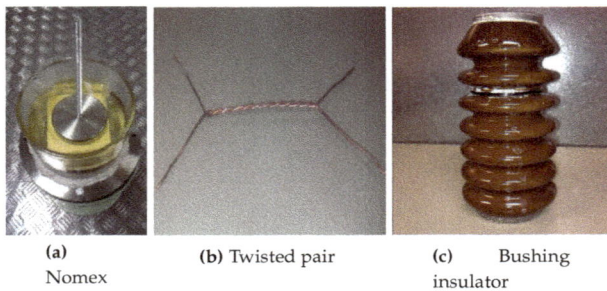

| (a) Nomex | (b) Twisted pair | (c) Bushing insulator |

Figure 5. Test objects: (**a**) Nomex 4-3-4 submerged in oil, (**b**) twisted pair of copper wire type Pulse Shield SD, 23.5 × 12 cm with a central twist of 13.5 cm and (**c**) bushing insulator with an electrode in the top, 21 × 8 cm.

Finally, it is used a bushing insulator made of porcelain commonly used in medium voltage transformers, Figure 5c. To obtain PD over its surface it is contaminated with saline and the measurements are carry out once moisture has disappeared. This kind of polluted environments are a very common situation in overhead lines in coastal areas. Some of the agents that pollute these elements are dust, bird droppings, rain, ice and, near the sea provisions, damp and saline environment. In locations with a high probability of contamination, the possible activity of this type of discharge can be critical and measures should be taken. In locations with a high-probability of contamination, the possible activity of this type of discharges can be critical and their measure should be taken into account.

5. Experimental Setup

At the very beginning, the type of discharges are previously confirmed using the commercial TechImp PD Check, its software PD Base and a High-Frequency Current Transformer (HFCT) with the indirect circuit, according to IEC-60270 standard [23]. For the measurements with the antennas, RG-223 coaxial cables 5 m long, with BNC connectors are used. The cables are connected to a Tektronix DPO 7254 oscilloscope 5 GS/s and the set-up is represented in Figure 6.

For these tests, two antennas, one inside and one outside the tank are deployed to determine the influence of shielding in PD propagation both inside and outside. As is indicated in Section 2, two antennas of 10 cm in length with a ground plane will be used to have comparable sensors inside and outside. The antenna inside the tank is located 30 cm from the test object and the second is placed outside the tank at the same distance, to have the same time-delay between pulses. When the bushing insulator is used, it is placed over the tank and equidistant to the outer antenna, which is also located at 30 cm, to be consistent with the others' experiments. Figure 7 shows the arrangement of the setup, which is the same for all experiments, of the antennas and the tank for measurements with the insulator. As it is shown, both antennas are supported by a tripod that prevents movement.

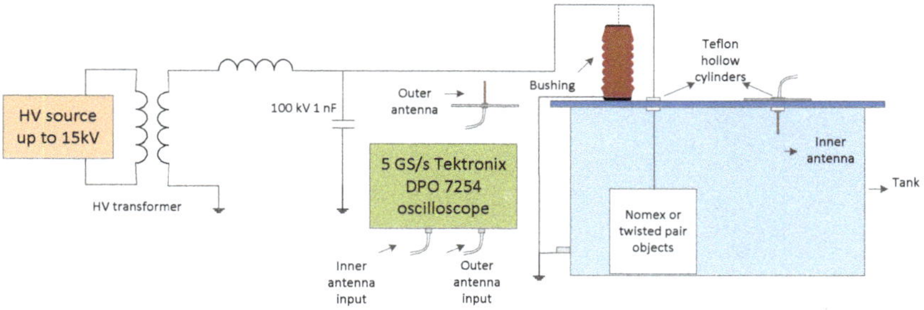

Figure 6. Experimental set-up scheme for the PD acquisitions for the three test objects.

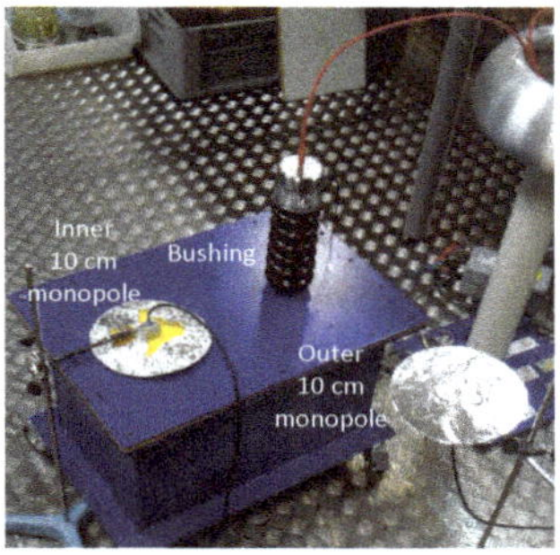

Figure 7. Layout of 10 cm monopole antennas with ground plane, inside and outside the tank, to measure surface PD from the bushing insulator at 12 kV.

6. Experimental Measurements and Results

6.1. Influence of the Tank in PD Measurements with Antennas

The measurement campaigns are made with 50 pulse acquisitions at 5 GS/s with a time window of 0.5 μs in the oscilloscope channels where the antennas are connected. For each pulse, the fast Fourier transform (FFT) was calculated using a rectangular window. Each frequency component n-th from the FFT was scaled to calculate the root-mean-square (RMS) and squared to obtain the power spectrum P_{f_n}, in V^2, of the signal at frequency f_n, being A_{f_n} the FFT amplitude, Equation (3) [8].

$$P_{f_n} = \left(\frac{A_{f_n}^2}{\sqrt{2}}\right) \tag{3}$$

In order to compare significant bands of frequency, several intervals are selected for each test object depending on the frequency components. Equation (4) is used to obtain the accumulative power in bands $\Delta P_{[f_L, f_H]}$ in V^2 in various intervals between a lower and higher frequency values, f_L and f_H, respectively.

$$\Delta P_{[f_L, f_H]} = \sum_{n=f_L}^{f_H} \left(\frac{A_{f_n}^2}{\sqrt{2}} \right) \qquad (4)$$

Equation (5) is used to calculate the signal-to-noise ratio (SNR) for the accumulative power when PD are acquired and when no voltage is applied. When SNR > 1 then the background noise power is lower than the PD power in the $[f_L, f_H]$ band.

$$SNR = \left(\frac{\Delta P_{[f_L, f_H]}^{PD}}{\Delta P_{[f_L, f_H]}^{noise}} \right) \qquad (5)$$

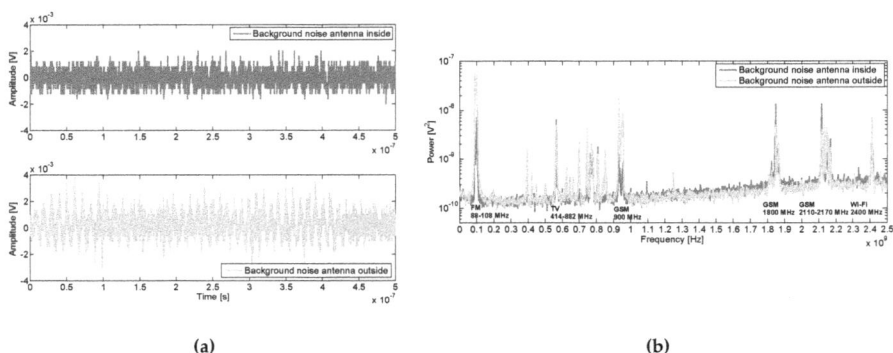

(a)   (b)

Figure 8. Background noise measured inside and outside the tank, (**a**) pulses and (**b**) spectrum with continuous sinusoidal noise from communications identified.

For the noise acquisition in the laboratory, no voltage was applied and the synchronization was performed with the AC 50 Hz line voltage of the oscilloscope. In Figure 8a a background noise pulse is represented for both inner and outer antennas. Additionally, the averaged spectrum of 50 pules of each antenna are compared in Figure 8b and the sinusoidal signals from communication systems such as:

- Modulated frequency (FM) radio
- Digital television (TV)
- Global System for Mobile communications (GSM)
- Wireless Fidelity (Wi-Fi)

are depicted. The frequency response inside the tank for FM, TV and GSM at 900 MHz are mitigated, however, the spectrum power is quite similar for both for GSM at 1800 MHz, 2110−2170 MHz and Wi-Fi sources.

Once the pulses from noise have been acquired, then the PD activity for each test object was measured. The voltage level was raised above the voltage ignition, v_i, for which PD activity start, up to set the applied voltage, v_a, Table 4. The measurement was carried out after 10 min to ensure a stable emission activity of PD and with a trigger level slightly above the noise level. The Phase-Resolved PD (PRPD) pattern for each PD source is registered as explained in the previous section and they are represented in Figure 9.

Table 4. Ignition and applied voltages for the test objects used.

Test Object	v_i (kV)	v_a (kV)
Nomex	9	13
Twisted pair	0.65	0.76
Porcelain insulator	10	12

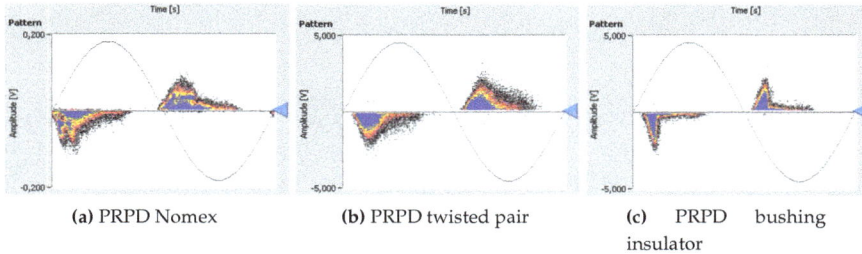

(a) PRPD Nomex   (b) PRPD twisted pair   (c) PRPD bushing insulator

Figure 9. PRPD patterns: (**a**) Internal PD in Nomex 4-3-4 submerged in oil, (**b**) surface PD in twisted pair and (**c**) surface PD in bushing insulator.

6.2. Internal PD in Nomex Inside the Tank

Figure 10a shows an internal PD pulse for Nomex measured by both antennas. Their spectrum compared to background noise are depicted in Figure 10b,c, inside and outside the tank, respectively. As might be expected, the signal for the inner antenna has more components in UHF than those measured outside the enclosure, which are mitigated due to the lost of its energy when the pulse is rebounding inside the tank [24].

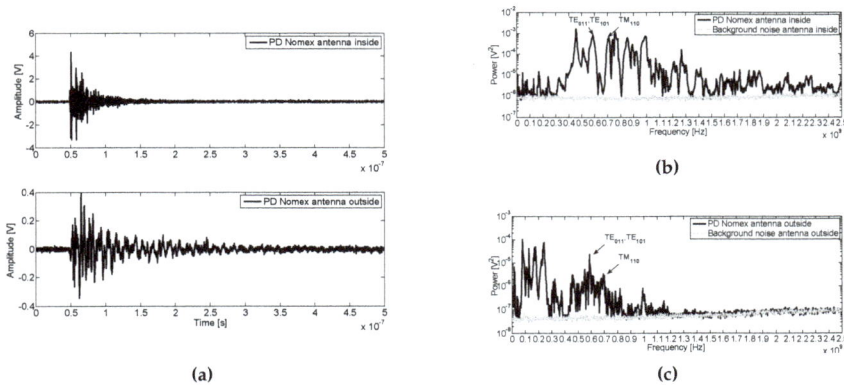

Figure 10. (**a**) Internal PD pulses in Nomex paper at 13 kV inside the enclosure acquired by the inner and outer antennas and their average power spectra compared with the noise inside (**b**) and outside (**c**).

As it is shown in Figure 10b,c, there is a power increase below the frequency of the first resonance mode TE_{011}, TE_{101} = 560 MHz, mainly at 450 MHz. All these components of the spectrum are due to the characteristics of the direct emission radiated of this type of internal PD at these frequencies [8], the frequency response of the antenna and the resonance characteristic of the enclosure.

For the power spectrum calculated for the pulses acquired with the antenna outside the tank, it can be seen that no significant power is received from the PD above 1.2 GHz, so the tank acts as a low pass filter. Furthermore, the content of power at 450 MHz is also received outside and mainly up to 400 MHz. Indeed, more power below 400 MHz is received outside the tank than inside. This can be due to the cable to ground of the enclosure which act as an antenna whose length radiate in these frequency ranges when the PD leave the cavity conductively through the ground path.

Finally, the first two resonance modes are marked, TE_{011}, $TE_{101} = 560$ MHz and $TM_{110} = 655$ MHz in Figure 10b,c verifying that they can be measured from outside the tank under these circumstances. Table 5 compares the first resonance modes for Nomex inside and outside the tank. As can be expected, in both cases, the peak power values are greater inside than outside. To compare which has the highest magnitude, the power inside is divided by the power outside (Ratio), so values greater than 1 mean that the mode has higher power inside the tank and viceversa.

Table 5. Peak values for first TE and TM modes inside and outside the tank for internal PD in Nomex.

Modes	Power (V^2)		
	Inside	Outside	Ratio
TE_{011}, $TE_{101} = 560$ MHz	8.1×10^{-4}	5.9×10^{-4}	1.4
$TM_{110} = 655$ MHz	2.4×10^{-5}	3.3×10^{-6}	7.3

Table 6 presents the accumulated power measured with the inner and outer 10 cm monopoles. In this case, three bands are analysed with their SNR: up to 300 MHz where there is the main contribution in power from the outer antenna; 300–1200 MHz, where there are the first resonance modes and; 1200–2500 MHz, where no PD radiation is received from outside. Focused on the 300–1200 MHz band, the ratio for the inner antenna is 166.5 while outside is 35.5, so it could be measured the the resonances inside the tank from outside with a high SNR. In Table 6, also the two first modes are included as example of comparison, with high values for the measure inside the tank. Up to 300 MHz, the ratio outside 214.3 is much greater than inside 2.6, so this components are not coming from inside the enclosure. From 1200–2500, the power is greater inside with a ratio of 6.8 than outside, 1.1, thus their low power can not come out the enclosure.

Table 6. Accumulated power and SNR for internal PD in Nomex inside the tank when measured with the inner and outer monopoles of 10 cm.

Antenna	Power (V^2)		
	13 kV	0 V	SNR
Inside ~300 MHz	3.4×10^{-4}	1.3×10^{-4}	2.6
Outside ~300 MHz	1.4×10^{-3}	6.3×10^{-6}	214.3
Inside 300–1200 MHz	6.6×10^{-2}	4×10^{-4}	166.5
Outside 300–1200 MHz	8.3×10^{-4}	2.3×10^{-5}	35.5
Inside 1200–2500 MHz	4.4×10^{-3}	6.4×10^{-4}	6.8
Outside 1200–2500 MHz	5.4×10^{-5}	4.9×10^{-6}	1.1

6.3. Surface PD in a Twisted Pair Inside the Tank

In Figure 11a surface PD pulses with the twisted pair inside the tank are shown from inside and outside. For the inner antenna, the pulse has high amplitude than outside the enclosure.

The spectrum of noise compare with PD obtained with both antennas are shown in Figure 11b,c. Inside the tank, the power is concentrated mainly between 500–1100 MHz with a maximum high-power peak value of 3×10^{-3} V^2. First resonance modes are marked on figures for these frequencies. Besides, it can be shown that there are other frequency components in two distinguished bands, the first up to 100 MHz and, the second, above 1100 MHz. In the second band, the power is concentrated between 1100–1400 MHz, 1500–1800 MHz and at 2100 MHz, with a peak value of 3×10^{-3} V^2 at 1700 MHz. As in the previous experiment, the power spectrum of the PD outside the tank is not increased above 1.2 GHz, receiving only the GSM and Wi-Fi emissions at 2.1 GHz and 2.4 GHz, respectively. Besides, an increase in power for the range of 500–1100 MHz due to PD activity in the tank is observed. Up to 150 MHz, there is a power increase from PD with a similar maximum power level as inside, taking values of 10^{-5} V^2. Note that, in the 150–500 MHz range, there is a power increase outside the tank that is not shown from inside due to the different scales used. This is because the pulse inside is 20 times greater than outside the tank, thus the high-scale needed to represent the inside pulse not allows enough resolution to identify low power components in the order of 10^{-6} V^2 because its peak power is at 5×10^{-3} V^2.

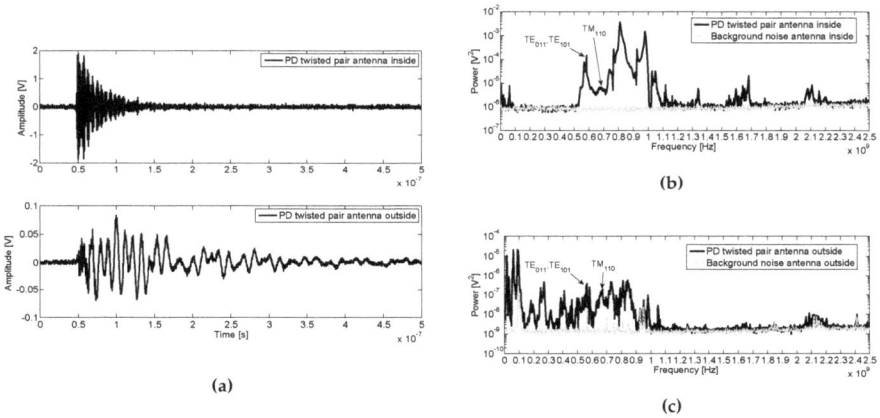

Figure 11. (**a**) Surface PD pulses in twisted pair at 760 V inside the enclosure acquired by the inner and outer antennas and their average power spectra compared with noise inside (**b**) and outside (**c**).

In Table 7 the first resonance modes for twisted par test object measured inside and outside the tank are compared. As expected, for both, the peak power values are greater inside than outside.

Table 7. Peak values for first TE and TM modes inside and outside the tank for surface PD in twisted pair.

Modes	Power (V^2)		
	Inside	Outside	Ratio
$TE_{011}, TE_{101} = 560$ MHz	1.5×10^{-4}	6.6×10^{-6}	22.7
$TM_{110} = 655$ MHz	3.8×10^{-7}	1.9×10^{-7}	2

Table 8 presents the accumulative power measured with the 10 cm monopoles. The three more significant intervals are: up to 500 MHz, where power from PD is seen outside and not from inside; 500–1200 MHz; and 1200–2500 MHz. Note that the intervals analysed are different within experiments depending on the frequency content nature in each test object. In this case, inside, the most representative band of power is 500–1200 MHz with a ratio of 158.6 which it is also measured from outside with a good SNR of 29.1 making it possible to measure the power content of the first resonances of the tank, excited by this type of PD, from outside and with a good SNR. Thus, discriminating the bands in which the kind of PD emits power, it is possible discern, in this case, if there is a concrete kind of discharges into the tank. Outside, the power contents of the first band up to 500 MHz has a higher SNR, 186.3, however this effect is only representative up to 100 MHz from inside Figure 11b. From 1.2 GHz, the tank do not allow to leave out radiation from inside.

Table 8. Accumulated power and SNR for surface PD in twisted pair inside the tank when measured with the inner and outer monopoles of 10 cm.

Antenna	Power (V^2)		
	760 V	0 V	SNR
Inside ~500 MHz	2.6×10^{-4}	2.1×10^{-4}	1.2
Outside ~500 MHz	1.7×10^{-4}	9×10^{-7}	186.3
Inside 500–1200 MHz	5.1×10^{-2}	$3.2 \cdot 10^{-4}$	158.6
Outside 500–1200 MHz	2.1×10^{-5}	7.3×10^{-7}	29.1
Inside 1200–2500 MHz	1.1×10^{-3}	6.9×10^{-4}	1.7
Outside 1200–2500 MHz	1.5×10^{-6}	1.2×10^{-6}	1.3

6.4. Surface PD in an Insulator on the Tank Cover

Figure 12a shows pulses for surface PD on the insulator supported on the cover tank. Spectra of noise and PD are represented in Figure 12b,c from inside and outside the enclosure, respectively. Inside and outside, this type of PD have a frequency distribution with a large content of power up to 300 MHz and between 300–1200 MHz as can be seen in, Figure 12b,c and in this second band is precisely where the first tank resonant modes are. It should be expected that external signals can not be measured inside the tank, but this is not so, and the distribution of power measured inside and outside is quite similar. Into the tank, resonances are observed and their power content is amplified for values above 1200 MHz. Below 300 MHz, the power is markedly amplified inside the tank compare to outside. This may be because the antenna is located in the hole of the cavity and the insulator above the tank, so the sensor can receive the pulse directly, as well as the noise level received inside has an order of magnitude lower than that received outside, thereby increase the SNR.

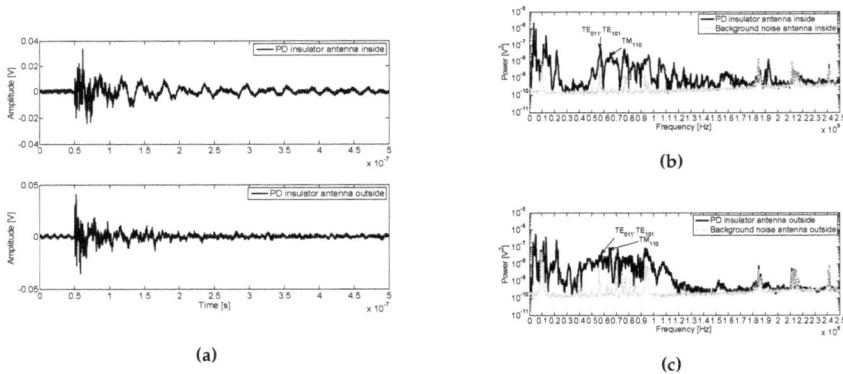

Figure 12. (a) Surface PD pulses on insulator at 12 kV outside the enclosure acquired by the inner and outer antennas and their average power spectra compared with the noise inside (b) and outside (c) the enclosure.

Table 9 represent the first resonance modes with each peak value of the tank measured inside and outside the tank. In this case, the first TE mode has higher power inside the tank than outside, while the first TM mode is lower inside, as expected. This trend it is not comply with the expectations for the

TE mode. It is because the power takes very low values in the order of 10^{-8} V^2 and the measurement error is significant.

Table 9. Peak values for first TE and TM modes inside and outside the tank for surface PD on insulator.

Modes	Power (V^2)		
	Inside	Outside	Ratio
$TE_{011}, TE_{101} = 560$ MHz	7.5×10^{-8}	3.3×10^{-8}	2.3
$TM_{110} = 655$ MHz	4.8×10^{-8}	8.8×10^{-8}	0.5

Table 10 shows the cumulative power for the spectrum calculated inside and outside the tank in three representatives bands for this test object: up to 300 MHz, wherein the PD power decays; from 300 MHz to 1200 MHz, where the first resonances are; and 1200–2500 MHz band in which PD have low power. Inside the tank, most of the power received is located up to 300 MHz, and it is noteworthy that the tank amplifies the value of the power in this band. This can be attributed to the ground system of the enclosure of the model tank, which gives a poor RF shielding comparing with the grounding in a real power transformer. In the band 300–1200 MHz, the SNR has the same order of magnitude for both, inside and outside, however it is greater the power of the PD received from outside, as expected. Finally, the power contained in the last interval is approximately equal from both sensors.

Table 10. Accumulated power and SNR for surface PD in insulator over the tank when measured with the inner and the outer 10 cm monopoles.

Antenna	Power (V^2)		
	12 kV	0 V	SNR
Inside ~300 MHz	7.1×10^{-6}	4.1×10^{-8}	171.5
Outside ~300 MHz	4.8×10^{-6}	3.0×10^{-7}	16
Inside 300–1200 MHz	2.9×10^{-6}	1.1×10^{-7}	27.1
Outside 300–1200 MHz	6.7×10^{-6}	2.1×10^{-7}	32.3
Inside 1200–2500 MHz	6.2×10^{-7}	2.8×10^{-7}	2.2
Outside 1200–2500 MHz	4.2×10^{-7}	2.4×10^{-7}	1.7

7. Discussion

Though in the previous section the important findings when comparing the spectrum of the three tests samples are mentioned, it is important to state any information that is found as a result comparing their spectra and SNRs at different frequencies. Thus, in Figure 13 are compared the PD spectra inside and outside the tank for the three test objects. The base values for the PD spectra inside for Nomex, Figure 13a and for the twisted pair Figure 13b is clearly lower than from outside. However, in the insulator case, Figure 13c, the spectrum takes similar shape and base magnitudes.

Table 11 shows the comparison of SNR for the three test objects in three different frequency bands: up to 500 MHz, 500–1200 MHz and 1200–2500 MHz. For the first band, the SNR is clearly greater than 1 for the three samples and from inside and outside expect inside the tank for the twisted pair, due to the PD pulses have great power than this effect attributable to the set-up. In the 500–1200 MHz band, where the first resonances are, it is shown that it is possible to measure this frequency components from both sides of the tank enclosure. Finally, the 1200–2500 MHz band is mitigated outside the enclosure with SNR closers to unit.

(a)

(b)

(c)

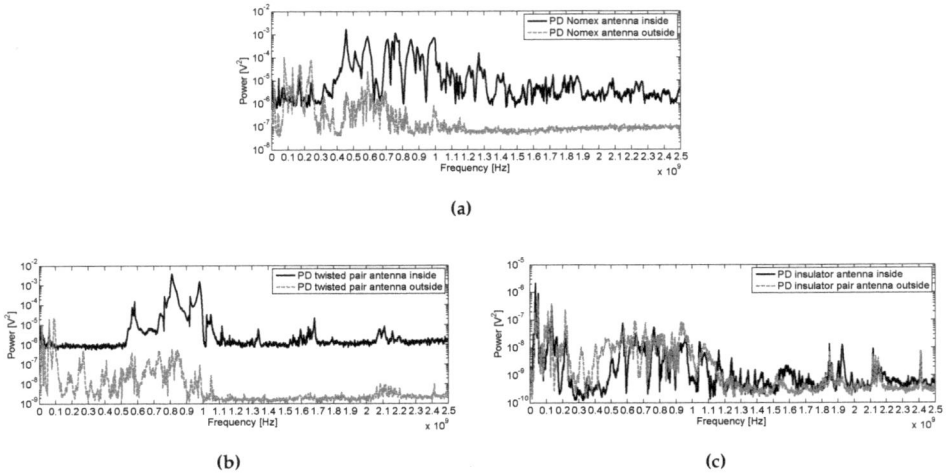

Figure 13. Average power spectra for PD measured from inside and outside the tank in (**a**) Nomex, (**b**) twisted pair and (**c**) insulator.

Table 11. SNRs for the three test objects measured from inside and outside.

Antenna	Object	SNR
Inside ~500 MHz	Nomex	51.5
	Twisted pair	1.2
	Insulator	127
Outside ~500 MHz	Nomex	131
	Twisted pair	186.3
	Insulator	17.4
Inside 500–1200 MHz	Nomex	204
	Twisted pair	158.6
	Insulator	30.8
Outside 500–1200 MHz	Nomex	13.7
	Twisted pair	29.1
	Insulator	27.8
Inside 1200–2500 MHz	Nomex	6.48
	Twisted pair	1.7
	Insulator	2.04
Outside 1200–2500 MHz	Nomex	1.06
	Twisted pair	1.3
	Insulator	1.38

In real power transformers, one of the most important problems to carry out PD detection with the UHF method is the necessity to locate a sensor inside the tank or in dielectric windows of its enclosure, such as the oil drain valve. These necessities are very problematic and usually require switching-off the electrical asset and remove some part of oil. Besides, real transformers have a star point connected to the neutral that is piped outside the tank through the bushing to avoid the galvanic connection between the active part of the windings and the tank. Thus, the tank shields the impulses. However, the model tank used in this work does not have a real grounded system that made the tank be really like a screen for the pulses. Thus, the model tank acts as an emitter at VHF and the outer antenna can register its effect.

8. Conclusions

Partial discharge measurements performed inside the model of a transformer tank show that the spectrum of the received signals are influenced by several factors. The first is the cavity, which depending on its dimensions allowing a certain resonant modes that have power for certain frequencies. The second is the frequency response of the antenna. The third depends on the PD type. When the EM emission comes out, the tank acts as a low-pass filter with a cut-off frequency of about 1.2 GHz, so that the outer antenna is not able to receive power from the PD above this frequency. Furthermore, it was found that a higher power content may appear outside than inside, below 300 MHz, and this is attributed to the ground cable of the enclosure which behaves as an antenna that amplify the PD radiation below 300 MHz. When the insulator is located outside the tank, similar power below 300 MHz is measured inside than outside, this could be due to the insufficient EM enclosure of the tank model. In this case, similarly than with the other test objects, resonance modes of the tank are excite and the cumulative power has the same order of magnitude both inside and outside in the 500–1200 MHz range. These results suggest that taking into account only the power spectrum of the PD, it is not trivial to identify the kind of PD source and additional studies in future must be carried out. In this sense, it will be intended to have a voltage reference to represent the PRPD pattern to identify the type of source in each case. However, this requires the use of another hardware to carry out the measurements in order to capture both, fast pulses in the order of *ns*, such as PD, synchronizing with the AC voltage reference in the order of *ms*, so a synchronized acquisition system with uncouple channels, with different frequency sample will be required. In future research, it would be necessary to deploy artificial defects with similar PRPD distributions as those described in this work in a real power transformer to measure its PD activity in UHF with various sensor arrangements. Future research will be a step forward in this research area and will allow for applying the knowledge acquired in this paper for application in the field.

Acknowledgments: This work was supported by Basal Project FB0008. The authors would like to thank the Department of Electrical Engineering at Universidad Carlos III de Madrid for all tests that were done in High-Voltage Research and Test Laboratory (LINEALT), as well as to the Department of Signal Theory and Communications for the same institution by the support in the measurements with the network analyser.

Author Contributions: The presented research was developed by the authors with the contribution of each one in a specific task. Ricardo Albarracín defined the tests to be performed and carried out the experimental measurements. Besides, he did the post-processing and analysis of the results. Jorge Alfredo Ardila-Rey supported Ricardo Albarracín in test object design and PD measurements in the laboratory. Finally, Ricardo Albarracín, Jorge Alfredo Ardila-Rey and Abdullahi Mas'ud contribute in a brainstorming to the analysis of the results.

Conflicts of Interest: The authors declare no conflict of interest.

References

1. Judd, M.; Yang, L.; Hunter, I. Partial discharge monitoring of power transformers using UHF sensors. Part I: sensors and signal interpretation. *IEEE Electri. Insul. Mag.* **2005**, *21*, 5–14.
2. Markalous, S.; Tenbohlen, S.; Feser, K. Detection and location of partial discharges in power transformers using acoustic and electromagnetic signals. *IEEE Trans. Dielectr. Electr. Insul.* **2008**, *15*, 1576–1583.
3. Robles, G.; Albarracín, R.; Vázquez, J.L. Antennas in Partial Discharge Sensing System. In *Handbook of Antenna Technologies*; Chen, Z.N., Ed.; Springer Singapore: Singapore, 2015; pp. 1–47.
4. IEC-62478. *High-Voltage Test Techniques—Measurement of Partial Discharges by Electromagnetic and Acoustic Methods. Proposed Horizontal Standard*, 1st ed.; International Electrotechnical Commission (IEC): Geneva, Switzerland, 2016.(submitted)
5. Coenen, S.; Tenbohlen, S.; Markalous, S.; Strehl, T. Sensitivity of UHF PD measurements in power transformers. *IEEE Trans. Dielectr. Electr. Insul.* **2008**, *15*, 1553–1558.
6. Robles, G.; Fresno, J.M.; Martínez-Tarifa, J.M. Separation of Radio-Frequency Sources and Localization of Partial Discharges in Noisy Environments. *Sensors* **2015**, *15*, 9882–9898.

7. Albarracín, R.; Robles, G.; Martínez-Tarifa, J.; Ardila-Rey, J. Separation of sources in radiofrequency measurements of partial discharges using time-power ratio maps. *ISA Trans.* **2015**, *58*, 389–397.

8. Robles, G.; Albarracín, R.; Martínez-Tarifa, J. Shielding effect of power transformers tanks in the ultra-high-frequency detection of partial discharges. *IEEE Trans. Dielectr. Electr. Insul.* **2013**, *20*, 678–684.

9. Mirzaei, H.; Akbari, A.; Zanjani, M.; Gockenbach, E.; Borsi, H. Investigating the partial discharge electromagnetic wave propagation in power transformers considering active part characteristics. In Proceedings of the IEEE 2012 International Conference on Condition Monitoring and Diagnosis (CMD), Bali, India, 23–27 September 2012; pp. 442–445.

10. Balanis, C.A. *Advanced Engineering Electromagnetics*, 2nd ed.; Wiley: Hoboken, NJ, USA, 2012.

11. Robles, G.; Sánchez-Fernández, M.; Albarracín-Sánchez, R.; Rojas-Moreno, M.; Rajo-Iglesias, E.; Martínez-Tarifa, J. Antenna Parametrization for the Detection of Partial Discharges. *IEEE Trans. Instrum. Meas.* **2013**, *62*, 932–941.

12. Balanis, C.A. *Antenna Theory: Analysis and Design*, 3rd ed.; Wiley-Interscience: Hoboken, NJ, USA, 2005.

13. Judd, M.; Farish, O.; Hampton, B. Broadband couplers for UHF detection of partial discharge in gas-insulated substations. *IEE Proc. Sci. Meas. Technol.* **1995**, *142*, 237–243.

14. Tang, Z.; Li, C.; Cheng, X.; Wang, W.; Li, J.; Li, J. Partial discharge location in power transformers using wideband RF detection. *IEEE Trans. Dielectr. Electr. Insul.* **2006**, *13*, 1193–1199.

15. López-Roldán, J.; Tang, T.; Gaskin, M. Optimisation of a sensor for onsite detection of partial discharges in power transformers by the UHF method. *IEEE Trans. Dielectr. Electr. Insul.* **2008**, *15*, 1634–1639.

16. Kaneko, S.; Okabe, S.; Yoshimura, M.; Muto, H.; Nishida, C.; Kamei, M. Detecting characteristics of various type antennas on partial discharge electromagnetic wave radiating through insulating spacer in gas insulated switchgear. *IEEE Trans. Dielectr. Electr. Insul.* **2009**, *16*, 1462–1472.

17. Fabiani, D.; Cavallini, A.; Montanari, G. A UHF Technique for Advanced PD Measurements on Inverter-Fed Motors. *IEEE Trans. Power Electron.* **2008**, *23*, 2546–2556.

18. Stutzman, W.L.; Thiele, G. *Antenna Theory and Design*; Wiley: Hoboken, NJ, USA, 1998.

19. Prevost, T.; Oommen, T.V. Cellulose insulation in oil-filled power transformers: Part I—History and development. *IEEE Electr. Insul. Mag.* **2006**, *22*, 28–35.

20. Cheng, D.K. *Fundamentals of Electromagnetism Engineering*; Addison Wesley Iberoamericana: Mexico DF, Mexico, 1998.

21. Jardine, A.K.; Lin, D.; Banjevic, D. A review on machinery diagnostics and prognostics implementing condition-based maintenance. *Mech. Syst. Signal Process.* **2006**, *20*, 1483–1510.

22. Gill, P. *Electrical Power Equipment Maintenance and Testing, Second Edition*; Power Engineering (Willis), Taylor & Francis: Boca Raton, FL, USA, 2008.

23. IEC-60270. *High-Voltage Test Techniques—Partial Discharge Measurements*, 3rd ed.; International Electrotechnical Commission (IEC): New Delhi, India, 2000.

24. Pozar, D. *Microwave Engineering*; Wiley: Hoboken, NJ, USA, 2005.

sensors

MDPI

Article

Automated Low-Cost Smartphone-Based Lateral Flow Saliva Test Reader for Drugs-of-Abuse Detection

Adrian Carrio [1],*, Carlos Sampedro [1], Jose Luis Sanchez-Lopez [1], Miguel Pimienta [2] and Pascual Campoy [1]

[1] Computer Vision Group, Centre for Automation and Robotics (UPM-CSIC), Calle José Gutiérrez Abascal 2, Madrid 28006, Spain; carlos.sampedro@upm.es (C.S.); jl.sanchez@upm.es (J.L.S.-L.); pascual.campoy@upm.es (P.C.)

[2] Aplitest Health Solutions, Paseo de la Castellana 164, Madrid 28046, Spain; mp@aplitest.com

* Author to whom correspondence should be addressed; adrian.carrio@upm.es; Tel.: +34-913-363-061; Fax: +34-913-363-010.

Academic Editor: Gonzalo Pajares Martinsanz

Received: 31 August 2015; Accepted: 16 November 2015; Published: 24 November 2015

Abstract: Lateral flow assay tests are nowadays becoming powerful, low-cost diagnostic tools. Obtaining a result is usually subject to visual interpretation of colored areas on the test by a human operator, introducing subjectivity and the possibility of errors in the extraction of the results. While automated test readers providing a result-consistent solution are widely available, they usually lack portability. In this paper, we present a smartphone-based automated reader for drug-of-abuse lateral flow assay tests, consisting of an inexpensive light box and a smartphone device. Test images captured with the smartphone camera are processed in the device using computer vision and machine learning techniques to perform automatic extraction of the results. A deep validation of the system has been carried out showing the high accuracy of the system. The proposed approach, applicable to any line-based or color-based lateral flow test in the market, effectively reduces the manufacturing costs of the reader and makes it portable and massively available while providing accurate, reliable results.

Keywords: smartphone; drugs-of-abuse; diagnostics; computer vision; machine learning; neural networks

---

1. Introduction

Most rapid tests or qualitative screening tests on the market are chromatographic immunoassays. They are used to detect the presence or absence of a substance (analyte) in an organic sample. The result is obtained in a few minutes and without the need of specialized processes or equipment. The use of this kind of test is an aid in the rapid diagnose of different diseases (*i.e.*, HIV, hepatitis, malaria, *etc.*) or certain physiological conditions (pregnancy, drugs-of-abuse, blood glucose levels, cholesterol, *etc.*).

In the particular case of drug-of-abuse detection, tests are commonly based on the principle of competitive binding: drugs that may be present in the organic sample (*i.e.*, urine, saliva, *etc.*) compete against a drug conjugate, present in the test strip, for the specific binding sites of the antibody. During the test procedure, the sample migrates along the test strip by capillary action.

If a substance present in the sample is available in a concentration lower than the cutoff level, it will not saturate the binding sites of the particles coated with the antibody on the test strips. The coated particles will be then captured by the immobilized conjugate of each substance (drug), and a specific area in the strip will be visibly colored. No coloration will appear in this area if the concentration of the substance is above the cutoff level, as it will saturate all of the binding points of the specific antigen for such a substance. An additional control area is usually disposed in the test strip and colored upon effective migration through the initial test area, to confirm the validity of the test result.

Most of the results obtained using the commercially-available test kits are interpreted visually by a human operator, either by the presence or absence of colored lines (Figure 1) or by comparison of the changes in color of particular areas of a test strip against a pattern (Figure 2). Some of the problems arising from the use of rapid/screening tests are the following:

- Interpretation: This is done by direct visual inspection, and thus, the results interpretation may vary depending on the human operator (training, skills, lighting conditions, *etc.*). Normally, under well-lit environments, there are less interpretation errors than when this kind of test is used under poor light conditions, as this may affect the ability of the operator to judge the result correctly (*i.e.*, testing for drugs during a roadside control in the middle of the night).

- Confirmation is required: Any results obtained should be confirmed using a technique with a higher specificity, such as mass spectrometry, specially when presumptive positive results are obtained.

- Conditions under which the tests are performed: Rapid tests are used in conditions where there is no availability of specialized equipment.

- Dispersion and structure of the data: Test results and subject data are scattered and stored, usually in paper-based format.

- Processing and analysis of data: Data for decision-making are gathered and analyzed manually by human operators, being prone to human errors.

While most lateral flow tests are intended to operate on a purely qualitative basis, it is possible to measure the color intensity of the test lines in order to determine the quantity of analyte in the sample. Computer vision has been proven as a useful tool for this purpose, as capturing and processing test images can provide objective color intensity measurements of the test lines with high repeatability. By using a computer vision algorithm, the user-specific bias is eliminated (ability to interpret), which may affect the result obtained.

Smartphones' versatility (connectivity, high resolution image sensors and high processing capabilities, use of multimedia contents, *etc.*) and performance condensed in small and lightweight devices, together with the current status of wireless telecommunication technologies, exhibit a promising potential for these devices to be utilized for lateral flow tests interpretation, even in the least developed parts of the world.

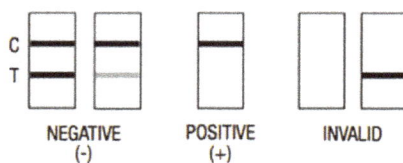

(**a**) Line-based test interpretation

(**b**) Negative   (**c**) Positive   (**d**) Invalid

Figure 1. (a) Line-based test interpretation samples; (**b–d**) Strip samples.

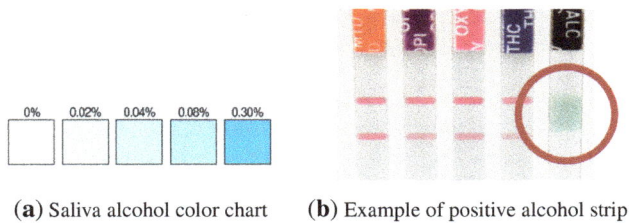

(**a**) Saliva alcohol color chart   (**b**) Example of positive alcohol strip

Figure 2. (**a**) Color chart to obtain the relative blood alcohol concentration by comparison with the colored area in the test strip; (**b**) test sample including an alcohol strip.

In this paper, we present a novel algorithm to qualitatively analyze lateral flow tests using computer vision and machine learning techniques running on a smartphone device. The smartphone and the tests are contained on a simple hardware system consisting of an inexpensive 3D-printed light box.

The light box provides controlled illumination of the test during the image capture process while the smartphone device captures and processes the image in order to obtain the result. Test data can be then easily stored and treated in a remote database by taking advantage of the smartphone connectivity capabilities, which can help to increase the efficiency in massive drug-of-abuse testing, for example in roadside controls, prisons or hospitals.

This approach effectively reduces the manufacturing costs of the reader, making it more accessible to the final customer, while providing accurate, reliable results. To the authors' knowledge, the interpretation of lateral flow saliva tests for drug-of-abuse detection using computer vision and machine learning techniques on a smartphone device is completely novel.

The remainder of the paper is organized as follows. Firstly, a review of the state of the art for hand-held diagnostic devices, in general, and for drug-of-abuse lateral flow readers, in particular, is presented. Secondly, the saliva test in which the solution was implemented is introduced. Thirdly, the software and hardware solutions proposed are described. Fourthly, the methodology for validating the results is discussed. Then, the results of the evaluation through agreement and precision tests are shown. Finally, conclusions are presented.

2. State of the Art

Martinez *et al.* [1] used paper-based microfluidic devices for running multiple assays simultaneously in order to clinically quantify relevant concentrations of glucose and protein in artificial urine. The intensity of color associated with each colorimetric assay was digitized using camera phones. The same phone was used to establish a communications infrastructure for transferring the digital information from the assay site to an off-site laboratory for analysis by a trained medical professional.

A lens-free cellphone microscope was developed by Tseng *et al.* [2] as a mobile approach to provide infectious disease diagnosis from bodily fluids, as well as rapid screening of the quality of water resources by imaging variously-sized micro-particles, including red blood cells, white blood cells, platelets and a water-borne parasite (*Giardia lamblia*). Improved works in this field were published by Zhu *et al.* [3], who also developed a smartphone-based system for the detection of *Escherichia coli* [4] on liquid samples in a glass capillary array.

Matthews *et al.* [5] developed a dengue paper test that could be imaged and processed by a smartphone. The test created a color on the paper, and a single image was captured of the test result and processed by the phone, quantifying the color levels by comparing them with reference colors.

Dell *et al.* [6] presented a mobile application that was able to automatically quantify immunoassay test data on a smartphone. Their system measured both the final intensity of the capture spot and the progress of the test over time, allowing more discriminating measurements to be made, while showing great speed and accuracy. However, registration marks and an intensity calibration pattern had to be included in the test to correctly process the image, and also, the use of an additional lens was required for image magnification.

Uses of rapid diagnostic test reader platforms for malaria, tuberculosis and HIV have been reported by Mudanyali *et al.* [7,8]. Smartphone technology was also used to develop a quantitative rapid diagnostic test for multi-bacillary leprosy, which provided quantifiable and consistent data to assist in the diagnosis of MBleprosy.

A smartphone-based colorimetric detection system was developed by Shen *et al.* [9], together with a calibration technique to compensate for measurement errors due to variability in ambient light. Oncescu *et al.* [10] proposed a similar system for the detection of biomarkers in sweat and saliva. Similar colorimetric methods for automatic extraction of the result in ELISA plates [11] and proteinuria in urine [12] have also been reported. However, none of those works presented developments on colored line detection.

In the area of drug-of-abuse detection, accurate confirmatory results are nowadays obtained in a laboratory usually by means of mass spectrometry techniques. These laboratory-based solutions are expensive and time consuming, as the organic sample has to be present in the laboratory in order to perform the analysis. In contrast, screening techniques provide *in situ*, low-cost, rapid presumptive results with a relatively low error rate, with immunoassay lateral flow tests currently being the most common technique for this type of test. Nonetheless, as most lateral flow tests operate on a purely qualitative basis, obtaining a result is usually subject to the visual interpretation of colored areas on the test by a human operator, therefore introducing subjectivity and the possibility of human errors to the test result. Hand-held diagnostic devices, known as lateral flow assay readers, are widely used to provide automated extraction of the test result.

VeriCheck [13] is an example of an automated reader for lateral flow saliva tests consisting of the use of a conventional image scanner and a laptop. However, this solution is far from portable, as it consists of multiple devices and requires at least a power outlet for the scanner.

DrugRead [14] offers a portable automated reader solution on a hand-held device. However, our system offers a low-cost, massively available solution, as it has been implemented on a common smartphone device.

3. Saliva Test Description

The proposed image processing methodology can be applied to any line presence or absence-based or color interpretation-based immunoassay tests on the market. For the results and validation presented, the rapid oral fluid drug test DrugCheck SalivaScan [15], manufactured by Express Diagnostics Inc. (Minneapolis, MN, USA), was used. DrugCheck SalivaScan is an immunoassay for rapid qualitative and presumptive detection of drugs-of-abuse in human oral fluid samples.

The device is made of one or several strips of membrane incorporated in a plastic holder, as shown in Figure 3. For sample collection, a swab with a sponge containing an inert substance that reduces saliva viscosity is used. A saturation indicator is placed inside the collection swab to control the volume of saliva collected and to provide the indication to start the reaction (incubation) of the rapid test. The test may contain any combination of the parameters/substances and cutoff levels as listed in Table 1.

Additionally, the test may include a strip for the detection of the presence of alcohol (ethanol) in oral fluid, providing an approximation of the relative blood alcohol concentration. When in contact with solutions of alcohol, the reactive pad in the strip will rapidly turn colors depending on the concentration of alcohol present. The pad employs a solid-phase chemistry that uses a highly specific enzyme reaction. The detection levels of relative blood alcohol concentration range between 0.02% up to 0.30%.

Table 1. Calibrators and cutoff levels for different line-based drug tests.

Test	Calibrator	Cutoff Level (ng/mL)
Amphetamine (AMP)	D-Amphetamine	50
Benzodiazepine (BZO)	Oxazepam	10
Buprenorphine (BUP)	Buprenorphine	5
Cocaine (COC)	Benzoylecgonine	20
Cotinine (COT)	Cotinine	50
EDDP (EDDP)	2-Ethylidene-1,5-dimethyl-3,3-diphenylpyrrolidine	20
Ketamine (KET)	Ketamine	50
Marijuana (THC)	11-nor-Δ9-THC-9 COOH	12
Marijuana (THC)	Δ9-THC	50
Methadone (MTD)	Methadone	30
Methamphetamine (MET)	D-Methamphetamine	50
Opiates (OPI)	Opiates	40
Oxycodone (OXY)	Oxycodone	20
Phencyclidine (PCP)	Phencyclidine	10
Propoxyphene (PPX)	Propoxyphene	50
Barbiturate (BAR)	Barbiturate	50

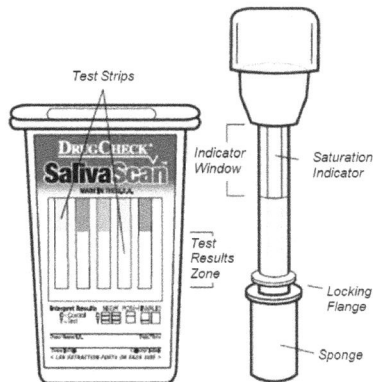

Figure 3. DrugCheck SalivaScan test. During the test procedure, the collection swab (**right**) will be inserted into the screening device (**left**).

3.1. Test Procedure

The first step in the test procedure consists of saturating the saliva test sponge. For this, the donor sweeps the inside of the mouth (cheek, gums, tongue) several times using a collection swab and holds it in his or her closed mouth until the color on the saturation indicator strip appears in the indicator window.

The collection swab can then be removed from the mouth and inserted into the screening device. Once the specimen is dispersed among all strips, the device should be set and kept upright on a flat surface while the test is running. After use, the device can be disposed of or sent to a laboratory for confirmation on a presumptive positive result.

3.2. Interpretation of Results

In the case of non-alcohol strips, interpretation is based on the presence or absence of lines. Two differently-colored lines may appear in each test strip, a control line (C) and a test line (T), leading to different test results, as shown in Figure 1. The areas where these lines may appear are called the control region and the test region, respectively.

Negative results can be read as soon as both lines appear on any test strip, which usually happens within 2 min. Presumptive positive results can be read after 10 min. Three possible results may be obtained:

1. Positive: Only one colored line appears in the control region. No colored line is formed in the test region for a particular substance. A positive result indicates that the concentration of the analyte in the sample exceeds the cutoff level.
2. Negative: Two colored lines appear on the membrane. One line is formed in the control region and another line in the test region for the corresponding substance. A negative result indicates that the analyte concentration is below the cutoff level.
3. Invalid result: No control line is formed. The result of any test in which there is no control line during the specified time should not be considered.

The intensity of the colored line in the test region may vary depending on the concentration of the analyte present in the specimen. Therefore, any shade of color in the test region should be considered negative.

In the case of saliva alcohol strips, the interpretation of the results should be made by comparing the color obtained in the reagent strip against a printed color pattern that is provided with the test (Figure 2). Alcohol strips must be read at 2 min, as pad color may change, and again, three possible results may be obtained:

1. Positive: The test will produce a color change in the presence of alcohol in the sample. The color intensity will range, being light blue at a 0.02% relative blood alcohol concentration and dark blue near a 0.30% relative blood alcohol concentration. An approximation of the relative blood alcohol concentration within this range can be obtained by comparison with the provided color pattern (Figure 2).
2. Negative: If the test presents no color changes, this should be interpreted as a negative result, indicating that alcohol has not been detected in the sample.
3. Invalid: If the color pad is already colored in blue before applying the saliva sample, the test should not be used.

4. System Description

In the following section, a description of the saliva test reader system is presented. Firstly, image acquisition aspects are discussed, including a description of the light box device used for illumination normalization purposes. Secondly, the computer vision algorithms used in the image processing stage are described. Finally, the machine learning algorithms used for lateral histogram classification are presented.

4.1. Image Acquisition

Image acquisition is an extremely important step in computer vision applications, as the quality of the acquired image will condition all further image processing steps. Images must meet certain requirements in terms of image quality (blur, contrast, illumination, *etc.*). The positions of the camera (a mobile device in our case) and the object to be captured (here, the test) should remain in a constant relative position for the best results. However, contrary to traditional image processing applications, a mobile device is hand-held and, therefore, does not have a fixed position with respect to the test, which can lead to motion blur artifacts. Furthermore, mobile devices are used in dynamic environments, implying that ambient illumination has to be considered in order to obtain repeatable results regardless of the illumination conditions.

In order to minimize all image acquisition-related problems, a small light box was designed to keep the relative position between the smartphone and the test approximately constant, while removing external illumination and projecting white light onto the test with an embedded electronic lighting system. The light box, with dimensions of $150 \times 70 \times 70$ mm, is shown in Figure 4. It is very portable, weighting only 300 g, and it can manufactured at a low cost with a 3D printer.

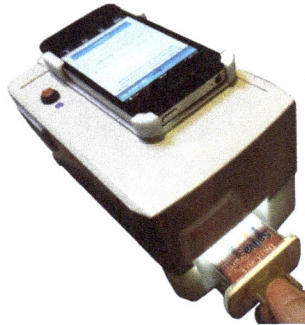

Figure 4. Light box with embedded electronic lighting system, which minimizes the relative movement between the smartphone and the test and the effects of external illumination changes.

In order to acquire an image, the saliva test is inserted into the light box; the lighting system is activated using a mounted button, and the smartphone is attached to the light box. The smartphone application has an implemented timer, which allows one to measure the elapsed time and to provide the user with a result as soon as it is available. The test reader provides a result by using a single captured image.

Three smartphone devices were selected for capturing and processing the test images, taking into account their technical specifications and the mobile phone market share: Apple iPhone 4, 4S and 5.

For the purpose of implementing and testing the computer vision and machine learning algorithms, a total of 683 images, containing a total of 2696 test strips, were acquired with the mentioned iPhone models.

4.2. Image Preprocessing and Strip Segmentation

Even with an elaborated approach for the image capture procedure, further image processing stages have to deal with image noise and small displacements and rotations of the test within the image, which can be caused by many factors. Just to highlight a few, smartphones might have a loose fit in the light box fastening system; the in-built smartphone cameras come in a variety of resolutions and lenses; furthermore, there might be slight differences in the brightness of the saliva strip's material.

Once the image has been acquired, the first step is to localize in the image the region corresponding to the strips. For this purpose, this area is manually defined in a template image, which is stored in a database. This template image is only defined once during the implementation process and is valid for all tests sharing the same format, independent of the number of strips and the drug configuration.

For defining the area corresponding to the strips in the actual processed image, a homography approach based on feature matching is used. For this purpose, the first step, which is done off-line, consists of calculating keypoints in the template image and extracting the corresponding descriptors. In this approach and with the aim of being computationally efficient, ORB (oriented FAST (Features from Accelerated Segment Test) and rotated BRIEF (Binary Robust Independent Elementary Features)) features are computed, and their descriptors are extracted. In [16], it is demonstrated how ORB is two orders of magnitude faster than SIFT. This is crucial in this kind of device for real-time processing.

ORB uses the well-known FAST keypoints [17,18] with a circular radius of nine (FAST-9) and introduces orientation to the keypoint by measuring the intensity centroid (oFAST). Then, the BRIEF descriptor [19] is computed, which consists of a bit string description of an image patch constructed from a set of binary intensity tests, using in this case a learning method for extracting subsets of uncorrelated binary tests (rBRIEF). The combination of oFAST keypoints and rBRIEF features conforms the final ORB descriptor and makes ORB features rotation invariant and robust to noise. Once the ORB keypoints and their descriptors have been extracted from the template image, they are stored in the smartphone.

When a new image is captured from the device, the first step consists of extracting the ORB keypoints and their descriptors and matching them with the ones extracted from the template (Figure 5), which will provide the homography between both images, that is the transformation that converts a point in the template image to the corresponding point in the current image.

The matching process is divided into three steps:

- First, a brute force matching is computed. In this process, the descriptors are matched according to their Hamming distance, selecting for the next stage the ones that have the minimum Hamming [20] distance between them.
- Second, a mutual consistency step is done for removing those matches that do not correspond uniquely to their counterparts in the other image.
- Finally, with the point pairs from the previous step, a homography transformation is computed. In this step, a random sample consensus (RANSAC) [21] method is used for removing the ones that do not fit the rigid perspective transform, which are called outliers.

Once the homography between the template image and the current image has been computed, this transformation is applied to the selected four points in the template image that define the region of interest (ROI) of the strips (the area within the green border in Figure 5). This feature matching-based homography approach successfully deals with image noise, small input image displacements and rotations and different image resolutions.

Figure 5. Matching process between oriented FAST and rotated BRIEF (ORB) descriptors in the template image (**left**) and original image (**right**). White lines denote the matched points. In green is depicted the searched region of interest that contains the strips in the actual processed image.

The next stage in the strip segmentation process is to localize the colored area of each strip in the image. The drug strip configuration of the test is known in advance through a unique batch number provided by the test manufacturer, so it is only required to check that the number of detected strips in the image matches the test configuration and to localize these strips in the image. As the interpretation method for the tests containing an alcohol strip is specific to this drug, images containing an alcohol strip will be processed differently. According to this, the process for segmenting and localizing the colored area of each strip is as follows:

- If the image contains an alcohol strip, a thresholding procedure by the Otsu [22] method is applied on the R, G and B channels of the image ROI given by the homography. After that, two AND operations between Channels G and B, and G and R, are applied with the purpose of filtering the image.
- If the image does not contain an alcohol strip, the image is thresholded using the Otsu method in the R and B channels, and finally, an OR operation between these images is applied.

Finally, in both procedures, a morphological closing operation is applied to the resulting binary image of the previous steps. Then, in order to remove isolated pixels, a filter based on the number of pixels in each column is computed. For each of the columns, if the number of non-zero pixels is less than 10% of the total number of pixels in that column, the column is set to zero. An example of the result of the segmentation process on a test image is shown in Figure 6a.

The final step in the segmentation process consists of a post-processing stage, in which the contours of the previous binary image are computed. In this stage, several filters are applied to the computed contours:

- First, a position-based filter is applied. The purpose of this filter is to remove those contours whose centroid is located under the lower half of the computed ROI (area bounded in green in Figure 5) of the image. This is useful to filter out spurious contours, as the ones bounded in red in Figure 6a.
- Second, we apply a filter based on the area enclosed by each of the computed contours. Based on this, we remove those contours that satisfy: $A_i < 0.5 \cdot A_{maxcontour}$, where A_i is the area of the contour i and $A_{maxcontour}$ is the area of the largest contour. Again, the objective is to filter out small spurious contours, which do not correspond to the colored regions of each strip, which indicate the type of drug.

By applying the explained segmentation process, the colored region of each strip, which indicates the type of drug of each test strip, is finally localized within the ROI image, as shown in Figure 6b. For each of these extracted ROIs, its size is used together with the position of its bottom-middle point as parameters to automatically determine the region on which the lateral histogram will be computed, bounded in green in Figure 6b. By limiting the extraction of the lateral histogram to this area, the problems of pixel intensity variations due to the shadows of the edges of the strip is minimized.

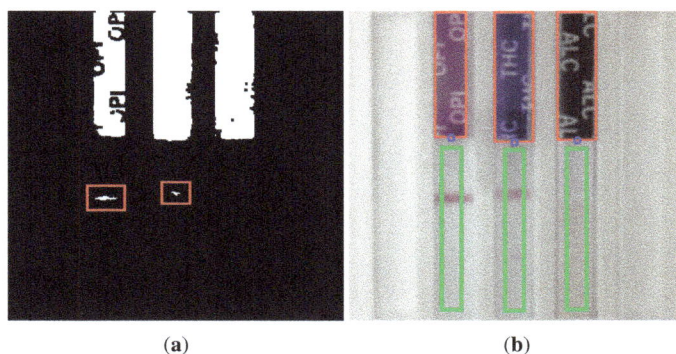

(a)   **(b)**

Figure 6. (**a**) Result of the segmentation process on a test image. Spurious contours, bounded in red, are filtered out during the post-processing stage; (**b**) Localization of the colored area of the strips after applying the segmentation process to a test image, bounded with a red rectangle. For each of these regions, its size is used together with the position of its bottom-middle point (depicted in blue) as parameters to automatically determine the region on which the lateral histogram will be computed, bounded with a green rectangle.

4.3. Lateral Histogram Extraction and Preprocessing

The regions of interest that represent each segmented strip in the previous section need to be processed before the classification step takes place. Due to the different way in which alcohol and non-alcohol strips should be interpreted, both require different preprocessing stages.

In the case of a non-alcohol strip, the preprocessing stage consists of four steps described below, whose mission is to simplify the information given to the classifier as much as possible to increase its efficiency. All of these steps are done for each non-alcohol segmented strip.

The first step is the lateral histogram extraction from the area extracted during the segmentation stage. The lateral histogram is computed for each of these areas, where the average pixel intensity value for each image row (strip transversal direction) is computed according to:

$$x_{HG}(i) = \frac{1}{n} \sum_{\forall j} s(i, j) \tag{1}$$

where $x_{HG}(i)$ is the lateral histogram extracted value in the i coordinate; $s(i, j)$ is the (i, j) pixel intensity value; i are image rows (strip transversal direction); j are image columns (strip longitudinal direction); and n is the number of image columns inside the lateral histogram area.

Due to differences of image acquisition, raw lateral histograms do not usually have the same amount of bins. Therefore, an adjustment in the lateral histogram number of bins is set to an arbitrary number $n_{bin} = 100$ through the use of quadratic interpolation. With this step, we ensure that the lateral histogram always has the same number of bins (set to $n_{bin} = 100$), independent of the image used.

Then, to minimize the effect of the light intensity changes in the same segmented strip and between two different images, a RANSAC technique is applied. The proposed model to be fitted is a linear model $y = a \cdot i + b$, where y is the lateral histogram value; i is the coordinate; and a and b are the estimated parameters. Then, the lateral histogram is recalculated:

$$x_{HR}(i) = x_{HN}(i) - (a \cdot i + b) \tag{2}$$

where $x_{HN}(i)$ is the lateral histogram value in the i coordinate after the adjustment to $n_{bin} = 100$.

As the lateral histogram peaks, which have the information of the test and control regions, are clearly separated, the fourth and last pre-processing step tries to exploit this information by lining up these peaks. The histogram x_{HR} is divided into two similar parts. The first part uses bins from one to $\frac{n_{bin}}{2} = 50$, while the last part uses bins from $\frac{n_{bin}}{2} = 50$ to $n_{bin} = 100$. The minima for each part, $h_{min1}(i)$ and $h_{min2}(i)$, respectively, are computed. The final lateral histogram h_x is the conjunction of both minimum values with a range of 15 bins in each direction. That means the final lateral histogram h_x has $n_{bin-final} = 62$ bins. Note that the first 31 bins will correspond to the control line (C), and the remaining bins will correspond to the test line (T).

In the case of an alcohol strip, a different processing is done: the first and the second steps are analogous, but with $n_{bin} = 62$. Then, a third step to reduce the light intensity differences between two different images is done. The average of only the first $n_{av} = 15$ values of the lateral histogram is calculated and then subtracted from the original lateral histogram.

$$h_x(i) = x_{HN}(i) - \frac{1}{n_{av}} \sum_{i=1}^{n_{av}} x_{HN}(i) \tag{3}$$

where h_x is the lateral histogram value adjusted to $n_{bin} = 62$ bins with no light intensity influence between two different images. This normalization is done only with respect to the first 15 values of the lateral histogram, because this part of the histogram has proven to account well for illumination changes along the strip. The normalization with respect to the mean of these 15 values helps to minimize the influence of differences in illumination along the strip.

4.4. Lateral Histogram Classification and Test Outcome

Three different supervised machine learning classifiers based on artificial neural networks (ANN) have been implemented for lateral histogram classification:

- A classifier for alcohol strip lateral histograms.
- A classifier for control lines in non-alcohol strip lateral histograms.
- A classifier for test lines in non-alcohol strip lateral histograms.

By dividing the classification task into three steps through the use of different classifiers, a better performance can be achieved due to its forced specialization. All three classifiers have the same kind of model, and only their structure (their size) and their parameters are different from one another.

Machine learning algorithms require a correct sample dataset in order to be trained. Supervised algorithms require examples of each of the desired classes to be learned in order to perform classification tasks.

The parameters of the classifier need to be set after the best structure is selected through a cross-validation algorithm. The available dataset is randomly, but uniformly divided into three sub-sets depending on its functionality. In our case, "training data" correspond to 70% of the available data and are used to adjust the parameters of the model; "validation data", 15%, are used to check the model parameters, ensuring the correct training; finally, "test data", 15%, test the performance of the classifier.

Five sequential algorithms have been considered for generating the classifiers. firstly, an unsupervised input data normalization is performed:

$$x_{normalized}(i) = \frac{h_x(i) - \overline{x}_{train}(i)}{std(x_{train}(i))} \tag{4}$$

where $h_x(i)$ is the preprocessed lateral histogram value in the coordinate i; the subscript *train* indicates the "training data" set; $\overline{x}_{train}(i)$ stands for the mean value; and *std* stands for the standard deviation.

Secondly, an unsupervised data mapping is done:

$$x_{mapped}(i) = 2 \cdot \frac{x_{normalized}(i) - min_{m,t}(i)}{max_{m,t}(i) - min_{m,t}(i)} - 1 \tag{5}$$

where $min_{m,t}(i)$ and $max_{m,t}(i)$ are the minimum and maximum values used for the mapping in the i coordinate and calculated with the "training data" set.

Then, a supervised ANN, multi-layer perceptron (MLP) [23], is first trained and afterwards executed to classify the data. This ANN is composed by several layers, called the "input layer", "hidden layers" and the "output layer", depicted in Figure 7. Each layer (except the "input layer", which is only the input to the ANN) is formed by several neurons. Each neuron has a single output; multiple inputs (all of the outputs of the neurons of the previous layer); a weight value associated with each input; and a bias value. The behavior of each neuron, *i.e.*, how its output is computed based on its inputs and internal parameters, is given by:

$$output = tansig\left(\sum_{\forall i} input(i) \cdot weight(i) + bias\right) \tag{6}$$

where $tansig(x)$ is the hyperbolic tangent sigmoid function.

After the input $\vec{x}_{NN} = \vec{x}_{mapped}$ is introduced to the ANN, it generates an output vector $\vec{y}_{NN}$. The number of neurons and its configuration (*i.e.*, the structure of the classifier) and the weights and biases of each neuron (*i.e.*, the parameters of the classifier) are calculated using the "training data" set.

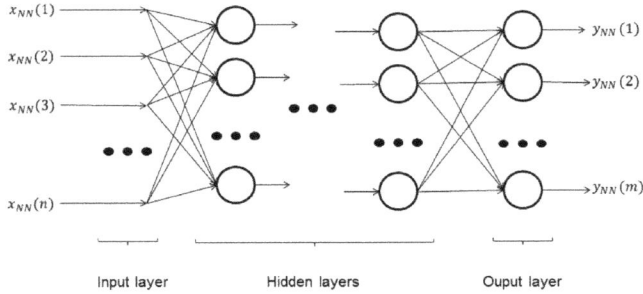

Figure 7. ANN-MLP structure diagram. Each neuron is represented by a circle, whose behavior is explained by Equation (6). The connections between the inputs, the neurons and the outputs are represented with arrows.

In the next step, an unsupervised output data reverse mapping is done:

$$y_{rev-mapped}(k) = \frac{y_{NN}(k)+1}{2(max_{r,t}(k) - min_{r,t}(k))} + min_{r,t}(k) \tag{7}$$

where $max_{r,t}(k)$ and $min_{r,t}(k)$ are the maximum and minimum values used for reverse mapping in the k coordinate of the output and calculated with the "training data" set.

Finally, an unsupervised thresholding operation generates the classification output. The maximum of the reverse mapped output is calculated: $y_{candidate} = max\left(\vec{y}_{rev-mapped}\right)$. Then, the thresholding operation is applied:

$$y_{class} = \begin{cases} k, & \text{if} y_{candidate} \geq y_{thres}(k) \\ \text{Undetermined}, & \text{otherwise} \end{cases} \tag{8}$$

This last step ensures that the algorithm gives only a trained output as a classification result if its confidence level is high enough, above a certain threshold. Otherwise, it will show a conservative behavior, outputting "undetermined" as the classification result.

4.4.1. Classifier for Alcohol Strip Lateral Histograms

The input to this classifier is the 62-bin preprocessed lateral histogram for alcohol strips. The output of the trained classifier is a value that indicates the result of the test. The alcohol strips used in these experiments allow for the detection of five alcohol levels in saliva, depicted in Figure 2a. However, only three output categories have been considered: "positive (1)", "negative (2)" or "undetermined (3)". The reason for this simplification can be found in the market demand, where saliva-based tests compete with other diagnostic technologies. Breath-based analyzers are generally used for measuring blood alcohol content in massive testing, such as roadside tests, where a high accuracy is expected and expensive equipment can be used. However, low-cost saliva-based alcohol tests are used in situations where only a binary output is necessary, such as detoxification clinics.

Due to the good separability between lateral histograms, these can be classified directly into the three output categories. The available complete dataset for this classifier is shown in Figure 8.

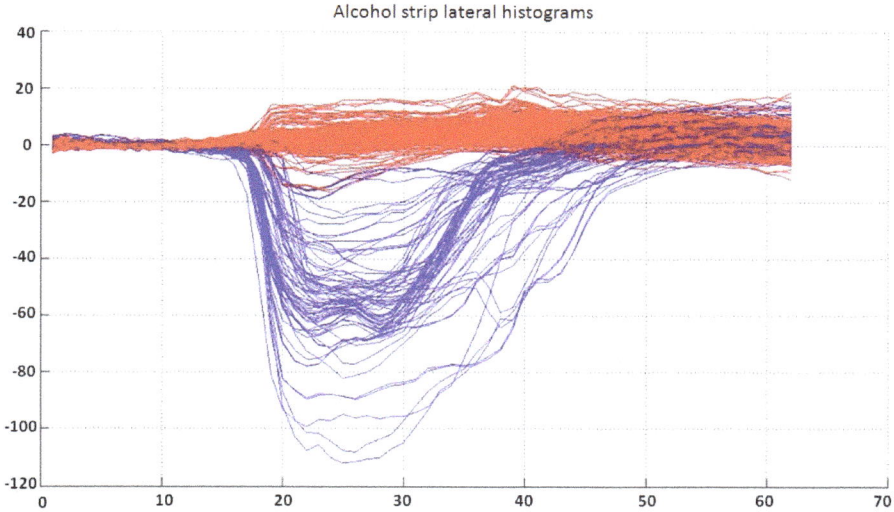

Figure 8. Lateral histograms used as the dataset for the classifier for the alcohol strip. Samples labeled as "negative" are represented in red, while "positive" samples are represented in blue. Note that the separability between classes is very good, and lateral histograms can be easily classified into three output categories ("positive", "negative" or "undetermined").

After an intensive training and output data analysis, it was determined that the best results were achieved using a classifier structure consisting of an ANN structure with 62 inputs, five neurons in a single hidden layer and two output neurons.

The evaluation of the classifier in the three data sub-sets after the structures and parameters are calculated is shown in Table 2. The performance of the classifier is excellent, showing no incorrect classifications.

Table 2. Evaluation of the classifier for alcohol strip lateral histograms. Note that confusion matrices have dimensions of 2×3, because "undetermined" samples were never considered as an input. "Undetermined" samples are just the classification output when a certain confidence level is not reached.

	Success	Confusion Matrix (P, N, U)
Training data (238 samples)	100%	$\begin{bmatrix} 52 & 0 & 0 \\ 0 & 186 & 0 \end{bmatrix}$
Validation data (50 samples)	100%	$\begin{bmatrix} 12 & 0 & 0 \\ 0 & 38 & 0 \end{bmatrix}$
Test data (50 samples)	100%	$\begin{bmatrix} 10 & 0 & 0 \\ 0 & 40 & 0 \end{bmatrix}$

4.4.2. Classifier for Control Lines in Non-Alcohol Strip Lateral Histograms

The inputs to this classifier are the first 31 bins of the preprocessed lateral histogram of a non-alcohol strip (inside a black box in Figure 9). The output of the trained classifier can be "valid (1)", "invalid (2)" or "undetermined (3)". The complete dataset available for this classifier is shown in Figure 9.

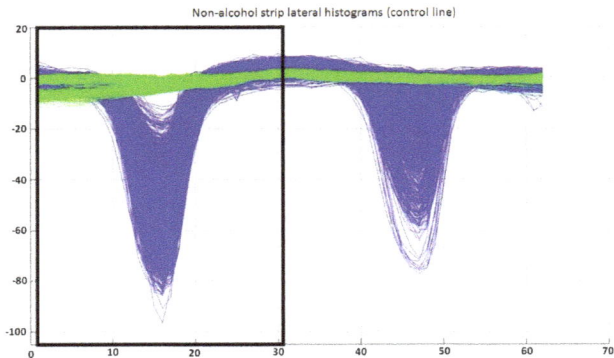

Figure 9. Lateral histograms used as the dataset for the classifier for control lines for non-alcohol strip lateral histograms. Samples labeled as "invalid" are represented in green, while "valid" samples are represented in blue. Note that only the first 31 bins of each lateral histogram are used by this classifier. It can also be observed that the separability between classes is very good, allowing the use of only three classes in this classifier ("valid", "invalid" or "undetermined").

After training and analyzing the output data, it was determined that the best results were achieved using a classifier structure consisting of an ANN structure with 31 inputs, a single neuron in a single hidden layer and two output neurons.

The evaluation of the classifier in the three data sub-sets after the structures and parameters are calculated is shown in Table 3. Again, the performance of the classifier is excellent, showing no incorrect classifications.

Table 3. Evaluation of the classifier for control lines in non-alcohol strip lateral histograms. Note that confusion matrices have dimensions of 2×3 because "undetermined" samples were never considered as an input. "Undetermined" samples are just the classification output when a certain confidence level is not reached.

	Success	Confusion Matrix (V, I, U)
Training data (1660 samples)	100%	$\begin{bmatrix} 1396 & 0 & 0 \\ 0 & 264 & 0 \end{bmatrix}$
Validation data (349 samples)	100%	$\begin{bmatrix} 293 & 0 & 0 \\ 0 & 56 & 0 \end{bmatrix}$
Test data (349 samples)	100%	$\begin{bmatrix} 293 & 0 & 0 \\ 0 & 56 & 0 \end{bmatrix}$

4.4.3. Classifier for Test Lines in Non-Alcohol Strip Lateral Histograms

The inputs to this classifier are the last 31 bins of the preprocessed lateral histogram of a non-alcohol strip (inside a black box in Figure 10). The output of the trained classifier can fall into six different categories: "very positive (1)", "positive (2)", "doubtful (3)", "negative (4)", "very negative (5)" or "undetermined (6)". Although there are only three possible output test results, "positive", "negative" or "undetermined", the internal use of a larger number of categories (see Figure 11) by the classifier allows one to have enhanced control of the treatment of doubtful cases, in order to adjust for a desired false positive or false negative ratio. The available complete dataset for this classifier is shown in Figure 10.

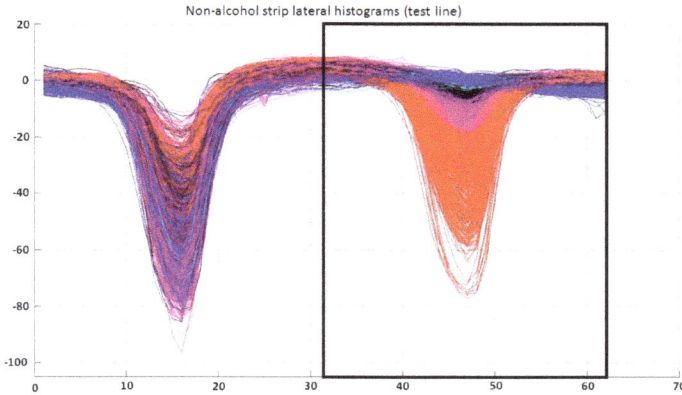

Figure 10. Lateral histogram samples used as training data for the classifier for test lines in non-alcohol strip lateral histograms. Samples labeled as "very negative" are represented in red; "negative" in magenta; "doubtful" in black; "positive" in cyan; and "very positive" in blue. Note that only the last 31 bins of each lateral histogram are used by this classifier. It can also be observed that the separability between classes is not very good, which justifies the use of six classes in this classifier ("very positive", "positive", "doubtful", "negative", "very negative" or "undetermined").

Once the training procedure was completed and the output data were analyzed, it was determined that the best results were achieved using a classifier structure consisting of an ANN structure with 31 inputs, two hidden layers with seven neurons each and five output neurons. The evaluation of the classifier in the three data sub-sets after the structures and parameters are calculated is shown in Table 4. The performance of the classifier is very good, showing only a few errors between adjacent labeled classes.

Table 4. Evaluation of the classifier for the test lines in non-alcohol strip lateral histograms. Note that confusion matrices have dimensions of 5 × 6 because "undetermined" samples were never considered as an input. "Undetermined" samples are just the classification output when a certain confidence level is not reached.

	Success	Confusion Matrix (VP, P, D, N, VN, U)					
Training Data (1369 samples)	100%	459	0	0	0	0	0
		0	122	0	0	0	0
		0	0	200	0	0	0
		0	0	0	258	0	0
		0	0	0	0	330	0
Validation data (293 samples)	91.809%	94	2	0	0	0	0
		5	19	2	0	0	0
		0	4	39	4	0	0
		0	0	2	49	3	0
		0	0	0	2	68	0
Test data (293 samples)	89.761%	93	2	0	0	0	0
		5	22	4	0	0	0
		0	1	35	8	0	0
		0	0	6	45	2	0
		0	0	0	2	68	0

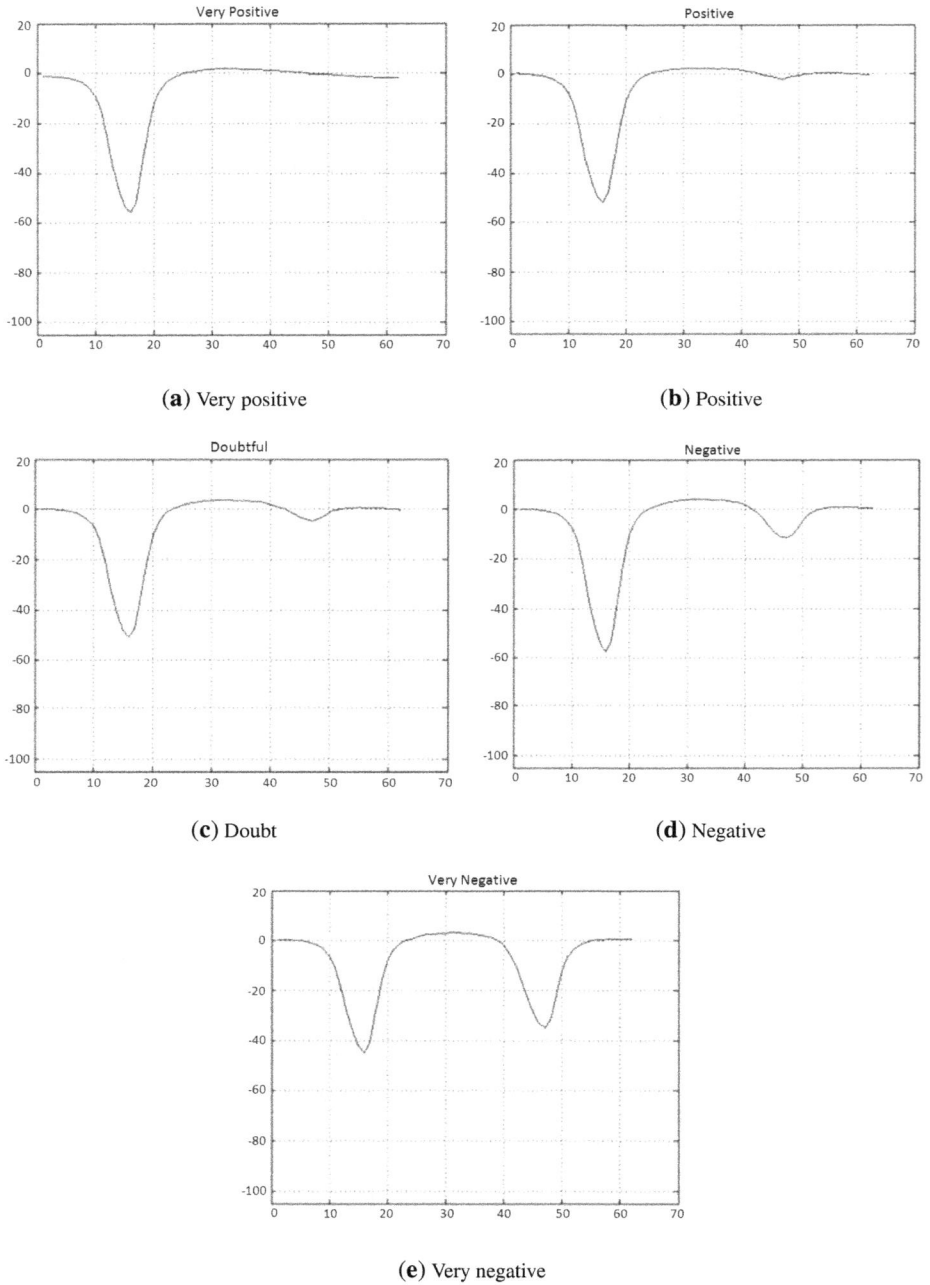

(**a**) Very positive

(**b**) Positive

(**c**) Doubt

(**d**) Negative

(**e**) Very negative

Figure 11. Lateral histograms representing a prototype of each labeled class used for the classifier for test lines in non-alcohol strips.

5. System Evaluation Methodology

A study was conducted by an external company to verify the performance of the application with regards to the two following aspects:

- Verify the agreement between visual results obtained by human operators and those obtained by the test reader (agreement).
- Check the repeatability of the results interpreted by the test reader (precision).

The following values should be established as an outcome of the testing:

1. Provide the number of cases (%) in which the results obtained by the test reader agreed with the interpretation made by an operator by visual interpretation, on a given representative sample size (N) (at least two or three independent operators should be considered, to minimize bias or the impact of subjectivity).
2. Provide the number of cases (%) in which the test reader is able to repeat the same result (positive, negative or invalid) when interpreting any test, given a representative sample size (N).

Ninety SalivaScan double-sided tests were used (detection of six drugs + alcohol in oral fluid) with the following configuration: amphetamine, ketamine, cocaine and methamphetamine on one side and opiates, marijuana and alcohol on the other side, with the cutoff levels indicated in Table 1. Three detection levels were considered: negative, cutoff and $3\times$ cutoff (saliva controls with three-times the cutoff concentration level).

In order to simulate the concentration on each detection level, positive and negative standard saliva controls were used. Such controls were sourced from Synergent Biochem Inc. (Hawaiian Gardens, CA, USA) [24].

The visual and automated results obtained for the alcohol (ALC) strip were not recorded, as the interpretation for the results is made by comparison of the color changes in the reagent pad against a pattern and not by the interpretation of the control and test lines.

The results interpreted by the test reader were extracted using an iPhone 4 (partial test) and an iPhone 4S (full test), both running iOS Version 6.1.3.

5.1. Agreement Test

In the agreement test, the results of 30 tests per detection level were considered. Two human operators made the visual interpretation of the results obtained on each test, and the second operator obtained a result using the test reader in addition to his or her own interpretation. Each operator would log the results independently from each other. The detailed steps followed on each test were the following:

1. The reaction on each test was started after adding the corresponding positive or negative control depending on the cutoff level being tested.
2. Operator 1 waited 10 min for incubation of the result, interpreted it and logged the test outcome.
3. Operator 2 interpreted the visual result, independently from Operator 1, and logged the test outcome.
4. Operator 2 interpreted the result using the test reader and logged the result.

A retest was performed at the negative and $3\times$ detection levels, using a different lot. On each level, 20 tests were performed.

5.2. Precision Test

Test devices from the agreement evaluation were randomly selected, three tests from each detection level (negative, cutoff, $3\times$). Test 1 corresponds to an (amphetamine (AMP), ketamine (KET), cocaine (COC), methamphetamine (MET)) configuration, while Tests 2 and 3 had an (opiates (OPI), THC, ALC) configuration. Each of these tests was processed 40 times repeatedly.

6. Results

6.1. Agreement Test Results

Agreement test results are summarized in Table 5. Results show that there are differences each time the visual interpretation is made by each human operator. Results also show how the disagreement between expert operators can be substantial (20% to 30%), which proves that interpreting the lines is not an easy task, especially in doubtful cases, and that specific training is necessary for correct interpretation. Usually, these interpretation differences occur when test lines are faint or remains of the reagent are present on the test strips. In such cases, it is normal that there are doubts about the result obtained, and this situation occurs normally when interpreting results at the cutoff level in contrast to the results obtained at the negative and 3× levels, which are expected to be easier to classify. It should be noted that the agreement on the THC strip in the levels of negative and 3× is very low, and not as expected (it should be near 100%), this was due to the fact that the results obtained on those levels showed a high number of faint test lines that caused doubts while the operators interpreted the results.

Table 5. Agreement test results summary. OP, operator. TR, test reader.

Agreement OP1 *vs.* OP2 Test/Retest	AMP	KET	COC	MET	OPI	THC	Average
Total Agreement (%)	93/100	93/100	100/100	89/95	100/80	87/100	94/96
Negative (%)	100/100	100/100	100/100	100/100	100/100	83/100	97/100
Cutoff (%)	80/-	90/-	100/-	67/-	100/-	97/-	89/-
3× (%)	100/100	90/100	100/100	100/90	100/60	80/100	95/92
Agreement OP1 *vs.* TR Test/Retest	**AMP**	**KET**	**COC**	**MET**	**OPI**	**THC**	**AVG**
Total Agreement (%)	87/100	92/100	100/95	91/100	98/95	94/98	93/98
Negative (%)	100/100	100/100	100/100	100/100	100/100	100/100	100/100
Cutoff (%)	60/-	93/-	100/-	73/-	93/-	85/-	84/-
3× (%)	100/100	83/100	100/90	100/100	100/90	96/95	97/96
Agreement OP2 *vs.* TR Test/Retest	**AMP**	**KET**	**COC**	**MET**	**OPI**	**THC**	**AVG**
Total Agreement (%)	87/100	92/100	100/95	89/95	98/75	69/98	89/94
Negative (%)	100/100	100/100	100/100	100/100	100/100	71/100	95/100
Cutoff (%)	60/-	97/-	100/-	67/-	93/-	60/-	79/-
3× (%)	100/100	79/100	100/90	100/90	100/50	76/95	93/88
OP1 and OP2 *vs.* TR Test/Retest	**AMP**	**KET**	**COC**	**MET**	**OPI**	**THC**	**AVG**
Total Agreement (%)	90/100	96/100	100/95	96/100	96/95	72/98	91/98
Negative (%)	100/100	100/100	100/100	100/100	93/100	70/100	94/100
Cutoff (%)	70/-	100/-	100/-	87/-	93/-	60/-	85/-
3× (%)	100/100	86/100	100/90	100/100	100/90	86/95	95/96

There was a large number of results obtained in the rapid test in which the test reader (TR) judged the result as negative, while the operators (OPs) judged such results as positives (OPs positive–TR negative). This was caused by the fact that the test lines in some strips were very faint, which induced the operators to judge the results incorrectly as positive when these should have been labeled as negative, as determined by the test reader.

The cases when the operators judged the result as negative and the test reader interpreted it as positive can be explained by the development of faint test lines.

There are discrepancies in the agreement between operators (OP1 *vs.* OP2) when interpreting the results on the OPI strip, especially at the 3× detection level; this may be due to the appearance of stains or color remaining on the reagent strip.

The agreement of Operator 2 is substantially lower than that of Operator 1. It can be seen that Operator 2 had some doubts while interpreting the results for the OPI strip at the level of 3×.

As can be seen, all errors correspond to the strips tested where it was expected to obtain positive results (3×). At such levels, the test reader interpreted the result as positive, while the operators judged the result as negative. This discrepancy may probably be associated with the presence of color remaining on the strip, which might give the impression to the operators of a test line, while such lines did not have the typical characteristics to be considered as test lines.

The cases in which both operators did not agree on the results obtained by the test reader can be classified in the categories shown in Table 6, as well as the results as a percentage of the total number of strips evaluated.

Table 6. Disagreement and adjusted agreement test results summary.

Both Operators *vs.* Test Reader	AMP	KET	COC	MET	OPI	THC	Total
Disagreements OPs negative–TR positive count (%)/retest count (%)	0 (0)/0 (0)	0 (0)/0 (0)	0 (0)/2 (5)	0 (0)/0 (0)	0 (0)/2 (5)	3 (3)/1 (3)	3 (1)/5 (2)
Disagreements OPs positive–TR negative count (%)/retest count (%)	9 (10)/0 (0)	4 (4)/0 (0)	0 (0)/0 (0)	4 (4)/0 (0)	2 (2)/0 (0)	20 (22)/0 (0)	39 (7)/0 (0)
Disagreements TR error count (%)/retest count (%)	0 (0)/0 (0)	0 (0)/0 (0)	0 (0)/0 (0)	0 (0)/0 (0)	2 (2)/0 (0)	2 (2)/0 (0)	4 (1)/0 (0)
Total disagreements count/retest count	9/0	4/0	0/2	4/0	4/2	25/1	46/5
Total adjusted agreement (%)/retest (%)	100/100	100/100	100/95	100/100	98/95	94/98	99/98
Total test count/retest count	89/40	89/40	89/40	89/40	89/40	89/40	534/240

Excluding the cases in which the operators indicated positive when the test reader indicated negative and counting them as interpretation errors attributable to the operators, the total adjusted agreement (excluding cases when OPs positive and TR negative) has been computed and is shown in Table 6.

6.2. Precision Test Results

The results indicated in Table 7 were obtained once the tests were processed repeatedly using the test reader for each detection level. The cases in which the test reader interpreted the result as "negative" when the operator interpreted it as "positive" correspond to strips showing very faint test lines, especially at "cutoff Test 1" and "3× Test 1" on the strips KET and OPI.

Table 7. Precision test results.

Detection Level	No. Tests	No. Strips per Side	Total Strips [1]	No. of Times That TR Gave The Same Result for a Given Strip [2]	Precision (%)
Negative test 1/2/3 Precision for negative	40/40/40	4/2/2	156/78/78 312	156/78/78 312	100/100/100 100
Cutoff test 1/2/3 Precision for cutoff	40/40/40	4/2/2	160/80/80 320	149/78/71 298	93/98/89 94
3× test 1/2/3 Precision for 3×	40/40/40	4/2/2	156/80/80 316	137/80/80 297	88/100/100 94
Total	360		948	907	96

[1] Total number of strips correctly interpreted; invalid strips or errors excluded; [2] in the case of different results, the maximum value was taken.

7. Conclusions

An innovative, low-cost, portable approach for the rapid interpretation of lateral flow saliva test results in drug-of-abuse detection based on the use of commonly-available smartphone devices has been presented and evaluated. A small inexpensive light box is used to control image quality parameters during the acquisition. This solution reuses an existing smartphone, and there is no additional equipment needed, besides the light box, which costs a fraction of the price for similar products on the market, with prices ranging around 3000 EUR/device. In order to segment the strips, images are first pre-processed to correct for small displacements and/or rotations with an ORB feature-based matching and homography strategy, and the strips corresponding to the different substances to be detected are segmented using color features and morphological operations. Finally, a lateral histogram containing the saliva test lines' intensity profile is extracted. Lateral histograms are then classified with a machine learning-based procedure, including unsupervised data normalization and classification using a multilayer perceptron artificial neural network (MLP-ANN). The implemented solution can be adapted to any line-based or color-based lateral flow test on the market. System agreement and precision tests were run for system evaluation, showing great agreement between the visual results obtained by human operators and those obtained by the test reader app, while showing high repeatability. The objective of the work is to demonstrate that the test reader is able to obtain the same result (or better) than a trained operator, therefore reducing the subjectivity of the analysis by standardizing the test interpretation conditions (illumination, test to image sensor distance, *etc.*) and by using a deterministic algorithm to obtain the results. In this sense, any operator independent of his/her level of experience can rely on the results obtained by the test reader. The system is automatic, allowing one to systematize the collection and analysis of the data in real time, removing the risks of manual results' management and allowing for centralized processing. The mentioned features can be very useful in places where there is no qualified staff and rapid detection is needed, such as on-site detection performed by a police officer or a first-aid operator.

Author Contributions: All authors have contributed to the development of the work presented in this paper, and they are its sole intellectual authors, all collaborating tightly in several parts for the final success of the system. Carlos Sampedro has been responsible for the computer vision part. Jose Luis Sanchez-Lopez has been responsible for the machine learning algorithms. Adrian Carrio has been responsible for the software and system integration. Miguel Pimienta has staged the problem and final objectives of the system. He has developed the light box, also providing the samples, and has been responsible for the external evaluation process. Pascual Campoy has acted as the general supervisor of the technical development of the project.

Conflicts of Interest: The system has been developed under a private contract between the company Vincilab Healthcare S.L. (Madrid, Spain) and the Universidad Politécnica de Madrid (UPM). The resulting procedure and system are patent pending, and the exploitation of their results is shared by the two companies split from the original one, named Vincilab Healthcare S.L. and Aplitest Solutions S.L. (Madrid, Spain), both retaining all of the industrial rights.

References

1. Martinez, A.W.; Phillips, S.T.; Carrilho, E.; Thomas, S.W.; Sindi, H.; Whitesides, G.M. Simple telemedicine for developing regions: Camera phones and paper-based microfluidic devices for real-time, off-site diagnosis. *Anal. Chem.* **2008**, *80*, 3699–3707. [CrossRef] [PubMed]
2. Tseng, D.; Mudanyali, O.; Oztoprak, C.; Isikman, S.O.; Sencan, I.; Yaglidere, O.; Ozcan, A. Lensfree microscopy on a cellphone. *Lab Chip* **2010**, *10*, 1787–1792. [CrossRef] [PubMed]
3. Zhu, H.; Yaglidere, O.; Su, T.W.; Tseng, D.; Ozcan, A. Cost-effective and compact wide-field fluorescent imaging on a cell-phone. *Lab Chip* **2011**, *11*, 315–322. [CrossRef] [PubMed]
4. Zhu, H.; Sikora, U.; Ozcan, A. Quantum dot enabled detection of Escherichia coli using a cell-phone. *Analyst* **2012**, *137*, 2541–2544. [CrossRef] [PubMed]
5. Matthews, J.; Kulkarni, R.; Gerla, M.; Massey, T. Rapid dengue and outbreak detection with mobile systems and social networks. *Mob. Netw. Appl.* **2012**, *17*, 178–191. [CrossRef]
6. Dell, N.L.; Venkatachalam, S.; Stevens, D.; Yager, P.; Borriello, G. Towards a point-of-care diagnostic system: Automated analysis of immunoassay test data on a cell phone. In Proceedings of the 5th ACM Workshop on Networked Systems for Developing Regions, New York, NY, USA, 28 June 2011; pp. 3–8.
7. Mudanyali, O.; Dimitrov, S.; Sikora, U.; Padmanabhan, S.; Navruz, I.; Ozcan, A. Integrated rapid-diagnostic-test reader platform on a cellphone. *Lab Chip* **2012**, *12*, 2678–2686. [CrossRef] [PubMed]

8.   Mudanyali, O.; Padmanabhan, S.; Dimitrov, S.; Navruz, I.; Sikora, U.; Ozcan, A. Smart rapid diagnostics test reader running on a cell-phone for real-time mapping of epidemics. In Proceedings of the Second ACM Workshop on Mobile Systems, Applications, and Services for HealthCare, Toronto, ON, Canada, 6–9 November 2012.

9.   Shen, L.; Hagen, J.A.; Papautsky, I. Point-of-care colorimetric detection with a smartphone. *Lab Chip* **2012**, *12*, 4240–4243. [CrossRef] [PubMed]

10. Oncescu, V.; O'Dell, D.; Erickson, D. Smartphone based health accessory for colorimetric detection of biomarkers in sweat and saliva. *Lab Chip* **2013**, *13*, 3232–3238. [CrossRef] [PubMed]

11. De la Fuente, J.B.; Garcia, M.P.; Cueli, J.G.; Cifuentes, D. A new low-cost reader system for ELISA plates based on automated analysis of digital pictures. In Proceedings of the IEEE Instrumentation and Measurement Technology Conference, Sorrento, Italy, 24–27 April 2006; pp. 1792–1794.

12. Velikova, M.; Lucas, P.; Smeets, R.; van Scheltinga, J. Fully-automated interpretation of biochemical tests for decision support by smartphones. In Proceedings of the 25th International Symposium on Computer-based Medical Systems, Roma, Italy, 20–22 June 2012; pp. 1–6.

13. DrugCheck VeriCheck. Available online: http://www.drugcheck.com/dc_vericheck.html (accessed on 22 November 2015).

14. Securetec DrugRead. Available online: http://www.securetec.net/en/products/drug-test/drugread-device.html (accessed on 22 November 2015).

15. DrugCheck SalivaScan. Available online: http://www.drugcheck.com/dc_salivascan.html (accessed on 22 November 2015).

16. Rublee, E.; Rabaud, V.; Konolige, K.; Bradski, G. ORB: An efficient alternative to SIFT or SURF. In Proceedings of the IEEE International Conference on Computer Vision, Barcelona, Spain, 6–13 November 2011; pp. 2564–2571.

17. Rosten, E.; Drummond, T. Machine Learning for High-Speed Corner Detection. In *Computer Vision–ECCV 2006*; Springer: Berlin, Germany, 2006; pp. 430–443.

18. Rosten, E.; Porter, R.; Drummond, T. Faster and better: A machine learning approach to corner detection. *IEEE Trans. Pattern Anal. Mach. Intell.* **2010**, *32*, 105–119. [CrossRef] [PubMed]

19. Calonder, M.; Lepetit, V.; Strecha, C.; Fua, P. Brief: Binary Robust Independent Elementary Features. In *Computer Vision–ECCV 2010*; Springer: Berlin, Germany, 2010; pp. 778–792.

20. Hamming, R.W. Error detecting and error correcting codes. *Bell Syst. Tech. J.* **1950**, *29*, 147–160. [CrossRef]

21. Fischler, M.A.; Bolles, R.C. Random sample consensus: A paradigm for model fitting with applications to image analysis and automated cartography. *Commun. ACM* **1981**, *24*, 381–395. [CrossRef]

22. Otsu, N. A threshold selection method from gray-level histograms. *IEEE Trans. Systems Man Cybern.* **1979**, *9*, 62–66.

23. Cybenko, G. Approximation by superpositions of a sigmoidal function. *Math. Control Signals Syst.* **1989**, *2*, 303–314. [CrossRef]

24. Synergent Biochem Inc. Homepage. Available online: http://www.synergentbiochem.com (accessed on 22 November 2015).

Article

Implementation of Context Aware e-Health Environments Based on Social Sensor Networks

Erik Aguirre [1], Santiago Led [1], Peio Lopez-Iturri [1], Leyre Azpilicueta [2], Luís Serrano [1] and Francisco Falcone [1,*]

[1] Electrical and Electronic Engineering Department, Public University of Navarre, Pamplona 31006, Spain; erik.aguirre@unavarra.es (E.A.); santiago.led@unavarra.es (S.L.); peio.lopez@unavarra.es (P.L.-I.); lserrano@unavarra.es (L.S.)

[2] School of Engineering and Sciences, Tecnologico de Monterrey, Monterrey 64849, Mexico; leyre.azpilicueta@itesm.mx

* Correspondence: francisco.falcone@unavarra.es; Tel.: +34-948-169-741; Fax: +34-948-169-720

Academic Editor: Gonzalo Pajares Martinsanz
Received: 23 December 2015; Accepted: 24 February 2016; Published: 1 March 2016

Abstract: In this work, context aware scenarios applied to e-Health and m-Health in the framework of typical households (urban and rural) by means of deploying Social Sensors will be described. Interaction with end-users and social/medical staff is achieved using a multi-signal input/output device, capable of sensing and transmitting environmental, biomedical or activity signals and information with the aid of a combined Bluetooth and Mobile system platform. The devices, which play the role of Social Sensors, are implemented and tested in order to guarantee adequate service levels in terms of multiple signal processing tasks as well as robustness in relation with the use wireless transceivers and channel variability. Initial tests within a Living Lab environment have been performed in order to validate overall system operation. The results obtained show good acceptance of the proposed system both by end users as well as by medical and social staff, increasing interaction, reducing overall response time and social inclusion levels, with a compact and moderate cost solution that can readily be largely deployed.

Keywords: social sensors; wireless body area networks; deterministic radio planning; back office

1. Introduction

Population ageing is unprecedented, without parallel in human history—and the twenty-first century will witness even more rapid ageing than did the century just past [1]. This global phenomenon is affecting the whole world, although with different evolution rates of the process, depending on regions or countries. In any case, the main goal for the future is to ensure people everywhere will be ageing actively, making possible their participation in social activities without any restrictions.

Hence, to reach the above challenge, an embedded Social-Health care action or strategy where citizen empowerment should be the solution's central point is mandatory [2]. This strategy must take into account all personal and context factors such as personal health, work and economic situation, social networks, *etc.* that affect active aging and assisted living processes. This Social-Health care strategy should be based on massive use of Information and Communication Technologies (ICT) for making possible a service deployment with several main features as cost-effective, plug and play operation, minimal user's intervention, and helpful for the largest number of citizens.

Adopted ICT Social-Health care solutions could be included in two possible scenarios [3]. On the one hand, indoor monitoring scenarios, so-called home monitoring is defined where the user's state is controlled at his or her own residence. These scenarios provide more real information about the user due to the fact that the remote control process is achieved at a comfortable and regular environment.

On the other hand, outdoor monitoring scenarios that include all situations where the user is away from home (office, walking on street, gym, *etc.*) and is also continuously monitored. These scenarios provide mobility to the user while the control process is performed. In both scenarios, the devices (for medical or behavioral monitoring), are portable or wearable, general purpose, and user-friendly. Moreover, they are equipped with short range (Bluetooth, Zigbee, IrDA, *etc.*) and/or large range (GSM-GPRS, UMTS, *etc.*) wireless technologies.

Currently, the use of Wireless Sensor Networks (WSN) continues to grow in a wide variety of application fields [4–6], such as agriculture and farming [7], infrastructure state monitoring [8], location and guiding [9], vehicular communications [10] and healthcare monitoring [11,12], to name a few. The future trend seems to be the increase of the number of wireless nodes in order to collect more information from the surrounding environment, bringing the Internet of Things (IoT) to our daily life.

Among WSNs, Mobile Ad-hoc Networks (MANET) have attracted the attention of many research groups around the world, becoming popular due to the unique characteristics they provide: wireless mobile devices with limited resources that can exchange information with each other without any fixed infrastructure. Thus, MANETs provide easier deployment, system maintenance and upgrade. These characteristics make MANETs an adequate solution to solve efficiently many applications, as is the case of the telemedicine and healthcare system presented in this work. In fact, in the literature can be found several works of MANETs applied to healthcare environments, such as hospital and big in-building environments [13], tele-emergency projects [14], tele-care and telemedicine systems [15,16], real time medical data acquisition and patient monitoring systems [17–19], and more specific applications such as tele-cardiology [20], emergency telemedicine system in disaster areas [21] and monitoring combat soldiers [22].

In Spain, the Social-Sanitary System term (in Spanish "*Sistema Socio-Sanitario*") makes reference to the public or private healthcare system focused on guaranteeing the best population' state regarding general aspects as life quality, well-being, and social integration. Meanwhile, the Sanitary term makes reference to the healthcare system part responsible for solving physiological and psychological problems suffered by the population. The Social term includes all the aspects associated with population social problems such as loneliness, integration into the society, energy poverty, population ageing, among others; problems that require specific healthcare professionals like social workers or assistants. The Social-Sanitary System concept is deeply adopted by Spanish population.

Social problems suffered by population can be detected and monitored in the same way as clinical diseases. Thus, it is required to know the user's social state by means of specific devices and sensors responsible for acquiring outstanding parameters associated to daily activity, habits, and in general any information that makes possible to work a social problem out. In this sense, environment measurements are combined with information related with user behavior, providing additional insight into the socio-sanitary context of the user; e.g., water consumption, electricity and gas consumption, operation time of HVAC (heating, ventilating and air conditioning) system, open/close status of doors and windows, among many others. Thus, water consumption can provide information about frequency the user goes to the toilet or takes a shower. A very low consumption can indicate the user suffers some degenerative disease or poverty situation. Electricity/gas consumption together with operation time of HVAC system can indicate the user is in energy poverty situation; if the user does not have enough resources for operating HVAC system during a suitable time period, the consumption values will be very low compared to average value. Because all these problems have a social nature and they must be treated by social workers and assistants, the sensors used to acquire their associated information are called social sensors throughout this paper.

In general, social sensors are environmental acquisition devices that require minimum user intervention, and for this reason, they are placed at previously established locations; ceiling for temperature/humidity sensors, doors and windows for open–close detection, sink pipe for water consumption sensors, *etc.* Social sensors exhibit reduced size, wireless communication capabilities, and minimum current consumption allowing them to be battery powered and highly operation autonomous; this feature allows the reduction of maintenance service costs. Moreover, social sensors must ensure the lowest visual and structural impact when embedded within the planned service scenario. Personal health devices show features similar to social sensors. However, they do have greater variety concerning form factor and location within the monitoring scenario. Some health devices like weigh scale, blood pressure monitor, and glucose meter could be used only a few times every day and they do not require to be worn by the user all the time. Thus, these devices could be portable sensors always located at the same living space, mainly bathroom and bedroom; when a biomedical measurement must be taken, the user goes to the living space where the medical device remains and uses it. Other health devices like electrocardiogram monitors require a real time monitoring process and they must be worn by the user. These devices must be wearable sensors and their location inside the monitoring scenario is not fixed, but changes as the user moves.

With regards to ICT, there is no implementation difference between health and social sensors because they require similar functional blocks: analog front-end (AFE), A/D conversion, microprocessor, power supply, and wireless communication. The resources provided by these blocks are conveniently used according to the sensor type, intelligent-processing needs, autonomy requirements, *etc.* This solution is currently being used for ICT Social-Health care service in Navarra (in Spanish "*Navarra-ASISte-TIC*", NASISTIC) project deployment [22] as a proof of concept of an integral Social-Healthcare system monitored by Red Cross in Navarra, and focused mainly on providing high-quality services to vulnerable citizens like the elderly or people suffering mental illnesses. Moreover, Navarra region shows an exceptional scenario for this deployment because it includes populations living in a medium-size city, Pamplona, as well as sparse citizens living in rural and mountain surroundings.

The novelty of the proposal presented within this work is focused in the development of Integrated Socio-Sanitary Services based on Information and Communication Technologies and with minimal user interaction, providing integral user care given by inherent interdependence between Health and Social Environment. In the context of elderly persons, a main target of the system (with elements such as dependence, loneliness, assisted living, *etc.*), this combined rollout is of particular importance, because degradation of health state can be caused by the degradation of the social context (energetic poverty, poor nutrition, cleanliness, *etc.*) and *vice versa*. In this way, deployments such as NASISTIC are of special interest in order to provide integral control of elderly people, people in risk of social exclusion, chronic patients, among others.

In this work, the implementation of Social Sensor devices and their application to e-Health monitoring within the framework of an urban scenario, given by project NASISTIC in Navarra, Spain, will be analyzed. Different considerations of physical implementation of the devices, considering multiple signal capability as well as usability factors (such as ergonomics and power consumption) will be analyzed. The usability and performance of the devices are strongly dependent on the behavior and influence of wireless transceivers, which are analyzed by deterministic techniques, providing assessment in the configuration of individual nodes as well as in the location and number of transceivers as a function of the scenario. The developed sensors will then be tested in a living lab configuration, which serves as the base for a future full-scale deployment.

The paper is structured as follows: Section 2 describes the architecture and functionality of the Social Sensor Devices, based on a multi signal Input/Output controller in which different sensors can potentially be implemented. Section 3 analyzes the impact of the wireless channel behavior, which is a key parameter to evaluate overall performance of the Social Sensor devices in terms of mobility, security and quality of the exchanged information, as well as on the design of individual nodes and layout of node network architecture. In Section 4, initial tests within a Living Lab framework are described, providing insight in complete system operation and the potential full-scale deployment. Finally, Section 5 provides conclusions and final remarks of the work.

2. Related Work

Different solutions based on multiple sensor platforms and configurations have been proposed in order to implement remote health-monitoring, diagnostics and medical services, with special interest in the past decade [23–25], due to their capability of reducing overall costs and increasing quality of life metrics. Solutions have been provided in the context of application of Information and Communication Technologies in order to allow system interoperability and enhance medical service, by means of solutions such as electronic patient records or digital image records, to name a few. A step further is achieved by adding mobility to previous e-Health scenarios, by means of integration of mobile terminals, connected to Public Land Mobile Networks (PLMN), allowing initial tele-monitoring capabilities. Other solutions, such as advanced care and alert portable telemedical monitor (AMON) project [26] employed GSM transmission links to communicate a wrist wearable device, with capability of measuring multiple signals, such as Electrocardiograms (ECG), Electromyograms (EMG), location or skin conductance. Cordless phones have also been employed as transmitter device, to which a specialized sensor unit was connected, implementing a real time wireless physiological monitoring system [27]. Prior to the popularization of Smart Phones, several solutions employed Personal Digital Assistants (PDA), with basic control functions of multiple sensor devices connected to sensor boards [28].

One of the main drivers in order to increase system interactivity is the use of compact size biomedical sensors, which can communicate in real time with medical specialists or social services, or can perform data-logging functions in order to store and send the relevant information later. In this sense, communication systems shift from PLMN based systems to Wireless Personal Area Network/Body Area Network devices, with predominance in the use of 802.15.4 standards, such as Bluetooth or ZigBee [23]. The main benefit in this approach is the inherent capability of connecting multiple devices, small size and reduced energy consumption as compared with PLMN/Wireless Local Area Network (WLAN) solutions.

Another relevant aspect in the adoption of remote health monitoring solutions is the advance in the implementation of sensors as well as the capability of embedding them in truly wearable configurations. In this sense, it is possible to measure multiple bio-physical parameters, such as glucose levels, blood pressure, oxygen saturation, respiration rate, ECG, and EMG, which can be combined with environmental parameters and user movement and location. The use of the combined information of these multiple sensors and sources leads to truly context aware Ambient Assisted Living (AAL) scenarios, with multiple solutions reported [23]. A following step in system integration is embedding sensors in textiles within user garments, in order to increase ergonomics as well as in depth user monitorization. Different solutions have been proposed, such as the MagIC vest or the MyHeart instrumented shirt, in which multiple sensor elements have been included within the clothing [29–33]. As a further step, specific software development frameworks, such as SPINE, have been proposed and multiple Body Sensor Network applications have also been proposed, such as rehabilitation, Gait analysis, emotional stress detection or handshake detection, to name a few [34]. Table 1 presents a comparison of different remote health monitoring systems.

Table 1. Health monitoring system comparison.

System	Transmission System	Architecture	Application Scenario	Detected Variables	Reference
LiveNET	PDA Connectivity	Centralized	Detection of Epilepsy Seizure, Parkinson symptom detection, Soldier Health Monitoring	3D Accelerometer, ECG, EMG, galvanic skin conductance	[25]
AMON	Wrist bracelet, GSM	Operator Based	Estimation of Patient Health Conditions	Blood Pressure, Blood Saturation, Skin Temperature, ECG	[26]
LifeGuard	Bluetooth to a base station	Centralized	Multiparameter Wearable Monitoring System	ECG, respiration rate, heart rate, oxygen saturation, body temperature, blood pressure, body movement.	[28]
Real Time Wireless Physiological Monitoring System	Low Power Cordless Phone To Base Station	Centralized	System Aid in Nursing Centers and Hospitals	Blood Pressure, Heart Rate and Temperature	[27]
Brain Injury Monitoring System	Bluetooth to Home PC	Centralized	Monitoring of Brain Injured Infants	Blood Saturation, Heart Rate, Respiration, Body Movement.	[29]
MyHeart	Communication to Data Logger	Off-line	Wearable System for Heart Disease Monitoring. Sensors knitted or embedded in garment.	ECG, Activity Sensor.	[30]
Wearable Health Care System (WEALTHY)	Bluetooth/GPRS	Centralized or Datalogger	Application to clinical patients during rehabilitation, elderly people, patients with chronic diseases	ECG, EMG, thoracic and abdominal respiration rate, body position, movement	[31]
MagIC	Bluetooth	Centralized	Woven textile sensors in a washable vest	ECG, respiration rate, motion level.	[32]
Medical Remote Monitoring of Clothes (MERMOTH)	RF Link to PDA	Data Logger	Wearable and Stretchable Sensing Garment	ECG, respiratory inductance plethysmography, skin temperature, activity.	[33]
NASISTIC	Bluetooth/2G-4G Connection	Locally Distributed/Centralized	Social Sensor Node Deployments	Combination of Biophysical signals (ECG) with user habits.	

3. Social Sensor Devices

In order to analyze the wireless communication in a generic social-sanitary monitoring system and emulate the operation of any acquisition sensor, a specific evaluation system has been developed. The system consists of two basic Bluetooth wireless modules: transmitter and receiver. The first one represents the health/social sensor, and it is placed at different locations inside the monitoring scenario in order to simulate the behavior and communication process of a real system. The second one represents the gateway device responsible for gathering the information acquired by sensors. Although this device could be implemented in a real smartphone or tablet platform, most of indoor scenarios include a set-top box for receiving data. For this reason, the receiver in the evaluation system is placed on a fixed location while the transmitter is moved around the scenario. Several wireless transceiver technologies have been analyzed and tested, mainly in the Wireless Body Area Network (WBAN)/Wireless Personal Area Network (WPAN) context. The election of Bluetooth allows inherent integration of mobile terminals such as smartphones within the Social Sensor scenario, enabling a wide variety of applications and functionalities which can be used by the end user, as well as by medical staff, social working staff and technical staff, increasing overall interaction and service provision.

Transmitter and receiver are *Ad-Hoc* evaluation modules connected to a PC or Laptop via serial port, which allow the execution of two relevant tasks. On the one hand, the configuration of data chunk size and transmission period are used by sensor module. By modifying these transmission parameters, it is possible to emulate the communication behavior of any acquisition sensor; from devices which transmit huge amounts of information in a continuous way being the case of real-time/event-driven electrocardiogram monitors, to devices with low data size and transmission rate requirements just like blood pressure monitors or humidity sensors. On the other hand, this allows the reception of link quality measurements. The receiver module provides information about received power in form of Received Signal Strength Indication (RSSI) levels and communication channel quality by means of Bit Error Rate (BER) values. These parameters are really useful in the radio propagation study. Figure 1 shows implemented evaluation modules as well as a final social sensor hub device.

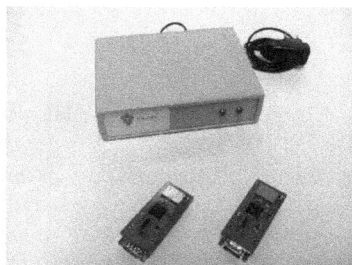

Figure 1. Evaluation module of the implemented Social Sensor device and a Home Hub.

The evaluation module contains several functional blocks (Figure 2). These blocks have been designed and implemented according to very low power consumption and reduced form factor requirements. Although the module is an evaluation board focused on providing versatility for modeling any acquisition sensor, the design restrictions mentioned previously allow the evaluation module to get close the features of portable and wearable real sensors. The description of the main functional blocks is the following:

- Wireless communication: This block manages all the Bluetooth communication tasks like device search, connection establishment, flow control, and data transmission, among others. Bluegiga's WT12 module is used for implementing this block. The module includes the whole protocol stack, application profiles, a virtual machine for execution of software code, and several General

Purpose Input/Output (GPIO) terminals. Concerning wireless communication, the evaluation module is a Bluetooth class 2 device with maximum transmission power of 3 dBm and data rates up to 3 Mbps. The application data exchange is based on Serial Port Profile (SPP), which allow the emulation of a serial wired RS-232 communication between transmitter and receiver.

- Microcontroller: This block is based on a 16-bit ultra-low power microprocessor, which achieves WT12 module management, acquired data processing, and serial wired communication. The microcontroller receives RSSI and BER parameters from wireless block and transmits them to Laptop in order to be displayed and/or stored. Any sensor must include a microcontroller block according to the device's features and processing requirements; weigh scales and open/close sensors could require 8-bit low resources processors while electrocardiogram monitors could be based on 32-bit processors with high CPU clock frequency as well as memory, communication ports, and processing resources.

- Power supply: This block provides to evaluation module the voltage required for operating correctly. Based on high efficiency DC-DC regulators together with Low Voltage-Low Power (LV-LP) integrated circuits, the block generates 3.3 V voltage supply by means of two 1.5 V AAA batteries. The main features of these batteries are reduced size and weight, high capacity, and short-time pulse current support. This power supply configuration is representative of most health and social sensors. Although the use of coin cell batteries in order to reduce device's form factor could be appropriate, this type of battery is not widely used with Bluetooth classic technology because it does not provide enough capacity.

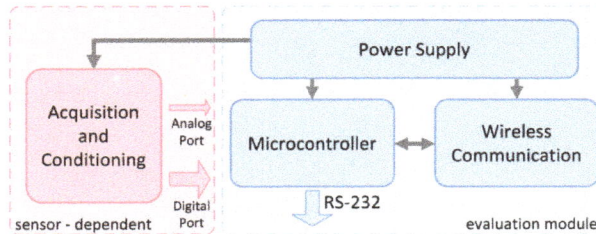

Figure 2. Evaluation Module architecture.

The evaluation module is depicted in Figure 3. Although the module has been specifically designed for performing radio propagation analysis, it shows features in fact close to real health/social sensors. On the one hand, the module does have reduced size (72.5 mm × 33.0 mm) and weight (24 g) as is required in portable and wearable devices. On the other hand, the module shows high operation autonomy thanks to use LV-LP integrated circuits together with optimized software implementation; in addition, a microcontroller and wireless communication block presents Deep Sleep states that allow reducing the power consumption. All these features make possible to have an evaluation module with features similar to commercial health and social sensors.

(a)   (b)

Figure 3. Evaluation module for the Social Sensor devices: (**a**) bottom view; and (**b**) top view.

Figure 4 shows a general architecture representative of any device or sensor. According to the device/sensor type, this architecture will include the required software module. The architecture layers are following:

- Transport layer: This layer implements the communication technology used by devices and sensor. In case of NASISTIC project, all devices use wireless Bluetooth technology. This layer also includes the data protocol required for exchange information between devices and set-top box. In this sense, two different protocols are used depending on the type of device or sensor. They are the following:

  - JAMP (*JSON Agent Management Protocol*) is a canned data protocol implemented in the social sensors that have been developed specifically for the project.
  - A manufacturer protocol implemented in medical devices. This is a binary data protocol defined by the manufacturer so it cannot be modified.

- Application layer: This layer can contain different software modules according to the device's features and functionality. In any case, there is an essential module that performs all the tasks related to measurement gathering; it is the so-called acquisition module. There are other optional software blocks such as User Interface and storage modules. All these modules and the interaction between application and transport layers are managed by a Kernel.

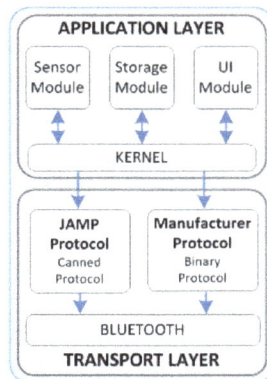

Figure 4. Software architecture for the sensor node devices.

In general, the gateway software makes the communication between the sensors and the back-end system easy. The software helps to deal with different transport technologies and protocols in order to gather data from a source (medical devices and social sensors) and deliver it to a destination (back-end system). The software architecture shown in Figure 5 defines the following main elements.

- Core: This is the most important element and includes all the software modules required for the management of layers, and the communication from one layer to each other.
- Plugins: This element represents all the software modules that can be connected to the platform in order to add features. The following are the main plugins.

  - Agents: These plugins are used to wrap connections between source and destination. The agent must be provided with a socket for performing a connection; thus, the socket makes the communication between the device/sensor and the back-end system possible. In addition, the agent provides its own services and methods: installation, open/close a connection, abort a connection, *etc.*
  - Transports: These plugins provide the different sockets required for agents. Obviously, the software architecture is able to use any implemented transport plugin, although NASISTIC software includes the following: Internet transport based on HTTP and TCP/IP protocols, and Bluetooth transport based on bluecove library.
  - Manager: This plugin registers the communication source and destination. These are kept in mind by the client application together its associated active agents. The manager also allows the client to establish a communication with any installed source/destination, or listen to incoming communication through any running transport.

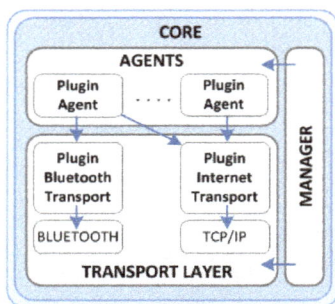

Figure 5. Software architecture for the gateway node device.

The back-end's architecture, depicted in Figure 6, is formed by several logical elements and abstraction layers. The following are the main blocks.

- The Spring Framework and Spring MVC (*Model-View-Controller*) allow the development of flexible and highly connected web applications.
- View layer: This layer provides the user interface according to the client request by means of the ZK library. This library provides an AJAX web framework in order to create a user interface through JAVA programming. In this way, it is possible to develop an environment with RIA (*Rich Internet Application*) features. The view layer also performs the data transmission to final client. The format used in the transmitted information is JSON (*JavaScript Object Notation*).
- Controller layer: It validates the data received from client, and selects the appropriate view for showing them.
- Data access layer: This layer is based on DAO (*Data Access Object*) elements and consists of two main logical blocks: the Java Persistence API focused on the management of persistence and objects mapping, and the Spring-ORM (*Object Relational Mapping*) module, which performs ORM container tasks.

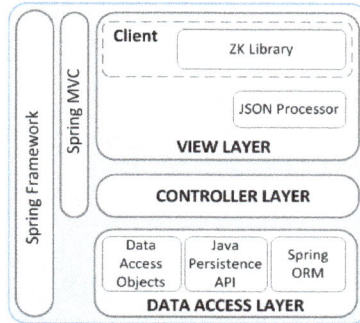

Figure 6. Back-end software architecture.

The architecture also includes a web-service layer for showing the measurements acquired by sensors. Afterwards, this information can be analyzed and exploited by the own back end system. This layer implements a REST (*Representational State Transfer*) web service. This simple technology has been selected for several reasons: simplicity of implementation in both client and server side, scalability in number of clients, and reduction of interaction latency between client and server, among others.

A real monitoring scenario for evaluation purposes based on social and health sensors is depicted in Figure 7. The scenario is a home monitoring service where the user's health and environment status is continuously acquired and received by a set-top box. The user's information is later transmitted by this gateway device to a remote call center in order for it to be displayed, stored, or analyzed by specialist staff. The system consists of multiple acquisition sensors; some of them (portable health and social sensors) are placed at fixed locations, and others (wearable health sensors) are worn by the user. Concerning wireless communication and processing capacity features, these acquisition sensors could be implemented with no particular restriction with the evaluation module's hardware introduced in this paper. Consequently, the real sensors should implement only a specific acquisition and conditioning block according to the type of measurement that will be taken. Some acquisition sensors that could be included in the monitoring service are described below.

- Blood pressure monitor: This health sensor is generally located in the same living space (bedroom or bathroom) and it is only used when the user must take a blood pressure measurement. This acquisition process is performed several times per day.
- Weigh scale: The performance of this sensor is similar to blood pressure monitor. It stays in the bathroom and provides weight measurements several times per day.
- Electrocardiogram monitor: This wearable sensor usually acquires the user's electrocardiographic signal continuously, and for this reason, it is not located at a fixed position. Instead, the user wears it all the time and the acquired electrocardiographic data are transmitted to the set-top box in real-time mode.
- Temperature/Humidity sensor: This social sensor is placed in every living space for taking temperature and humidity measurements frequently (one or two measurements per minute).
- Gas sensor: This sensor is usually placed in the kitchen and takes gas measurements frequently in order to detect possible gas leaks.
- Open/close detector.

Obviously, a real scenario could include other sensors like movement detector, water consumption monitor, glucometer, pill dispenser/reminder, among others; the monitoring service will include health and social sensors according to the control requirements of the user.

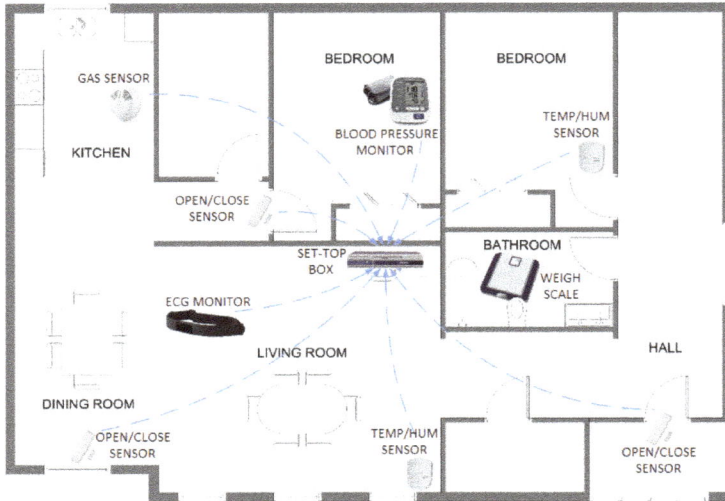

Figure 7. A home monitoring scenario in which a Social Sensor network is deployed.

In general, social sensors are characterized by enabling autonomous operation, and not requiring any specific user intervention. Sensors acquire the environment information periodically, and establish wireless communication with set-top box each time a relevant measurement is available. An important feature shown by these sensors is the possibility of configuring functional parameters like acquisition rate and transmission threshold; in this way, the sensors' power consumption can be optimized while maintaining the transmission of outstanding environment data. The scenario includes social sensors that require different responses by the monitoring service. On the one hand, sensors like temperature and humidity monitors that do not demand any fast intervention on user's environment if the acquired data exceed a fixed level. The system only must send the information to the call center and switch on the air conditioning unit. However, other sensors trigger a fast intervention on user's environment when the measurement value is over the threshold because it can imply a hazardous situation to the health; gas detector is a representative example of this type of sensor.

Concerning health sensors, electrocardiogram monitor operates generally in an autonomous way. When the user switches on the sensor, it starts the acquisition and transmission of electrocardiographic signal; the sensor's operation is completely transparent to the user. Other sensors (weigh scale, blood pressure monitor, glucose meter, *etc.*) require the user to take specific measurement steps: sensor switching on, blood pressure cuff placement, start button, wireless communication establishment, among others. In any case, all health information acquired by sensors is transmitted to the set-top box.

Independently of the sensor type, all of them must establish Bluetooth wireless communication with set-top box when there is outstanding data to transmit. Usually, the sensor will establish connection any time it must transmit information, although the sensor can also maintain an established connection permanently and transmit the information when is required. While the sensor does not have measurements to be transmitted, it can remain in low power consumption states (Park, Sniff and Hold are specific Bluetooth low power states) but the communication stays alive. Thus, set-top box receives continuously acquired data from all social and health sensors. Once the set-top box has received outstanding information, it is transmitted to the remote call center to be displayed and analyzed by specialist staff. This large-range communication can be achieved through data service of mobile phone network, although it will be carried out using Internet access with wired technologies, mainly ADSL services. This type of Internet access is widely available in society so use of these communication resources in order to send the information associated to the social-sanitary monitoring service is reasonable.

4. Characterization and Impact of Wireless Channel Behavior

Once the Social Sensor device has been designed and implemented and with the aim of testing system viability, wireless channel behavior has been estimated using an in-house developed 3D Ray Launching code and empirical measurements. The relevance of this test is given by the fact that wireless channels, especially in complex indoor environments lead to large signal degradation and hence, overall poor system operation. In this sense, the impact of the scenario in transceiver channel as a function of device location as well as on the obstacle density of the scenario under analysis will be tested. Estimation of propagation losses provides an estimation of received power level, which in turn can be compared with sensitivity thresholds. This way, coverage areas can be determined, which also depend on the information transmitted (*i.e.*, required bit rate and user mobility). In this way, the required node configuration can be implemented, in terms of antenna election, node placement and required number of nodes to be deployed. The simulation method has been widely tested in the literature [35–39] providing good results with a low computational cost.

The in-house developed 3D Ray Launching code is based on Geometrical Optics (GO) and the Uniform Theory of Diffraction (UTD). The principle method is that a bundle of rays are launched from the transmitter point with a horizontal and vertical angular resolution within a solid angle. Several transmitters can be placed within the indoor scenario. It is important to emphasize that the whole scenario is divided into a grid of cuboids, thus the parameters of the rays propagating along the space are stored in each cuboid for later computation. Parameters such as frequency of operation, radiation patterns of the antennas, number of multipath reflections and cuboid resolution are introduced. Electromagnetic phenomena such as reflection, refraction and diffraction have been also taken into account.

The scenario where a campaign of measurements has been deployed, depicted in Figure 8, corresponds to a typical office environment in the R&D Building of the Public University of Navarre. The scenario has been modeled three dimensionally and embedded in the simulator. All the objects and furniture within the environment have been considered, like the tables, chairs, shelves and windows, considering their material properties in terms of conductivity and dielectric constant. The size of the scenario is 13 m × 7 m × 4.2 m. In order to achieve a compromise between accuracy of the results and simulations computational time, the cuboids size and the considered number of reflections have been fixed to 0.1 m × 0.1 m × 0.1 m and 5, respectively.

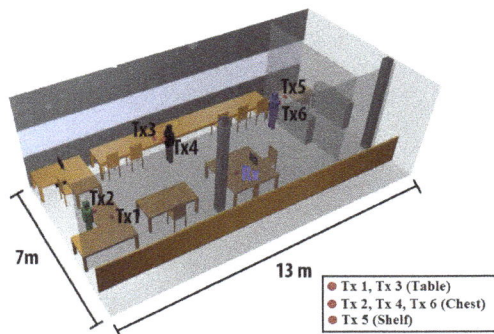

Figure 8. Schematic representation of the considered scenario with the three different positions of the human body and six different transmitter antenna points. The position of the receiver is also shown.

In order to emulate a real situation where a wireless communication system is working with a patient in a typical indoor environment, different positions of the person and the transmitter antenna have been considered. Figure 5 shows the fixed position of the receiver, which is on the surface of a table, emulating a fixed receiver device that is in charge of recollecting the data of the patient. Three different positions of the person have been considered, as depicted in Figure 8. Besides, for each person position, two different cases for the transmitter antenna have been analyzed. The first case was with the transmitter antenna above a surface (Transmitters 1, 3 and 5) emulating a fixed transmitter nearby the person, for example, a bascule. The second case (Transmitters 2, 4 and 6) was with the transmitter devices on-body, specifically in the chest of the person, emulating an on-body device, like a pacemaker.

Real antennas have been considered for simulation, taking into account their radiation diagram pattern, polarization, transceivers gain, and transmitted power. The simulation parameters are shown in Table 2.

Table 2. Simulation transmitter antenna characteristics.

Parameters in the Ray Launching Simulation	
Frequency	2.43 GHz
Transmitter power	0 dBm
Antenna gain	−1 dBi
Horizontal plane angle resolution ($\Delta\Phi$)	1°
Vertical plane angle resolution ($\Delta\theta$)	1°
Reflections	5

Figure 9 shows simulation results for Transmitters 1 and 2. It shows the bi-dimensional planes of received power for the receiver antenna height (1 m). It should be pointed out that Transmitter 1 was collocated on the table and the Transmitter 2 was on-body, specifically in the chest. The radio channel complexity in both planes can be observed. On the one hand, in the case of Transmitter 1, there is a higher influence of the surface of the table in the received values, and, on the other hand, in the case of Transmitter 2, a great impact of the person is observed due to the scattering originated for the position of the antenna in the chest of the person. In addition, the influence of walls and furniture in the environment is also represented. Due to the use of transceiver elements employing Bluetooth communication, intra-system interference is not a relevant issue in the present case, due to inherent interference control in channel access of transmitters within the network. In any case, it is compulsory to meet with receiver sensitivity levels in order to guarantee adequate service as a function of transmission bit rate.

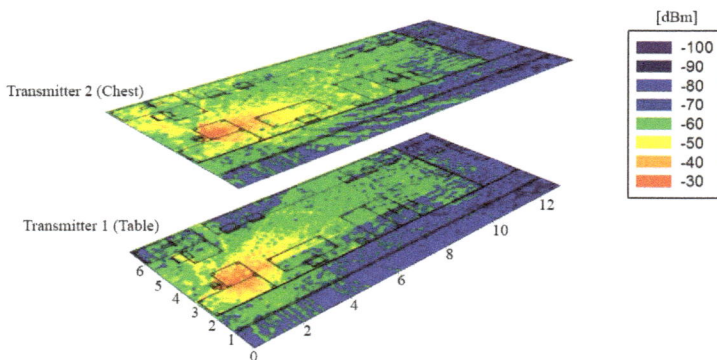

Figure 9. Bi-dimensional planes of Received Power (dBm) for the receiver antenna height (1 m) for two different transmitting cases, Transmitters 1 and 2.

As can be seen in Figure 9, the impact of multipath propagation in the environment is highly important. In order to represent this effect, the Power Delay Profiles (PDPs) in the receiver point for two different positions of the person have been depicted. Figure 10 represents the PDPs for Position 1 (green person in Figure 8) and Position 2 (black person in Figure 8) for both positions of the transmitter, above the table and on-body. It can be seen that multipath behavior is absolutely vital in this type of environment. Furthermore, not only does the position of the antenna (on-body or above a surface) have a great impact in electromagnetic propagation, but also the distance between transceivers: in the case of Position 2, there are higher values of multipath received values.

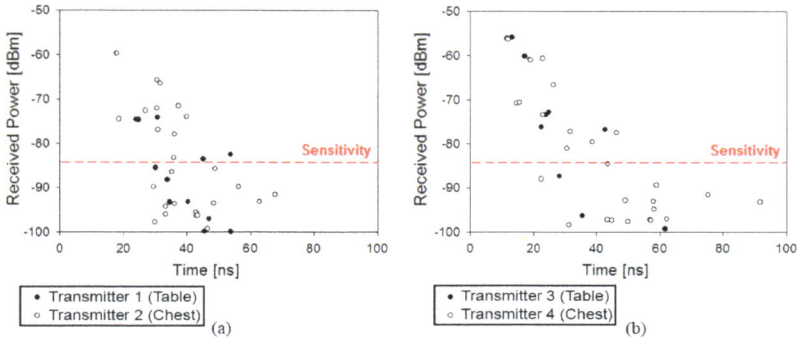

Figure 10. Comparison of Power Delay Profiles for both positions of the antenna (Transmitters 1 and 2): (**a**) Position 1 (green person in Figure 8); and (**b**) Position 2 (black person in Figure 8).

The delay spread can be a good indicator of the propagation dispersion when the complexity of the scenario is elevated. In Figure 11, a Delay Spread map is depicted for the case of Transmitters 1 and 2. Higher delays are visible nearby the transmitter antenna. This is due to the fact that higher values of power are received in these points and the contribution of the reflections is higher. Nevertheless, there are some critical points such as corners or around some obstacles where the delay spread could be elevated depending on the morphology of the environment. This is the reason why the Delay Spread and Power distribution changes completely when the position of the transmitter antenna changes. The impact of delay spread variation to end-user operation is given by the fact that larger fast fading losses can decrease overall received signal levels, with values below sensitivity thresholds and hence with large error rates. Moreover, reception of multiple propagation components increases bit error rate, given mainly by Inter Symbol Interference. In the case of Bluetooth transmitters, effective transmission bit rates are low compared to operation bandwidth, implying that conventional channel equalization techniques can be employed to mitigate potential information degradation. Care, however, should be taken if transmission rate is potentially increased (*i.e.*, by using new standards) or operational frequency is decreased.

Finally, the comparison between measurements and simulations for the aforementioned six points is depicted in Figure 12. Measurement results have been obtained with the aid of an Agilent 9912 portable spectrum analyzer, coupled to short vertical monopoles, tuned in the center of the 2.4 GHz band, consistent with Bluetooth specifications given by the employed transceivers. The spectrum analyzer data and the RSSI obtained from the device are compared and two different data rates are also considered. Good agreement between simulation and measurement data can be seen, even in the case of a complex scenario, obtaining a mean error of 0.18 dB and a standard deviation of 3.24 dB.

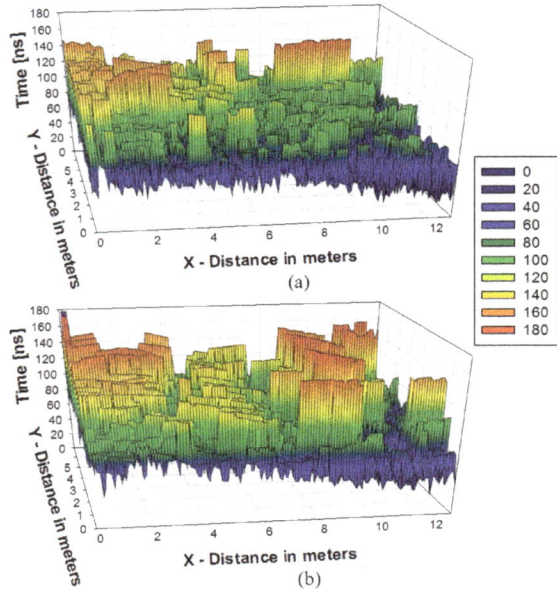

Figure 11. Delay Spread estimation at a bi-dimensional plane at 1 m height for Position 1 (green person in Figure 8): (**a**) Transmitter 1 (Table); and (**b**) Transmitter 2 (Chest).

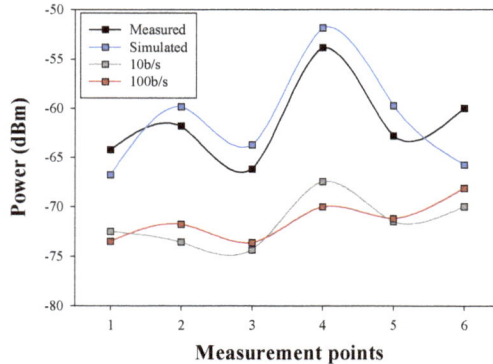

Figure 12. Comparison among simulated, Received Signal Strength Indication (RSSI) and measured power values.

The RSSI values usually have an error when they are compared with the spectrum analyzer caused by the kind of modulation used and therefore depending on the device used. In this case, an error of 10 dB is introduced and it is corrected in Figure 13. The existing error between corrected RSSI and the simulation is of 0.26 dB for 10 b/s and 0.07 dB for 100 b/s. The lower bit rate adjust better than the higher to the simulations power values, considering that the standard deviation in the first case is 4.1 dB and in the second case is 4.85 dB. The values that have been obtained are in all cases 15 dB above the sensitivity threshold of the employed transceivers, indicating that communication is feasible at any given location and position of the transceivers within the scenario under analysis, for the given transmitter to receiver linear radial. However, if location is changed, or if height is modified, sensitivity levels are not achieved. In the example scenario, the optimal configuration would be achieved with 3 transceivers (for the case of a relatively large room, as depicted in Figure 8), located in equidistant positions at a medium height (*i.e.*, equivalent to chest height of the user standing in the scenario).

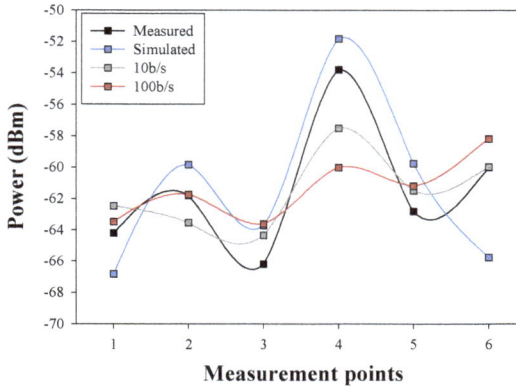

Figure 13. Comparison among simulated, RSSI and measured power values introducing RSSI power deviation correction.

5. Social Sensor System Design and Discussion

As previously stated, the solution implemented within this work has been developed under the framework of NASISTIC project, in the region of Navarra, in Spain. The NASISTIC project has been deployed in five households (dense urban area in the capital Pamplona, rural, mountain, people at risk of social exclusion and vulnerable older staff) in which medical devices such as Glucometer, Tensiometer, weigh scale, Thermometer, Control and Medication pulsioximeter as well as social sensors (humidity, temperature, presence, *etc.*), all based on the hardware-software approach described in this paper, are included. This uniformity in hardware has enabled rapid time-to-market development, reducing overall cost. A schematic description of the NASISTIC architecture, elements and back end is depicted in Figure 14.

Red Cross in Navarra has been the institution responsible for selecting the pilot users, addressing their training for the use of medical devices and taking the Call Center project. It has also been commissioned to conduct a survey of user satisfaction. In this sense, we can highlight:

- The proper functioning of the hardware proposed within the standard dimensions of a household in Spain (typically <90 m^2 as an average value).
- The need to simplify as much as possible the use of medical devices. On the contrary, the Social Care system is almost user transparent (plug and play function). This plug and play operation for medical devices is well accepted.
- The users showed interest in participating in the pilot deployment phase. From the medical point of view, the service provides them certain levels of self-control. The users have found of particular interest knowing their blood pressure on a daily basis as a security parameter related to cardiac health. From the social point of view, the interest of users has not been as high as with medical parameters. By contrast, in this case family members (second-users) have shown interest in issues such as temperature variation in the monitored homes, flood, *etc.*
- The large amount of stored data enables the development of additional projects related to social behavioral patterns, monitoring of chronic patients with social problems such as exclusion, loneliness, energy poverty, *etc.*

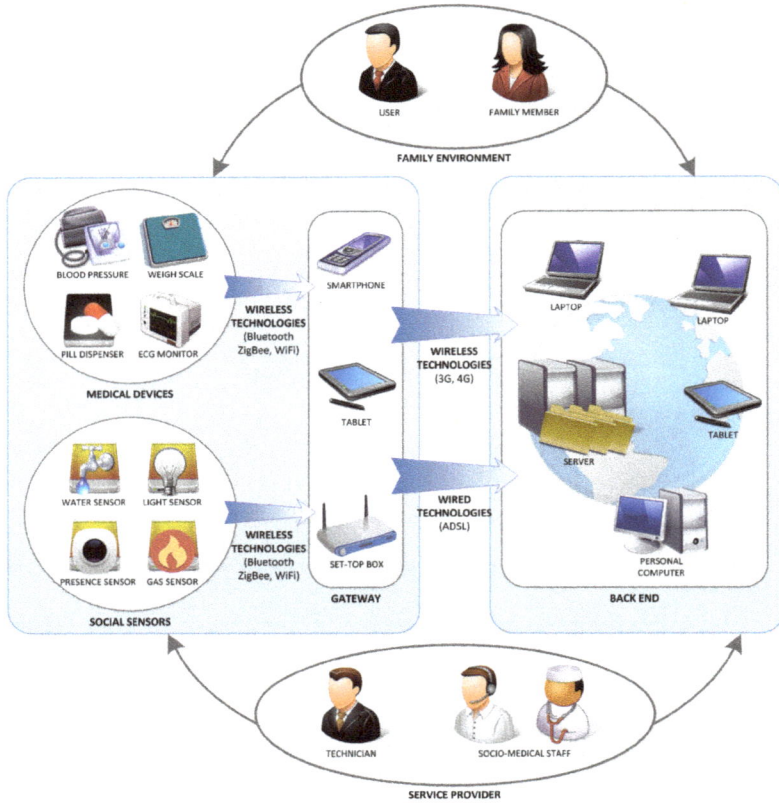

(a)

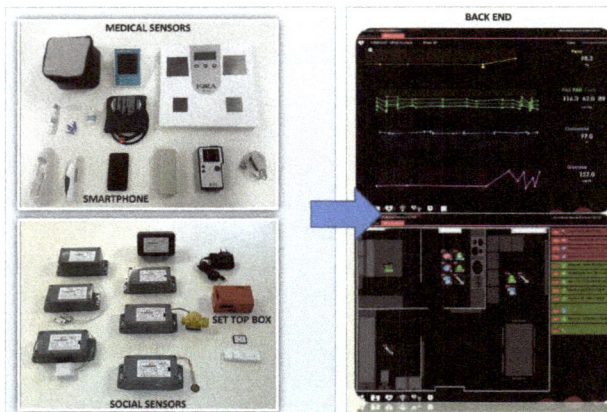

(b)

Figure 14. Overall view of the NASISTIC Social Sensor architecture, as well as an expanded view of the employed sensor test bed (**a**) generic architecture; (**b**) medical and social sensors together with application screenshots.

The acquisition devices gather the user's information and they can be divided into following types: medical devices and social sensors. Medical devices perform the acquisition of outstanding biomedical data in order to know the user's health state. Some examples of medical devices are blood pressure, weigh scale, pulsioximeter, glucometer, and pill dispenser, among others. Currently, most of these devices are portable and battery powered. Due to the fact that they are used only a few times a day, these portable devices are generally placed at fixed location in the user's residence; when a biomedical measurement is required, the user takes the medical device and uses it. However, some medical devices require continuous use in order to diagnose a disease suffered by the user. Usually, these devices are worn in the user's body. For this reason, essential features of these medical devices are reduced size and weight, battery powered, high operation autonomy, intelligence, and wireless communication; so-called wearable devices. An example of this type of device is an electrocardiogram monitor that performs continuous acquisition and automatic detection of outstanding cardiac episodes. Social sensors gather the user's environment information as well as data related to his/her behavior and daily activities. Some data acquired by social sensors are temperature, humidity, window and door state, gas detection, flood, and many others. The features of these sensors depend on both the acquired parameter and the location where they are placed, but in general, they are devices with very low form factor, battery powered, and high operation autonomy.

Medical devices and social sensors use short-range wireless technology in order to transmit the gathered information to a nearby Gateway system. The implemented wireless technology depends on several factors such as coverage range, data rate, power consumption, and security, among others. Currently, most systems use standard technologies such as Bluetooth, ZigBee, and WiFi, although proprietary communication technologies are also implemented in some cases for getting high levels of optimization, data rate and/or power consumption.

Gateway device performs bridge functions between acquisition devices and back-end system. Mainly, this device receives the acquired information and transmits it to the back-end. The Gateway device can also perform additional tasks: data pre-processing and temporary storage. The first one makes it possible to obtain outstanding information about user's state and system operation prior to the back-end system processing; it also allows the optimization of wirelessly transmitted data in order to reduce power consumption and cost. The second one prevents the system from losing data in case of communication or coverage failures.

The gateway device can be implemented in different platforms such as smartphone, tablet, or set-top box. All these platforms use short-range wireless technologies to communicate with acquisition devices. However, the data exchange between Gateway and back-end system can be performed with different technologies according to platform and availability of Internet access at user's household. Thus, mobile platforms (smartphones and tablets) will use 3G/4G technology provided by the mobile phone network; this guarantees a total coverage both inside and outside user's household during the monitoring process. In the case of set-top box platform, the technology depends on whether the user has ADSL or cable Internet service. If Internet service is provided, set-top box can use it directly for data transmission to the back-end system. Otherwise, the set-top box must incorporate its own 3G/4G technology by means of wireless dongle or similar device.

Finally, the back-end system achieves the user's data reception. It is made up of some servers and a database management system in order to store all the system's information: gathered data, agents, access profiles, authentication and credentials, *etc*. The back-end system provides a web service that allows the agents to access it with their personal devices.

Several agents can be involved in a social sensor system. On the one hand, user and family members can interact with acquisition sensors and Gateway device in order to start or modify system operation; the level of this interaction depends mainly on user/family member technical knowledge. The back-end system can also be accessed remotely by the user and family members to perform several tasks: management and visualization of gathered data, reception of medical/social staff notifications, question suggestions, among others. On the other hand, medical/social staff can also access the back-end system in order to analyze and visualize the user's information. Depending on this information and automatic alerts generated by the own system, the agent can know the user' state and even determine behavioral patterns that must be corrected. In these cases, the medical/social staff will make contact with the user or family member in order to notify behavioral advices, medication prescription, *etc.* Finally, technical workers are in charge of installing and configuring the system operation. These agents also have maintenance functions when some system malfunction has occurred. There will be problems whose solution will be performed remotely from the service provider, but other ones will require the movement of technical worker to the user's household; device malfunction, system configuration failure, and wrong device communication are common problems that cannot be solved by users or family members, so a technician must do it in person.

Centralized and secure management in the deployment of Socio-Sanitary services such as NASISTIC is a mandatory requirement. These back-end tools are designed for managing users with different roles (end-users, family users or secondary users, stakeholder users, *etc.*) as well as the management of information received from all monitored houses. Obviously, the end-users and their family users can access their own information. In the case of stakeholder users, access will be for all homes or end-users of their institution. In any case, these tools provide a complete configurability of roles.

Furthermore, in order to ease as far as possible the service configuration, tools like NASISTIC provide graphical aids such as either 2D or 3D drawings of housing for sensor placement, whether they are social sensors or health sensors, and tools for viewing the received data (time graphs, *etc.*). Moreover, besides a user-friendly configuration wizard, tools for configuring alerts from the received measurements are also implemented. For example, for a user, maximum and minimum blood pressure levels can be set (relative or absolute in %) or any other health sensor and alerts on social sensors like temperature (energy poverty, *etc.*) or moisture. These alerts generate messages (SMS, emails, WhatsApp messages, *etc.*) or phone calls to those in charge of the user who previously have been configured (family-users, physicians, stakeholder users or even end users).

In order to provide insight into the detailed operation of the system and to effectively test its suitability as an AAL system [40], examples of multiple signals available have been tested in a Living Lab environment, within the Public University of Navarra. As was previously mentioned, the evaluation module could emulate the functional operation of any monitoring sensor, being only required the implementation of analogue acquisition block. In this way, it is possible to emulate the home monitoring service's operation as the system was implemented in a real scenario. For this emulation, we consider four sensors: temperature/humidity sensor, electrocardiogram monitor, weigh scale, and open/close detector. Table 3 shows the main emulated features of sensors. The gateway device is also emulated with an evaluation module that gathers the acquired information and transmits it to Laptop via RS-232 serial communication. Laptop does not implement a display software application, so the raw data are stored in a stream file for later analyze and graphic representation.

Table 3. Features of emulated health and social sensors.

Sensor	Location	Accuracy	Transmission Rate	Range	Acquisition Rate
Temperature/Humidity	Living room	±0.5 °C ±1% RH	1 value/min	0 °C–65 °C	2 samples/s
Weigh scale	Bathroom	±100 gr.	2 value/day	0 gr–150 gr	2 sample/d
Electrocardiogram monitor	User's body	±45 µV/LSB	Real-time	−1V–2V	500 samples/s
Open/close	Hall door	-	-	0V–1V	-

As an example, Figure 15a,b shows measurements related to weigh scale and electrocardiogram monitor devices, respectively. The first one shows the user's weigh scale evolution throughout a week. This evolution, if it were to continue for too long, could be representative of a poor diet in ageing people or even suffering a neurodegenerative disease like Alzheimer's disease. The second figure shows five seconds of user's electrocardiographic signal. This health information is not usually analyzed by the user but instead a cardiologist in order to verify the heart activity and diagnose possible cardiac diseases, such as ventricular arrhythmias, atrial fibrillation, *etc.*

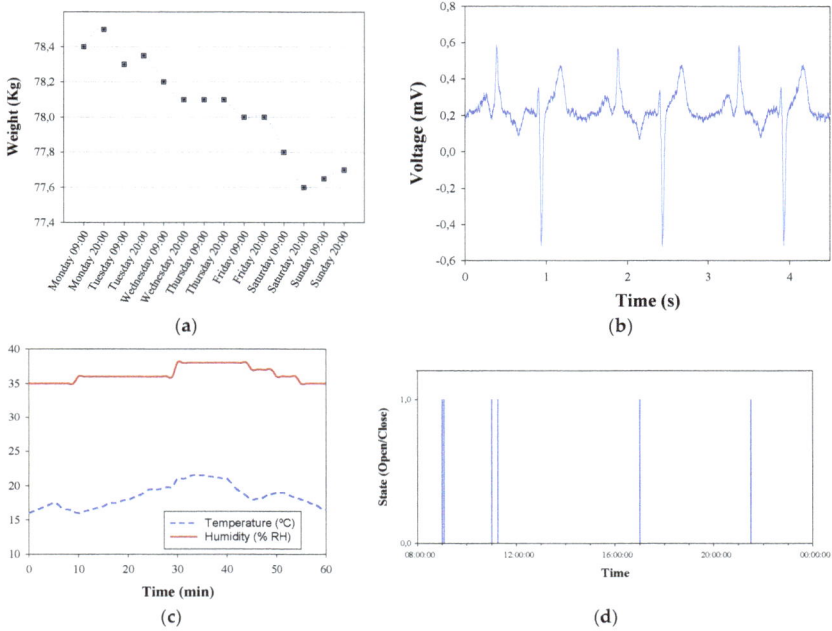

Figure 15. (a) Weight measurements obtained from the Social Sensor network devices. (b) Electrocardiographic signal obtained from the Social Sensor network devices. (c) Temperature and relative humidity measurements. (d) Hall door opening and closing events.

With regards to social sensors, temperature and relative humidity measurements are depicted in Figure 15c. As can be seen, temperature values range from 16 °C to 19 °C, and relative humidity from 35% to 38%, which could be usual in a home monitoring system. These measurements give the user and service provider relevant information about user's comfort and environment state during daily life conditions. Once again, ambient temperature and humidity values out of suitable comfort ranges could be representative of *Heating-Ventilation-Air-Conditioning* (HAVC) system malfunction and/or its inappropriate control by the user. Finally, Figure 15d shows the number of times the hall door is opened/closed and how many time it has remained in each state. Although it does not provide specific user's information, its analysis could be useful for inferring behavior aspects and daily activities. In this case, it is possible to know the time when the user has left from/arrived to home, or some people have visited the user; the information even allows knowing the time period the hall door has stayed opened.

As previously stated, the information gathered by the social sensors and the health sensors are processed by the back-end servers, where database processing as well user interfaces are implemented and managed. Figure 16 shows the implemented application layer, in which information such as monitoring of social sensors and health sensors, location of sensors within the household or detailed view of existing alarms can be obtained. The system is designed in order to provide a modular and easily accessible screen configuration to access relevant information, such as bio medical signal status and alarm monitors.

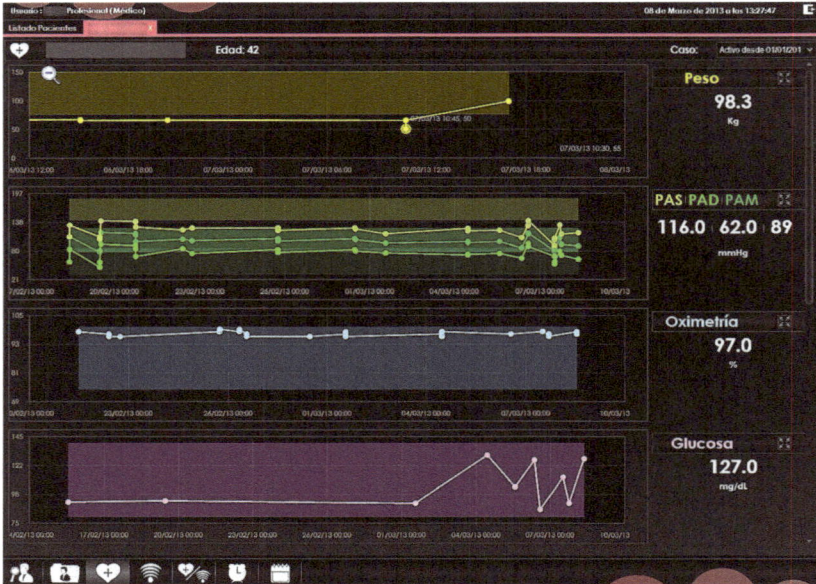

(a)

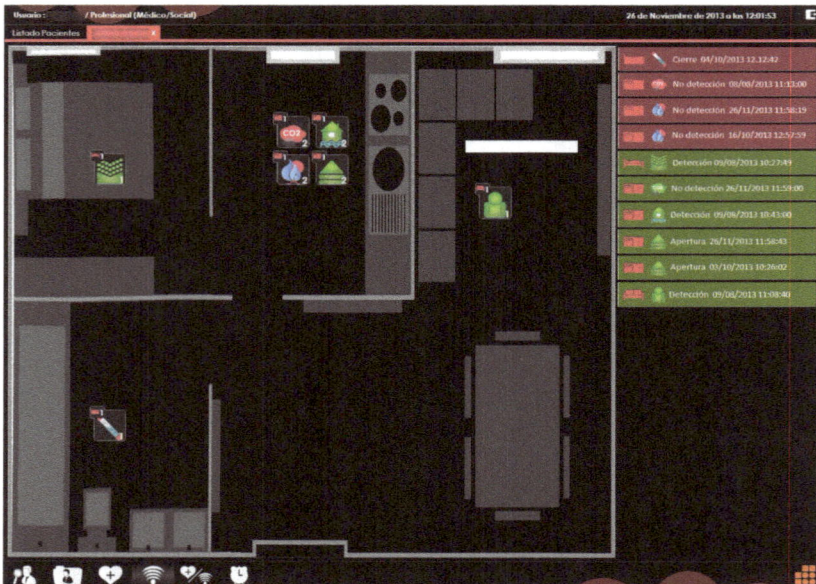

(b)

Figure 16. *Cont.*

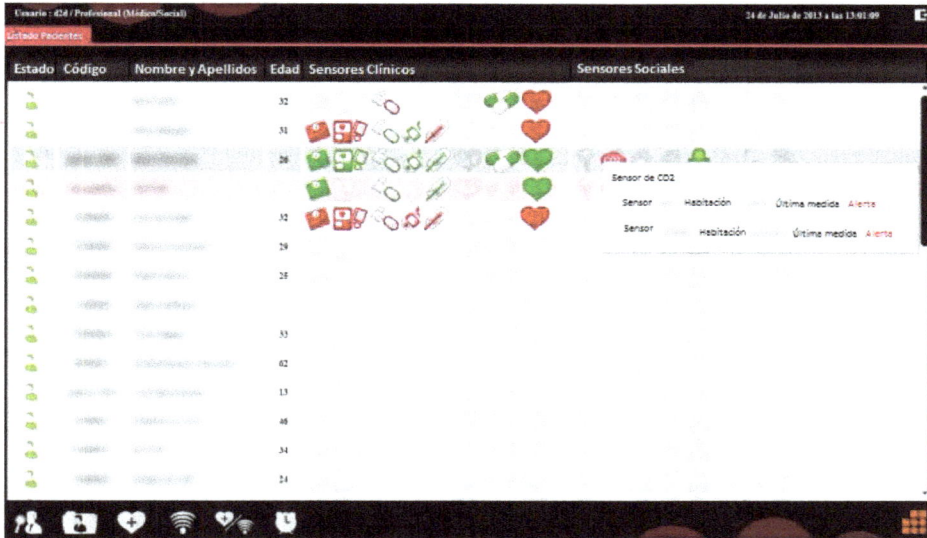

(c)

Figure 16. View of the application layer implemented in the back-end of the NASISTIC system. A view of different sensor signals, location map and message alerts are depicted (**a**) Sensor signals. (**b**) Location map of sensors. (**c**) Sensors and alerts associated to users.

The system has been developed in order to minimize the operation problems. In any case, the following are the main reasons for malfunction and detection mechanisms provided by the system to correct them.

- Device malfunction or misuse: At any time, a medical device or social sensor can show a wrong operation; even when that device has a proper operation, it can be misused by the user or caregiver. The service provider staff must detect these situations in order to solve the problem, by either replacing the broken device or retraining the user/caregiver on the correct use. Thus, the NASISTIC system includes a log message module that registers the main activity performed by devices. These messages are transmitted to the back-end so they can be viewed and analyzed by technical worker.

- False notifications: The system allows the configuration of alerts based on the received measurements in order to generate messages/notifications to those in charge of the user. During system operation, some occasional wrong measurement can be taken due to device malfunction/misuse or unusual environment conditions, which will generate false alerts. Obviously, there are acquisition parameters (e.g., temperature and humidity) that allow analyzing measurement trends in order to detect occasional out of range data. In this way, the system rejects the data and does not generate a false alert. Other acquisition parameters (e.g., gas, open-close, flood) require generating an alert in any case the measurement is out of range because it implies a hazardous situation. Currently, the NASISTIC system does not implement any strategy for distinguishing occasional out of range data with no risk for the user; all alerts lead to warning notifications.

Some problems were registered during the deployment of NASISTIC system. On the one hand, operation failures because of wrong system configuration. Devices configured with erroneous network address, sensors set with inappropriate operation features as device type, measurement identification, or date/time reference, and back end' server address not configured in the gateway device were the most usual mistakes. Problems related to erroneous configuration of sensor's features were detected by means of log messages, while those ones related to wrong setting of communication parameters were detected simply with no reception of measurements after a time the system was operating. In any case, the technician inspected the system in the user's residence and reconfigured it in order to guarantee the correct operation.

On the other hand, limited number of false alerts originated in medical device misuse; body temperature not registered correctly due to inadequate contact between skin and thermometer was the most relevant problem. These false alerts caused the corresponding calls to the user/caregiver in order to notify him or her about the event and remind him or her how to correctly use the medical device.

In general, the system has been developed with the aim of reaching most social monitoring situations. Certainly, the correct use of the system depends on several relevant factors. On the one hand, the user or caregiver/family member must have a basic knowledge of medical devices use. In those social situations where this requirement is not met, the monitoring system cannot be used; the number of these situations is limited but not zero. On the other hand, the user's household must be located in a region where Internet access and/or 3G wireless coverage are not limited. If this requirement is not fulfilled, the communication between Gateway device and back-end system is impossible and thus the user cannot receive follow up care. These situations are also limited, although users living in isolated regions or mountain villages can suffer from this problem.

In order to provide a holistic view of the implemented solution, an initial assessment from end users as well as from medical/social professionals involved in initial NASISTIC trials reveals the following statements:

- The proper operation of the proposed hardware is possible within the standard dimensions of a household in Spain (typically <90 m^2 as an average value). Although there are many factors involved in the wireless communication between devices, the use of Bluetooth Class 2 technology guarantees a suitable coverage range of approximately 10 m. This communication range, together with the set-top box usually being placed in the house's central room, makes it possible to use the proposed system in most households.

- The need to simplify as much as possible the use of sensors and devices. In the case of social system, the sensor's operation is almost user transparent because they only require being one time configured and switched on; this plug and play operation is well accepted. In the same way, the use of medical devices must be as simple as possible in order to guarantee the user/caregiver approval, and reduce possible measurement errors. It must be taken in account that end user likely show low technology knowledge, and any additional required intervention in device operation can make the user reject the system; even in the case of a medical device being used by the caregiver, a complex device operation can lead to wrong measurement acquisition, incorrect function selection, or data transmission fail. For this reason, and because most of the medical devices require some user intervention (press button, cuff placing, *etc.*), it is necessary to minimize the number of these actions. All these considerations have been taken into account in NASISTIC project development.

- The users and social/medical staff have shown significant interest in participating in the NASISTIC pilot deployment phase. With regards to users, personal interviews between them and Red Cross assistants have derived some first conclusions. On the one hand, they accept the use of proposed system, which is perceived as an additional element within the household. On the other hand, they have found of particular interest knowing some biomedical parameters (blood pressure mainly) on a daily basis as security data related to health state. From the social point of view, the interest of users has not been as high as with medical parameters; in contrast, in this case, family members (second users) have shown interest in issues such as temperature variation in the monitored homes, flood, *etc.* With regards to social/medical staff, the main results obtained are the following: response times are reduced due to real time interaction capabilities offered by the social sensors, and the general service provides them certain levels of self-control. Although specific survey forms have not been implemented, the daily personal contact between users and social/medical staff during the project deployment has made obtaining these first satisfaction data possible.

- The large amount of stored data enables the development of additional projects related to social behavioral patterns. Meanwhile, biomedical data allow the knowledge of user's health state and mood, while other activity data, such as watching TV, use of HVAC system, water consumption, or length of phone calls, can provide information about user's behavior and social relationships. All this information can be combined with user activity on the Internet and specific web social networks like Facebook, Twitter, *etc.* This makes broader monitoring applications focused on following up with users with chronic social problems such as exclusion, loneliness, energy poverty, *etc.* possible.

- Technical workers, engineers and architects involved in the deployment indicate that the proposed system is simple to integrate and maintain, reducing overall installation and operational costs. Thus, start-up system requires minimum intervention by technical worker because elements are pre-configured from factory or service provider; the worker only has to verify the correctness of system operation the first installation time. In addition, the system maintenance neither demands a great number of home interventions by engineers or technical workers unless some element must be repaired or replaced.

6. Conclusions

In this work, the implementation of Context Aware in order to aid in the assistance of end users within the framework of e-Health and m-Health scenarios has been described. A prototype of Social Sensor device has been designed and implemented in order to provide multiple-signal processing capabilities, while exhibiting flexible, low complexity and moderate cost features. The Social Sensor device is based on a Bluetooth transceiver, in which a dedicated microcontroller provides access to several signals, which can provide medical, environmental or behavioral information. In order to analyze the robustness of the system, detailed wireless channel analysis has been performed, providing insight on the behavior of the devices in real complex indoor scenarios, in which large signal variability can severely degrade overall performance. Estimations of received power levels have been compared with experimental results, indicating that received power levels comply with receiver sensitivity levels in the scenario for any potential transceiver location, implying adequate overall system service levels. Initial tests under Living Lab conditions, given by NASISTIC operational requirements have also been performed, showing the feasibility to successfully interchange multiple information within real time regime. Trial runs have shown the feasibility in data acquisition of multiple bio-physical parameters as well as user information, leading to a practical AAL platform. The proposed solution is currently being prepared for large-scale deployment because of the ease in installation and maintenance, positive adoption by end users and adequate results in terms of response time reduction and preventive actions that can be adopted.

The NASISTIC system is a remote health and behavior monitoring system, providing means to health professionals as well as social services to provide increased quality of living to vulnerable population segments, such as elderly people, and those suffering from some form of disability or mental illness. The difference between NASISTIC and other proposed systems is the integral approach in capturing biomedical signals as well as other signals that are related with user habits, such as usage of lights, heating, electrical appliances, water consumption or open/close indication of doors. In this way, combined action of health specialists and social services can be provided in order to enable higher degrees of autonomy to the users as well as increasing quality of life levels.

As future steps, complete system validation, integration of additional sensor elements, and the implementation of a distributed system architecture based on cloud integration will be performed.

The current implementation of NASISTIC system uses a back-end server for information storage. Although this first version is based on a centralized solution, the system is not limited to it. Suitable modifications on back-end element can lead to distributed implementations based on cloud solutions.

Additionally, complete user experience assessment will be obtained in order to identify usability, leading to further modifications and enhancements.

Acknowledgments: The authors wish to acknowledge funding of project NASISTIC, from the Government of Navarra, Spain; Project TEC2013-45585-C2-1-R, funded by the Spanish Ministry of Economy and Competiveness; collaboration and support from Red Cross Navarra; and D2D, AH and ID corporations, in Spain.

Author Contributions: Erik Aguirre, Peio López-Iturri, Leyre Azpilicueta and Francisco Falcone conducted the characterization of wireless propagation mechanisms in the Living Lab scenarios. Santiago Led and Luis Serrano participated in the development of the NASISTIC architecture and back end.

Conflicts of Interest: The authors declare no conflict of interest.

References

1. UN Department of Economic and Social Affairs. Available online: http://www.un.org/esa/population/publications/worldageing19502050/ (accessed on 25 February 2016).
2. Monteagudo, J.L.; Moreno, O. eHealth for Patient Empowerment in Europe, 2009. Available Online: http://ec.europa.eu (accessed on 25 February 2016).
3. Led, S.; Serrano, L.; Galarraga, M. Wearable Wireless Monitoring System Based on Bluetooth Technology: A Tutorial. In Proceedings of the 3rd European Medical and Biological Engineering Conference EMBEC, Prague, Czech Republic, 20–25 November 2005.
4. Xiao, Y.; Pan, Y. *Emerging Wireless LANs, Wireless PANs, and Wireless MANs*; John Wiley & Sons: New York, NJ, USA, 2009.
5. Dargie, W.; Poella, B.C. *Fundamentals of Wireless Sensor Networks Theory and Practice*; John Wiley & Sons: Chichester, UK, 2010.
6. Kim, Y.; Evans, R.G.; Iversen, W.M. Remote sensing and control of an irrigation system using a distributed wireless sensor network. *IEEE Trans. Instrum. Meas.* **2008**, *57*, 1379–1387.
7. Grosse, C.U.; Glaser, S.D.; Kruger, M. Initial development of wireless acoustic emission sensor Motes for civil infrastructure state monitoring. *Smart Struct. Syst.* **2010**, *6*, 197–209. [CrossRef]
8. Rahal, Y.; Pigot, H.; Mabilleau, P. Location estimation in a smart home: System implementation and evaluation using experimental data. *Int. J. Telemed. Appl.* **2008**. [CrossRef] [PubMed]
9. Tayal, S.; Tripathy, M.R. VANET-Challenges in selection of Vehicular Mobility Model. In *Second International Conference on Advanced Computing & Communication Technologies (ACCT)*; IEEE: Rohtak, India, 2012; pp. 231–235.
10. Li, Y.Z.; Wang, L.; Wu, X.M.; Zhang, Y.T. Experimental analysis on radio transmission and localization of a ZigBee based wireless healthcare monitoring platform. In Proceedings of the 5th International Conference on Information Technology and Applications in Biomedicine (ITAB'08), Nanjing, China, 30–31 May 2008; pp. 488–490.
11. Dagtas, S.; Pekhteryev, G.; Sahinoglu, Z.; Cam, H.; Challa, N. Real-time and secure wireless health monitoring. *Int. J. Telemed. Appl.* **2008**. [CrossRef]
12. Villanueva, F.J.; de la Morena, J.; Barba, J.; Moya, F.; Lopez, J.C. Mobile Ad-Hoc Networks for Large in-Building Environments. In Proceedings of the International Conference on Wireless Networks, Communications and Mobile Computing, Maui, HI, USA, 13–16 June 2005.
13. Husni, E.M.; Heryadi, Y.; Arifianto, M.S. The Smart Tele-emergency Project: A Mobile Telemedicine Unit Based on Mobile IPv6 and Mobile Ad-HOC Network for Sabah Areas. In Proceedings of the RF and Microwave Conference, RFM, Selangor, Malaysia, 5–6 October 2004.
14. Husni, E.M.; Heryadi, Y.; Woon, W.T.H.; Arifianto, M.S.; Viswacheda, D.V.; Barukang, L. Mobile Ad Hoc Network and Mobile IP for Future Mobile Telemedicine System. In Proceedings of the IFIP International Conference on Wireless and Optical Communications Networks, Bangalore, India, 11–13 April 2006.
15. Aziz, M.; Al-Akaidi, M. Security Issues in Wireless Ad Hoc Networks and the Application to the Telecare Project. In Proceedings of the 15th International Conference on Digital Signal. Processing, Cardiff, UK, 1–4 July 2007.

16. Yang, S.H.; Song, K.T. An Adaptive Routing Protocol for Health Monitoring with a Sensor Network and Mobile Robot. In Proceedings of the 36th Annual Conference on IEEE Industrial Electronics Society, IECON, Glendale, CA, USA, 7–10 November 2010.

17. Rashid, R.A.; Ch'ng, H.S.; Alias, M.A.; Fisal, N. Real time medical data acquisition over wireless Ad-Hoc network. In Proceedings of the Asia-Pacific Conference on Applied Electromagnetics, APACE, Johor Baru, Malaysia, 20–21 December 2005.

18. Singh, D.; Kew, H.-P.; Tiwary, U.S.; Lee, H.-J.; Chung, W.-Y. Global Patient Monitoring System Using IP-Enabled Ubiquitous Sensor Network. In Proceedings of the WRI World Congress on Computer Science and Information Engineering, Los Angeles, CA, USA, 31 March–2 April 2009.

19. Hu, F.; Jiang, M.; Wagner, M.; Dong, D.C. Privacy-Preserving Telecardiology Sensor Networks: Toward a Low-Cost Portable Wireless Hardware/Software Codesign. *IEEE Trans. Inf. Technol. Biomed.* **2007**, *11*, 6–10. [CrossRef]

20. Kim, J.C.; Kim, D.Y.; Jung, S.M.; Lee, M.H.; Kim, K.S.; Lee, C.K.; Nah, J.Y.; Lee, S.H.; Kim, J.H.; Choi, W.J.; *et al.* Implementation and performance evaluation of mobile ad hoc network for Emergency Telemedicine System in disaster areas. In Proceedings of the Annual International Conference of the IEEE Engineering in Medicine and Biology Society, EMBC, Minneapolis, MN, USA, 2–6 September 2009.

21. Egbogah, E.E.; Fapojuwo, A.O.; Shi, L. An energy-efficient transmission scheme for monitoring of combat soldier health in tactical mobile ad hoc networks. In Proceedings of the Military Communications Conference, MILCOM, Orlando, FL, USA, 29 October–1 November 2012.

22. NASISTIC Project Description. Available online: http://nasistic.com/index.php/nasistic (accessed on 25 February 2016).

23. Soh, P.J.; Vandenbosch, G.A.E.; Mercuri, M.; Schreurs, D.M.M.-P. Wearable Wireless Health Monitoring: Current Developments, Challenges, and Future Trends. *IEEE Microw. Mag.* **2015**, *16*, 55–70. [CrossRef]

24. Pantelopoulos, A.; Bourbakis, N.G. A Survey on Wearable Sensor-Based Systems for Health Monitoring and Prognosis. *IEEE Trans. Syst. Man Cybern. Part. C Appl. Rev.* **2010**, *40*, 1–12. [CrossRef]

25. Sung, M.; Marci, C.; Pentland, A. Wearable feedback systems for rehabilitation. *J. NeuroEng. Rehabil.* **2005**, *2*, 17–20. [CrossRef] [PubMed]

26. Anliker, U.; Ward, J.A.; Lukowicz, P.; Tröster, G.; Dolveck, F.; Baer, M.; Keita, F.; Schenker, E.B.; Catarsi, F.; Coluccini, L.; *et al.* AMON: A Wearable Multiparameter Medical Monitoring and Alert System. *IEEE Trans. Inf. Technol. Biomed.* **2004**, *8*, 415–427. [CrossRef] [PubMed]

27. Lin, B.S.; Chou, N.K.; Chong, F.C.; Chen, S.J. RTWPMS: A Real-Time Wireless Physiological Monitoring System. *IEEE Trans. Inf. Technol. Biomed.* **2006**, *10*, 647–656. [CrossRef] [PubMed]

28. Mundt, C.W.; Montgomery, K.N.; Udoh, U.E.; Barker, V.N.; Thonier, G.C.; Tellier, A.M.; Ricks, R.D.; Darling, R.B.; Cagle, Y.D.; Cabrol, N.A.; *et al.* A multiparameter wearable physiological monitoring system for space and terrestrial applications. *IEEE Trans. Inf. Technol. Biomed.* **2005**, *9*, 382–391. [CrossRef] [PubMed]

29. Tura, A.; Badanai, M.; Longo, D.; Quareni, L. A medical wearable device with wireless bluetooth-based data transmission. *Meas. Sci. Rev.* **2003**, *3*, 1–4.

30. Habetha, J. The MyHeart project—Fighting cardiovascular diseases by prevention and early diagnosis. In Proceedings of the 28th Annual International IEEE EMBS Conference, New York, NY, USA, 31 August–3 September 2006; pp. 6746–6749.

31. Paradiso, R.; Loriga, G.; Taccini, N. A wearable health care system based on knitted integral sensors. *IEEE Trans. Inf. Technol. Biomed.* **2005**, *9*, 337–344. [CrossRef] [PubMed]

32. Di Rienzo, M.; Rizzo, F.; Parati, G.; Brambilla, G.; Ferratini, M.; Castiglioni, P. MagIC System: A New Textile-Based Wearable Device for Biological Signalmonitoring Applicability in Daily Life and Clinical Setting. In Proceedings of the 27th Annual International IEEE EMBS Conference, Shanghai, China, 1–4 September 2005; pp. 7167–7169.

33. Luprano, J. *European projects on Smart Fabrics, Interactive Textiles: Sharing Opportunities and Challenges*; Workshop Wearable Technol. Intel.; Textiles: Helsinki, Finland, 2006.

34. Fortino, G.; Giannantonio, R.; Gravina, R.; Kuryloski, P.; Jafari, R. Enabling Effective Programming and Flexible Management of Efficient Body Sensor Network Applications. *IEEE Trans. Hum. Mach. Syst.* **2013**, *43*, 115–133. [CrossRef]

35. Led, S.; Azpilicueta, L.; Aguirre, E.; Martínez de Espronceda, M.; Serrano, L.; Falcone, F. Analysis and Description of HOLTIN Service Provision for AECG monitoring in Complex Indoor Environments. *Sensors* **2013**, *13*, 4947–4960. [CrossRef] [PubMed]

36. Iturri, P.L.; Nazábal, J.A.; Azpilicueta, L.; Rodriguez, P.; Beruete, M.; Fernández-Valdivielso, C.; Falcone, F. Impact of High Power Interference Sources in Planning and Deployment of Wireless Sensor Networks and Devices in the 2.4 GHz frequency band in Heterogeneous Environments. *Sensors* **2012**, *12*, 15689–15708. [CrossRef] [PubMed]

37. Aguirre, E.; Arpón, J.; Azpilicueta, L.; de Miguel, S.; Ramos, V.; Falcone, F. Evaluation of electromagnetic dosimetry of wireless systems in complex indoor scenarios within body human interaction. *Progr. Electromagn. Res. B* **2012**, *43*, 189–209. [CrossRef]

38. Azpilicueta, L.; Falcone, F.; Astráin, J.J.; Villadangos, J.; García, Z.I.J.; Landaluce, H.; Angulo, I.; Perallos, A. Measurement and modeling of a UHF-RFID system in a metallic closed vehicle. *Microw. Opt. Technol. Lett.* **2012**, *54*, 2126–2130. [CrossRef]

39. Nazábal, J.A.; Iturri, P.L.; Azpilicueta, L.; Falcone, F.; Fernández-Valdivielso, C. Performance Analysis of IEEE 802.15.4 Compliant Wireless Devices for Heterogeneous Indoor Home Automation Environments. *Int. J. Antennas Propag.* **2012**, *2012*. [CrossRef]

40. Lloret, J.; Canovas, A.; Sendra, S.; Parra, L. A smart communication architecture for ambient assisted living. *IEEE Commun. Mag.* **2015**, *53*, 26–33. [CrossRef]

MDPI

Article

Event-Based Control Strategy for Mobile Robots in Wireless Environments

Rafael Socas *, Sebastián Dormido, Raquel Dormido and Ernesto Fabregas

Departamento de Informática y Automática, Universidad Nacional de Educación a Distancia, Juan del Rosal 16, Madrid 28040, Spain; sdormido@dia.uned.es (S.D.); raquel@dia.uned.es (R.D.); efabregas@bec.uned.es (E.F.)
* Correspondence: rsocas@telefonica.net; Tel.: +34-629-578-386

Academic Editor: Gonzalo Pajares Martinsanz
Received: 26 October 2015; Accepted: 27 November 2015; Published: 2 December 2015

Abstract: In this paper, a new event-based control strategy for mobile robots is presented. It has been designed to work in wireless environments where a centralized controller has to interchange information with the robots over an RF (radio frequency) interface. The event-based architectures have been developed for differential wheeled robots, although they can be applied to other kinds of robots in a simple way. The solution has been checked over classical navigation algorithms, like wall following and obstacle avoidance, using scenarios with a unique or multiple robots. A comparison between the proposed architectures and the classical discrete-time strategy is also carried out. The experimental results shows that the proposed solution has a higher efficiency in communication resource usage than the classical discrete-time strategy with the same accuracy.

Keywords: event-based control; navigation algorithms; mobile robots; wireless communication

---

1. Introduction

A distributed networked control system (NCS) is composed of numerous subsystems (fixed or mobile) called agents, which are geographically distributed. In such a system, the individual subsystems exchange information over a communication network. The communication network generally uses wireless technology, which has a limited bandwidth and constraints in throughput and delays. In practice, communication, especially in wireless networks, uses digital network infrastructures, where the information is transmitted in discrete packets. These packets may be lost during communication. Moreover, the communication media are a resource that is usually accessed in an exclusive manner by the agents. This means that the throughput capacity of such networks and the number of elements is limited. In this kind of network, there are effects of non-linearities and constraints in closed loop control that may impact the performance and control stability of the system. Some effects are the loss of information when transmitting data, variable communication delays, signal sampling and quantization issues due to information packaging and constraints related to limited bandwidth [1]. On the other hand, it is important to decrease the input traffic to the network in a short period of time. This means that the input data to a network will be reduced. An important result of this kind of reduction is that the network can simply guarantee a predictable bandwidth to a control loop, and furthermore, it simplifies the analysis of network delay, which affects the control loop [2]. For these reasons, an important issue in the implementation of these systems is to define methods that use the limited network resources available for transmitting the state and control information more effectively.

To deal with this problem, the timing issue in NCS has been investigated by some researches [3]. In traditional approaches, the controllers are used under the assumption of perfect communication, and then, the maximum allowable transfer interval (MATI) between two subsequent message transmissions that ensures closed loop stability under a network protocol is determined. Examples of them are the try once discard (TOD) or round robin (RR) protocol. The MATI scheme is often done in a centralized manner; therefore, it is impractical for large-scale systems. Moreover, because the MATI is computed before the system is deployed, it must ensure performance levels over all possible system states. As a result, the MATI may be conservative in the sense of being shorter than necessary to assure a specified performance level. Consequently, the bandwidth of the network has to be higher than necessary to ensure the MATI is not violated.

Many others authors have achieved an important reduction of bandwidth utilization without significant loss of performance. Two approaches have been considered: model-based networked control systems (MBNCS) and event-based control. The MBNCS approach was introduced in [4,5], and it was considered in networks of coupled systems [6] using periodic communication. Another philosophy to deal this problem is based on an event-based feedback scheme in NCS [7–10]. In event-based systems, an agent broadcasts its state information only when it is needed. In this case, this means that some measure of the agent's state error is above a specified threshold (event threshold). The architecture is decentralized in the sense that an agent is able to produce broadcast state information using its locally-sampled data. Moreover, the selection of the event threshold only requires local information from the sensors of the agent, so that the design is decentralized.

Event-based control has been widely used for the stabilization of dynamical systems while reducing the number of measurements that the sensors need to send to the controller. The events based on state errors have been used extensively [11]. In [12–14], the same ideas have been extended to consider networked interconnected systems. A common characteristic in these works on event-based control is the use of a zero order hold (ZOH) model in the controller node and the assumption that the models being used are the same as the plants that they represent. In recent years, the research interest in the field of event-based control has been growing, and the volume of publications is increasing.

This work presents a new control strategy for mobile robots in wireless environments. The main idea is to develop architectures that manage the limited radio resources efficiently with a similar accuracy as the classical solutions. The event-based control strategies have been investigated to implement navigation algorithms and to compare them to the discrete-time implementations.

The paper is organized as follow. Section 2 shows an overview of the event-based control. In Section 3, the classical control strategies for wireless environments are described. The proposed control strategy is presented in Section 4. In Section 5, the navigation algorithms investigated in this work are presented. The experimental results are discussed in Section 6. Finally, the conclusions and future work are presented in Section 7.

2. Event-Based Control Overview

In contrast to the continuous-time or discrete-time control strategies, in event-based control systems [15], information exchanges among the sensors, the controller and the actuators are triggered with dependence on the system behavior, e.g., when the system variables exceed certain tolerance bounds. In other words, the activity of the controller and the communication between the components in the control system are restricted to time intervals in which the controller inevitably must act in a closed loop manner in order to guarantee the desired specifications of the closed loop system.

The basic structure of the event-based control strategy is depicted in Figure 1. It consists of the following elements: (1) the plant, (2) the event generator and (3) the control input generator [16]. The plant has input vector $u(t)$, output vector $y(t)$, state vector $x(t)$, disturbance vector $d(t)$ and noise in the sensors $v(t)$. The event generator determines the event times t_k at which information $\hat{x}(t_k)$ (estimated state vector) is sent towards the control input generator. The event generator block calculates the difference between the output of the system $y(t)$ and the reference signal $w(t)$. If this value crosses a certain value defined as the event threshold, an event is generated, and the estimated state vector $\hat{x}(t_k)$ is sent to the controller. The control input generator computes the continuous time input $u(t)$ of the plant based on the information obtained at time t_k and the command input $w(t)$. See [17] for further information.

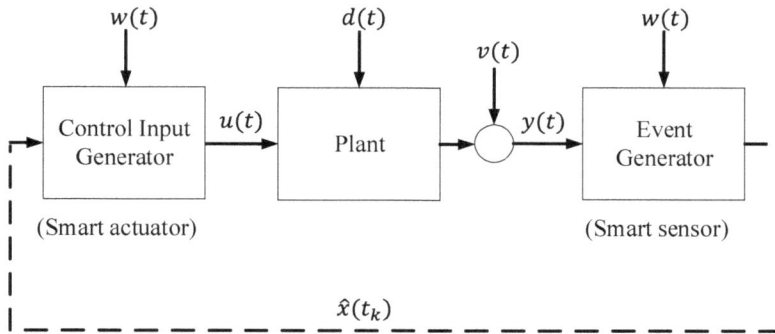

Figure 1. General scheme of an event-based control strategy.

In recent years, event-based control has motivated the interest of researchers, and multiple control schemes and new applications have been developed based on these ideas. In [18], an event-driven sampling method called the area-triggered method has been proposed. In this scheme, sensor data are sent only when the integral of differences between the current sensor value and the last transmitted one is greater than a given threshold. The proposed method not only reduces the data transmission rate, but also improves estimation performance in comparison with the conventional event-driven method. The work presented in [19] describes how greenhouse climate control can be represented as an event-based system in combination with wireless sensor networks, where low-frequency dynamics variables have to be controlled, and control actions are mainly calculated against events produced by external disturbances. The proposed control system allows saving costs related to wear minimization and prolonging the actuator life, but keeping promising performance results. Event-based sampling according to a constant energy of sampling error is analyzed in [20]. The defined criterion is suitable for applications where the energy of the sampling error should be bounded (e.g., in building automation or in greenhouse climate monitoring and control). The proposed sampling principle extends a range of event-based sampling schemes and makes the choice of a particular sampling criterion more flexible for the application requirements. Finally, a modified fault isolation filter for a discrete-time networked control system with multiple faults is implemented by a particular form of the Kalman filter in [21]. The proposed fault isolation filter improves the resource utilization with graceful fault estimation performance degradation.

3. Control Strategies in Wireless Environments for Mobile Robots

In a wireless environment, the typical architecture for mobile robots is a discrete-time control system, as is shown in Figure 2.

The elements of this system work in the discrete-time domain; the control signals ($u_c[n]$ and $u_r[n]$) and the sensor signals ($y_r[n]$ and $y_c[n]$) have a sampling period of $1/f_s$, where f_s is the sampling frequency of the system. The RF channels ($Ch_1(t)$ and $Ch_2(t)$) could use analog or digital modulations to transport the information between the elements. In both cases, this RF interface works in the continuous-time domain.

In this scheme, the controller sends the control signals $u_c[n]$ over a communication channel $Ch_1(t)$; this information is received in the robot $u_r[n]$ and acts over the actuators. The difference between $u_c[n]$ and $u_r[n]$ is the noise and the perturbations in the communication channel $Ch_1(t)$. The sensor signals $y_r[n]$ are sent towards the controller over another communication channel $Ch_2(t)$. Finally, the controller receives the sensor signals $y_c[n]$ and calculates the control signals $u_c[n]$, computing the reference signal $w[n]$ and the information from the sensors. As in channel $Ch_1(t)$, the noise and the perturbations in channel $Ch_2(t)$ produce two different signals $y_r[n]$ and $y_c[n]$. This system interchanges information between the controller and the mobile robot every $1/f_s$ seconds.

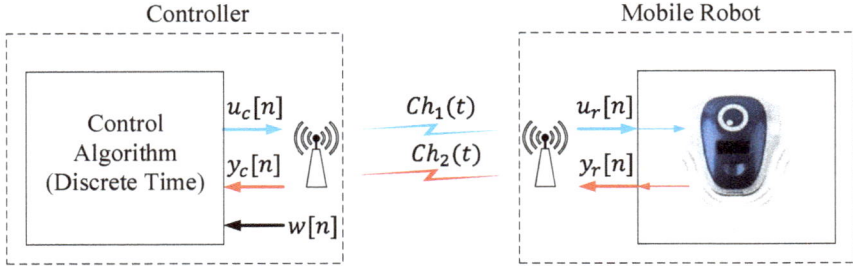

Figure 2. Wireless discrete-time control.

In this structure, the communication channels $Ch_1(t)$ and $Ch_2(t)$ are busy every $1/f_s$, because the system is periodically interchanging information between the controller and the robot. When the robot is in a steady state, it is not necessary to interchange information between the robot and the controller, but the channel is busy due to the system sending information over the channels each period of time $1/f_s$ unnecessarily. When a wireless infrastructure is used to connect the elements of the control system, it is mandatory to use the radio resources only for critical purposes due to these communication systems frequently having a limited spectrum, and only a few elements can send information at the same time. For this reason, in this work, other control methodologies has been investigated to improve the efficiency in the wireless communication channels.

4. Proposed Event-Based Control Strategy in a Wireless Environment

As was mentioned before, if the system is in a steady state, the event-based strategies have two main advantages *versus* the discrete-time control ones: (1) the communication resources are not used and (2) the controller does not need to compute new control signals. For these reasons, in this paper, a new event-based control strategy for mobile robots in wireless environments is proposed.

4.1. Control Architectures

Based on [22], a new approach for a wireless control system for mobile robots has been proposed. The architecture of the system is depicted in Figure 3.

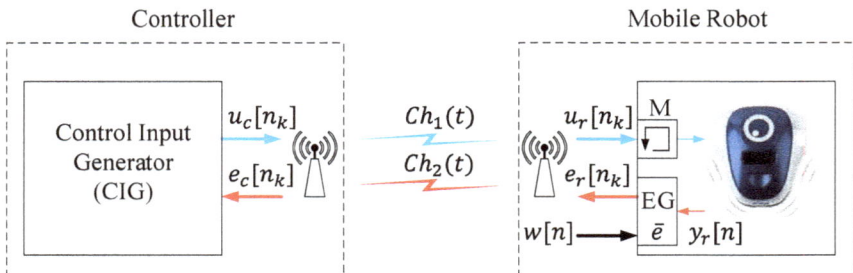

Figure 3. Proposed event-based control strategy.

In the robot, an event generator (EG) compares the sensor signals $y_r[n]$ to the reference signal $w[n]$. If the difference between both signals crosses the event threshold $\bar{e}$, an event k is generated, and the error signal $e_r[n_k]$ is sent to the controller over the radio interface. Different methodologies can be used to generate events. In [23], a review of these procedures has been analyzed. In a general way, the event threshold usually is a constant value, and it has to be set carefully. Otherwise, it can be defined as a function of the noise or as a function of other variables to get more accuracy in event generation;

see [24,25] for details. The error signal $e_r[n_k]$ goes to the controller via the communication channel $Ch_2(t)$. In this case, the channel is busy only when the EG generates events. When the robot is in a steady state, the signals $w[n]$ and $y_r[n]$ are very close, and no event is generated. In this case, the communication channel $Ch_2(t)$ is free. Every time an event is generated and the error signal $e_r[n_k]$ is sent to the controller, the robot receives the control signal $u_r[n_k]$; this signal is saved in a memory M. When an event k is generated, the memory is updated with new values, but in the period between events, the saved value in the memory is used to act on the robot.

When an event is generated, the control input generator (CIG) receives the error signal $e_c[n_k]$, and the control signal $u_c[n_k]$ is calculated; then, this information is sent to the robot via the communication channel $Ch_1(t)$. The communication channel $Ch_1(t)$ and the channel $Ch_2(t)$ are busy only when the events are generated, in other periods of time they are free. As in the discrete-time architecture; the differences between the error signals $(e_r[n_k], e_c[n_k])$ and the control signals $(u_c[n_k], u_r[n_k])$ are the noise and the perturbations in the communication channels $Ch_1(t)$ and $Ch_2(t)$.

If the proposed architecture is compared to the discrete-time strategy, it has two main advantages: (1) the communication channels are busy only when the events are generated and (2) the controller does not have to compute control signals when the robot is in the steady state. In this event-based architecture, the reference signal $w[n]$ is computed in the event generator (EG); for this reason, the communication channel $Ch_1(t)$ is used to send this information. In a general way, this information tends to be constant or it changes with a low frequency. In this way, the effects in the usage of the radio resources should be negligible.

The classical control algorithms can be implemented in this scheme in a simple way. Depending on the kind of algorithm to be implemented, the EG and the CIG will be defined in a different way, as is described in Section 5.

4.2. Mobile Robot Model

In this work, to check the proposed strategy, some navigation algorithms have been studied. The navigation algorithm will be defined for differential wheeled robots, although the architecture proposed can be adapted to another kind of robot in a simple way. The mobile robot that has been analyzed in this paper has the blocks presented in Figure 4. These robots have a dynamic model, which depends on the length between wheels L, the radio R and the angular speed of each wheel $speed_{left}$ and $speed_{right}$. In this model, the position in 2D (x and y) and the heading angle φ of the robot can be expressed by Equations (1)–(3).

$$\dot{x} = \frac{R}{2}(speed_{right} + speed_{left}) \cos\varphi \tag{1}$$

$$\dot{y} = \frac{R}{2}(speed_{right} + speed_{left}) \sin\varphi \tag{2}$$

$$\dot{\varphi} = \frac{R}{L}(speed_{right} - speed_{left}) \tag{3}$$

The position and orientation of the robot can be managed acting over its angular speeds $speed_{left}$ and $speed_{right}$. In the proposed system, these speeds will be calculated by the controller.

Other important blocks in this robot are the obstacle detection sensors, which use infra-red signals to detect the objects located in front of the robot. Generally, these vehicles have four of these sensors, two on the front obt_{front_left} and obt_{front_right} and two on the sides obt_{lat_left} and obt_{lat_right}. The value obtained from these sensors depends on the proximity of the obstacles: if the distance to the obstacle descends, the value presented by the sensor increases. Other important information that is obtained from these devices is the position of the detected objects, which can be calculated combining the data of the different sensors.

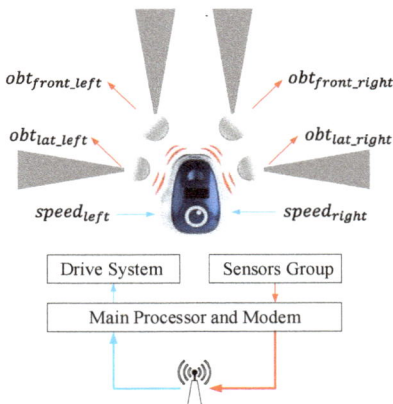

Figure 4. General structure of the differential wheeled robots.

5. Navigation Algorithms

In this section, some navigation algorithms will be defined to analyze the proposed architectures. In mobile robot navigation tasks, some well-known navigation algorithms are widely used in this kind of application. Go to goal, wall following and obstacle avoidance are the most commonly used in mobile robot environments [26–28]. In this work, some implementations of these algorithms are proposed using the presented event-based architecture. Moreover, a comparison with a discrete-time solution to solve the same problem is presented.

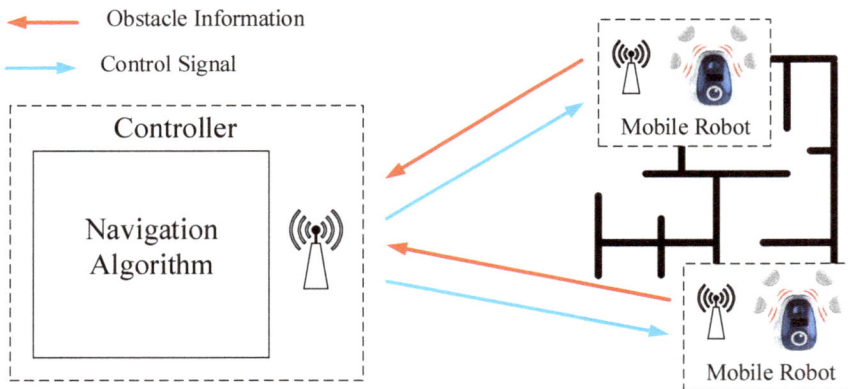

Figure 5. Navigation algorithm implementation based on centralized control.

The architecture used to check these algorithms is depicted in Figure 5. The system consists of a centralized controller and some mobile robots, which use an RF communication between them. The navigation algorithm is implemented in the controller. It sends the control signals (angular speeds) to the robots, and the mobile robots send to the controller the sensor signals (obstacle information). The communication between elements (robots and controller) occurs only when the robot sends sensor information to the controller. When this happens, the controller calculates the control signals and sends this information to the robot. In the discrete-time architecture, the communication between elements is periodical. In the event-based solution proposed in this work, the communication occurs only when the event condition is satisfied.

In the next sections, a wall following and an obstacle avoidance algorithms will be developed to analyze the behavior of the proposed strategy and to compare to a classical discrete-time solution.

5.1. Wall Following Algorithm

A wall following algorithm is one of the most common process used in mobile robot navigation. The basic algorithm is as follows:

(1) Find the closest wall.
(2) Move to a desired distance from the wall.
(3) Turn and start moving along the wall, staying the desired distance away.

Wall following algorithms have two variants of implementation: clockwise or counter-clockwise; in this case, the second option is selected. To implement this algorithm in the controller, a simple control law has been designed; it is presented in Equation (4).

$$
\begin{aligned}
&\text{if } (obt_{front\_left} > 0) \\
&\quad \{front\_obstacle = 1\} \\
&\text{else } \{front\_obstacle = 0\}
\end{aligned}
$$

$$
\begin{aligned}
sc_{left}(\%) &= \left(\frac{obt_{lat\_left}-50}{40}\right) \cdot 40 + 50 \\
sc_{right}(\%) &= 50 - (front\_obstacle) \cdot 50
\end{aligned}
$$

(4)

In this expression, obt_{front_left} and obt_{lat_left} are the values of the sensors, the range of which goes from 0% (obstacle is out of the range of the sensor) to 100% (the obstacle is touching the robot). The auxiliary variable $front\_obstacle$ is used to detect if an obstacle is in front of the robot. sc_{left} and sc_{right} are the angular speeds calculated in the controller, which have to be sent to the robot over the radio channel; the range of these speeds go from 0%–100%.

The control law in Equation (4) modifies the wheel speed of the robot to force it to stay at a constant distance (50% of the lateral sensor's range) from the wall with a constant speed (50% of the robot's maximum speed).

This navigation algorithm is the same for the discrete-time solution and for the event-based one. In the mobile robot, when the discrete-time solution is used, every sampling time $1/f_s$, the robot sends the sensor information (obt_{front_left}, obt_{front_right}, obt_{side_left} and obt_{side_right}) over the radio interface. After the controller receives this information, it computes the control signals (sc_{left} and sc_{right}), and they are sent to the robot over the same radio interface. Finally, in the robot, the received control signals are assigned by the drive system to the wheels ($speed_{left} = sc_{left}$ and $speed_{right} = sc_{right}$).

When the event-based proposal is considered, as was mentioned before, the algorithm in the controller is the same, but in the mobile robot, an event-based algorithm is implemented. The proposed event-based system for the mobile robot is depicted in Figure 6. The event-based algorithm in the robot depends on the event condition, which is presented in Equation (5), where w represents the reference signal, which is 50% for this algorithm. The event threshold $\bar{e}$ is the parameter that defines the accuracy of the system, the number of the events and its influence on the communication load in the RF interface.

$$
\begin{aligned}
&\text{if } ((obt_{front\_left} > 0) \text{ OR } (abs(obt_{lat\_left} - w) > \bar{e}) \\
&\quad \{event = true\} \\
&\text{else } \{event = false\}
\end{aligned}
$$

(5)

When the event condition is satisfied, an event is generated, and the robot sends to the controller the sensor information. At the same time, the controller, after computing the control signals, returns to the robot this information, which contains new speed values. Every time the event condition is true, the RF channel is busy with the information that the robot and the controller are interchanging; the rest of the time, the communication channel is free.

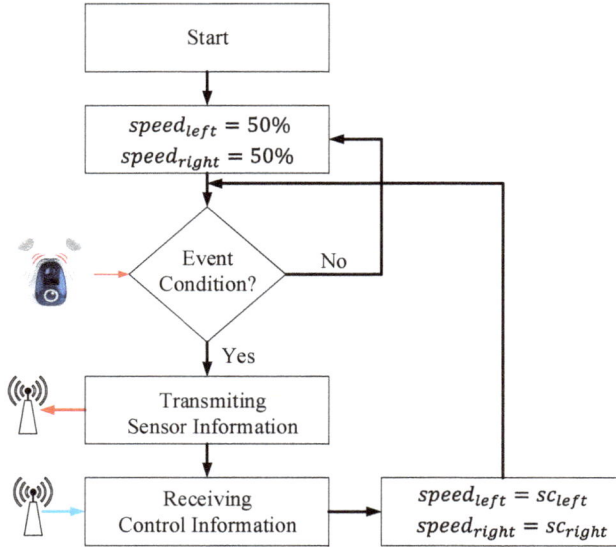

Figure 6. Event-based architecture for the mobile robots.

5.2. Obstacle Avoidance Algorithm

An obstacle avoidance algorithm makes the robot able to reach a destination without any collisions. To analyze the event-based architectures proposed in this work, a simple obstacle avoidance algorithm has been proposed. It is depicted in Figure 7, and the control law is given by the Equation (6).

$$
\begin{aligned}
&\text{if } (C_1) \ \{sc_{left} = 40\%, sc_{right} = 0\%\} \\
&\text{else if } (C_2) \text{ OR } (C_3) \text{ OR } (C_4) \ \{sc_{left} = 20\%, sc_{right} = 0\%\} \\
&\text{else if } (C_5) \text{ OR } (C_6) \text{ OR } (C_7) \ \{sc_{left} = 0\%, sc_{right} = 20\%\} \\
&\text{else if } (C_8) \ \{sc_{left} = 0\%, sc_{right} = 0\%\} \\
&\text{else } (C_9) \ \{sc_{left} = 10\%, sc_{right} = 10\%\}
\end{aligned} \tag{6}
$$

where the conditions C_1–C_9 are defined by Equations (7)–(15).

$$
\text{if } ((obt_{front\_left} > 0) \text{ AND } (obt_{front\_right} > 0)) \ \{C_1 = true\} \tag{7}
$$

$$
\text{if } (obt_{front\_left} > 0) \ \{C_2 = true\} \tag{8}
$$

$$
\text{if } (obt_{lat\_left} > 0) \ \{C_3 = true\} \tag{9}
$$

$$
\text{if } ((obt_{front\_left} > 0) \text{ AND } (obt_{lat\_left} > 0)) \ \{C_4 = true\} \tag{10}
$$

$$
\text{if } (obt_{front\_right} > 0) \ \{C_5 = true\} \tag{11}
$$

$$
\text{if } (obt_{lat\_right} > 0) \ \{C_6 = true\} \tag{12}
$$

$$
\text{if } ((obt_{front\_right} > 0) \text{ AND } (obt_{lat\_right} > 0)) \ \{C_7 = true\} \tag{13}
$$

$$
\text{if } ((obt_{front\_left} > 0) \text{ AND } (obt_{lat\_left} > 0) \text{ AND } (obt_{front\_right} > 0) \text{ AND } (obt_{lat\_right} > 0)) \\ \{C_8 = true\} \tag{14}
$$

$$if\ ((obt_{front\_left} = 0)\ AND\ (obt_{lat\_left} = 0)\ AND\ (obt_{front\_right} = 0)\ AND\ (obt_{lat\_right} = 0))$$
$$\{C_9 = true\} \tag{15}$$

$$y_c[n]\ \text{or}\ e_c[n_k]$$

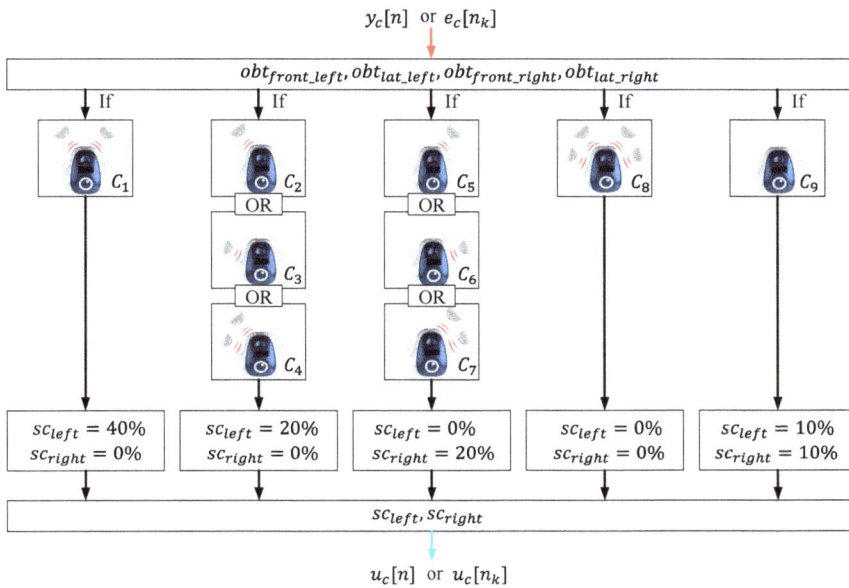

Figure 7. Block diagram of the obstacle avoidance algorithm.

This obstacle avoidance algorithm calculates the speed of each wheel as a function of the information in the obstacle detection sensors. Then, this information is sent to the robot via an RF interface. In this case, as in the case of the wall following algorithm, the control law is the same for the discrete-time implementation and for the event-based control strategy proposed in this work.

In the discrete-time implementation, every sampling period $(1/f_s)$, the robot sends to the controller the sensor information. At the same time, the controller sends to the robot the control signals, which contain the angular speeds to modify the trajectory of the robot.

For the event-based solution, the structure of the algorithm in the robot is the same as the one defined for the wall following algorithm depicted in Figure 6, with the following differences:

(1) the steady speeds are 10% for each wheel
(2) and the event condition is defined by Equation (16):

$$if\ ((obt_{front\_left} - w > \bar{e})\ OR\ (obt_{lat\_left} - w > \bar{e})\ OR\ ...$$
$$...(obt_{front\_right} - w > \bar{e})\ OR\ (obt_{lat\_right} - w > \bar{e}))$$
$$\{event = true\} \tag{16}$$
$$else\ \{event = false\}$$

In this case, the w is the reference signal, which for the obstacle avoidance algorithm is 0%, and $\bar{e}$ is the event threshold of the system.

6. Experimental Results

To check the proposed event-based control strategy in this work, a test laboratory with the low cost mOway differential wheel drive robots has been developed; see Figure 8.

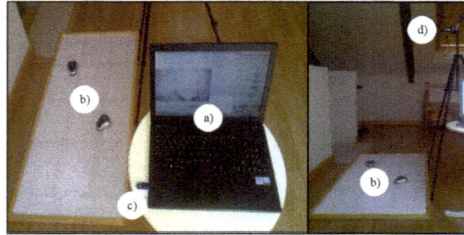

Figure 8. Components of the test laboratory: (**a**) controller; (**b**) mOway mobile robots; (**c**) RF interface; and (**d**) video camera.

The differential wheeled robot used in the experiments is depicted in Figure 9. The robots have four infra-red obstacle detection sensors, two in the front and the other two on the sides with a maximum range of 3 cm. The drive system of the robots permits angular speeds from 0 rad/s–10.9 rad/s. The dimensions regarding the dynamic model are: the distance between the wheel $L = 6.6$ cm and the wheel's radio $R = 1.6$ cm.

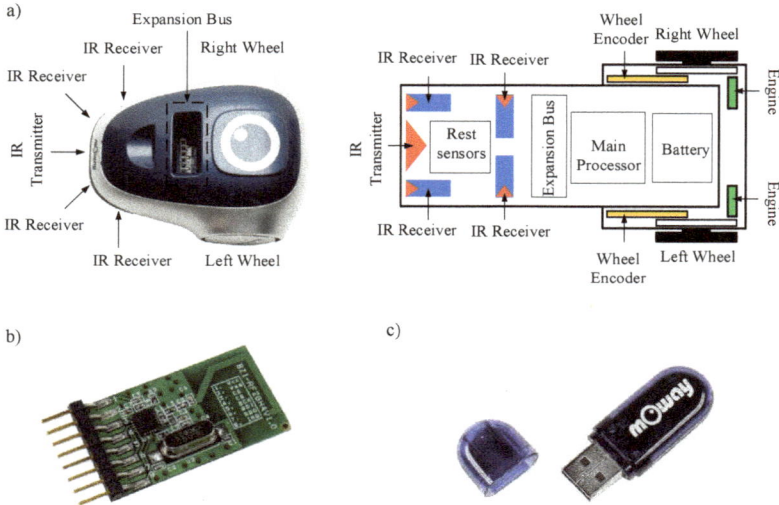

Figure 9. mOway mobile robot platform: (**a**) structure and components diagram; (**b**) RF module; and (**c**) RF USB interface.

The controller has been implemented in a laptop and the navigation algorithm in C++ with the Windows OS. In the robots, the discrete-time and the event-based solutions were programmed in the mOway World application. In both cases, the discrete-time and the event-based structures, the robot works with a sampling frequency of $f_s = 10$ Hz. A radio frequency link of 2.4 GHz has been used for communication between the robot and the controller. Finally, a video camera and the Tracker video processing tool were utilized to capture the real path of the robots.

The wireless communication systems (RF module and RF USB) are based on the nRF24L01 transceiver manufactured by Nordic Semiconductors. The RF interface works in the worldwide ISM frequency band at 2.400–2.4835 GHz and uses GFSK modulation. It has user-configurable parameters, like frequency channel, output power and air data rate. The air data rate for each channel is configurable to 2 Mbps, and 126 channels can be configured. The used bandwidth in the system depends on the traffic in the wireless network. The control system proposed in this work uses less control information than the classical discrete time solution. In this case, the bandwidth not utilized to send control information can be used to transmit other user information or to include more agents in the system.

The wall following and the obstacle avoidance navigation algorithms, which were defined in Section 5, have been checked in the test laboratory using the two control strategies, the proposed event-based control and the classical discrete-time control. The obtained results and a comparison between them are presented in the next sections.

6.1. Wall Following

The wall following algorithm described in Section 5.1 has been programmed in the test laboratory using the discrete-time architecture and the proposed event-based strategy. To check the response of both strategies, three experiments have been set: one with the discrete-time architecture and two with the event-based strategy. For the event-based solution, two event thresholds have been set, an event threshold of $\bar{e} = 10\%$ for the second experiment and $\bar{e} = 15\%$ for the third one. In all experiments, the robot has been navigating during 30 s in an environment with different objects as in Figures 10–12 show.

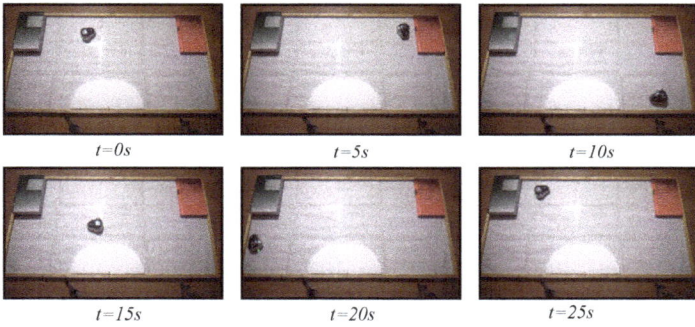

$t=0s$ $\qquad$ $t=5s$ $\qquad$ $t=10s$

$t=15s$ $\qquad$ $t=20s$ $\qquad$ $t=25s$

Figure 10. Experiment 1. Wall following algorithm using discrete-time architecture.

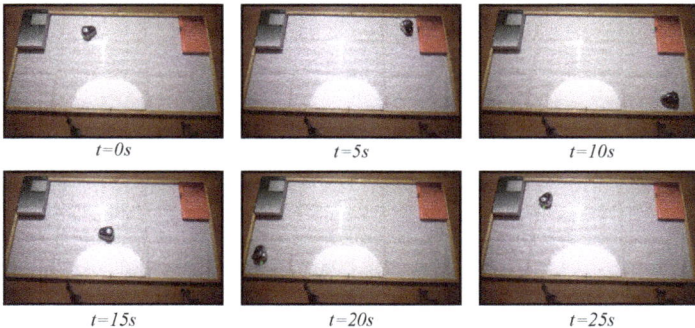

$t=0s$ $\qquad$ $t=5s$ $\qquad$ $t=10s$

$t=15s$ $\qquad$ $t=20s$ $\qquad$ $t=25s$

Figure 11. Experiment 2. Wall following algorithm using event-based architecture with $\bar{e} = 10\%$.

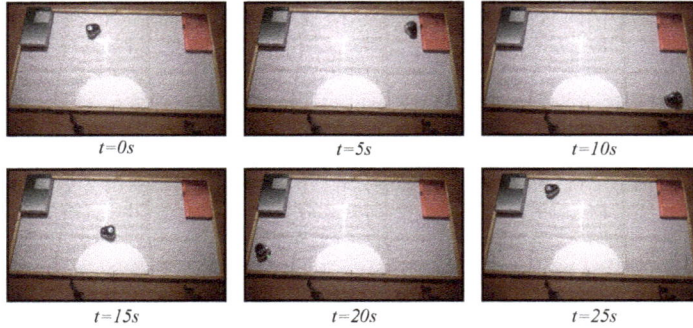

Figure 12. Experiment 3. Wall following algorithm using event-based architecture with $\bar{e} = 15\%$.

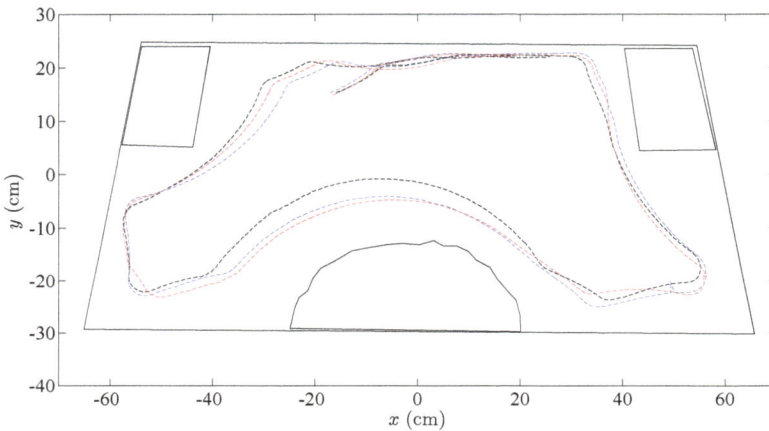

Figure 13. Trajectories of the robots controlled by the wall following algorithm. The black dashed line is the discrete-time strategy; the blue dashed line represents the event-based strategy with $\bar{e} = 10\%$; and the red dashed line is the event-based strategy with $\bar{e} = 15\%$.

In the three cases, the video of the experiments has been analyzed with the tracker tool, and the trajectories of the robots have been obtained. As shown in Figure 13, the three algorithms resolve the navigation problem in the correct way, and the trajectories of the robots are very similar.

The RF channel utilization has been analyzed for the three experiments. In this case, in each period, the robot transmits the information of its four obstacles sensors (the information of each sensor is coded by eight bits, in total 32 bits per each transmission). At the same time, the controller sends to the robot the speed of each wheel (the angular speed of each wheel is coded by eight bits, this means 16 bits per each transmission). Taking into account the previous considerations, the bandwidth utilized by the system is presented in Figure 14. The results show that the event-based solutions have a higher efficiency *versus* the discrete-time one. In the event-based case, the channel utilization also depends on the selected event threshold. If the event threshold increases, the efficiency in the RF channel increases, as well. The event threshold has to be selected carefully, because of if this parameter is too high, the channel utilization decreases, but the navigation problem cannot be resolved with the required accuracy.

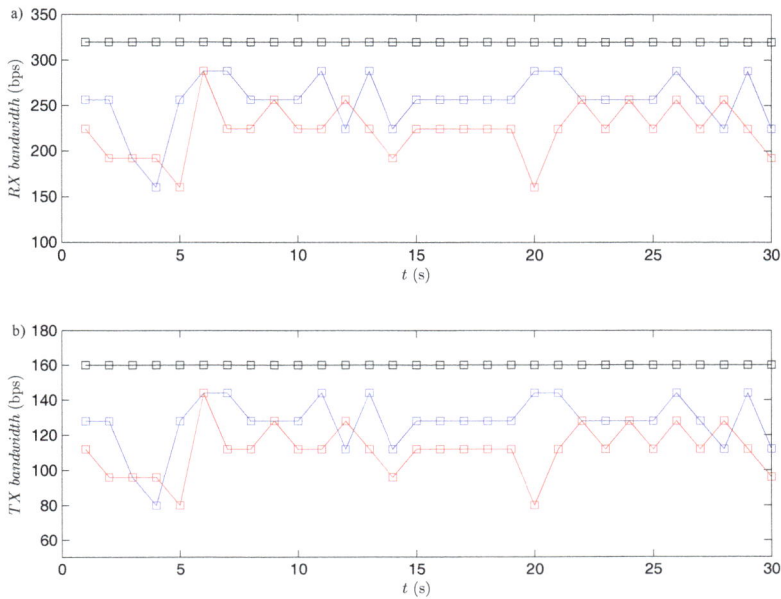

Figure 14. RF channel utilization. The black square is the discrete-time strategy; the blue square represents the event-based strategy with $\bar{e} = 10\%$; and and red dashed line is the event-based strategy with $\bar{e} = 15\%$. (**a**) Bandwidth used in reception (from the robots to the controller); (**b**) bandwidth used in transmission (from the controller to the robots).

To study the efficiency of the proposed solution, the number of accumulated transmissions of these experiments has been calculated. In Table 1, this parameter is presented for the three experiments.

Table 1. RF channel utilization in the wall following algorithm. Number of accumulated transmissions.

t (s)	Discrete-time	Event-based ($\bar{e} = 10\%$)	Event-based ($\bar{e} = 15\%$)
5	50	35	31
10	100	77	69
15	150	117	104
20	200	158	137
25	250	199	174
30	300	239	210

The results of Table 1 clearly show that the proposed event-based solution needs to use the RF channel less than a classical discrete-time one. The presented architecture has about 20.3% efficiency with $\bar{e} = 10\%$ and about 30% efficiency with $\bar{e} = 15\%$.

6.2. Obstacle Avoidance

To study the performance of the proposed architecture in an environment of multiple robots, the obstacle avoidance algorithm presented in Section 5.2 has been implemented in the test laboratory. In this case, two mOway robots have been used, and the two strategies, the discrete-time and the event-based, have been analyzed. Two experiments have been set up. In the fist one, the classical discrete-time solution has been used. In the second experiment, the proposed event-based solution has been programmed with an event threshold $\bar{e} = 0\%$ to check the worst case in the event-based solution.

As in the experiments of the wall following algorithms, in this case, the robots have been controlled by the navigation algorithm for 30 s.

The snapshots of these two experiments are depicted in Figures 15 and 16. The videos of the experiments were analyzed by the tracker tool to obtain the trajectories of the different robots; they are presented in Figure 17. In the two experiments, the navigation problem was resolved in a satisfactory way and without instabilities in the system.

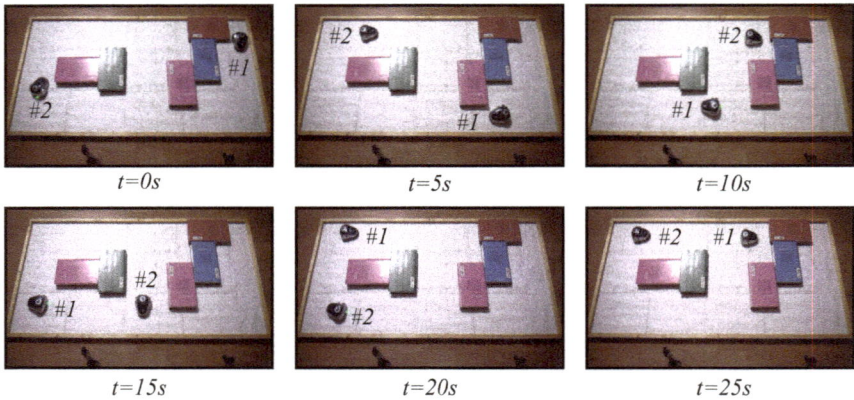

Figure 15. Experiment 1. The obstacle avoidance algorithm using a discrete-time strategy with a centralized controller and two robots. The symbols #1 and #2 represent Robot 1 and Robot 2, respectively.

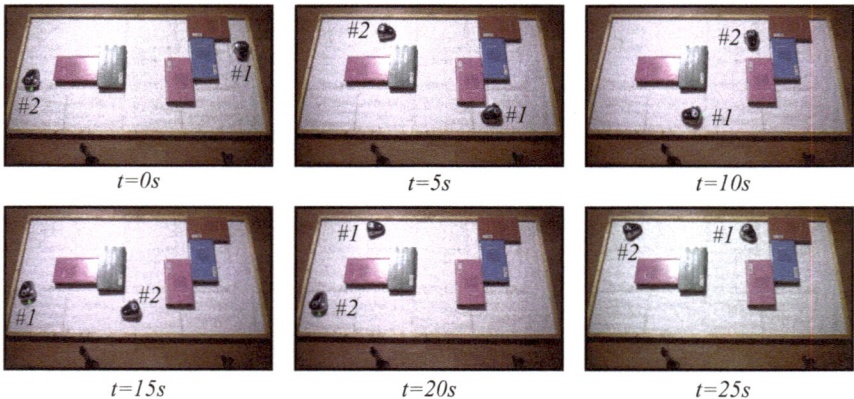

Figure 16. Experiment 2. The obstacle avoidance algorithm using an event-based strategy with a centralized controller and two robots. The symbols #1 and #2 represent Robot 1 and Robot 2, respectively.

As in the previous experiment, every period, the robots send to the controller the information of their obstacle sensors (each robot sends 32 bits), and the controller transmits to the robots the angular speeds (16 bits per each robot). The used bandwidth in both directions is presented in Figure 18. The results show that the event-based solution has a higher efficiency in communication resource utilization than the discrete-time one.

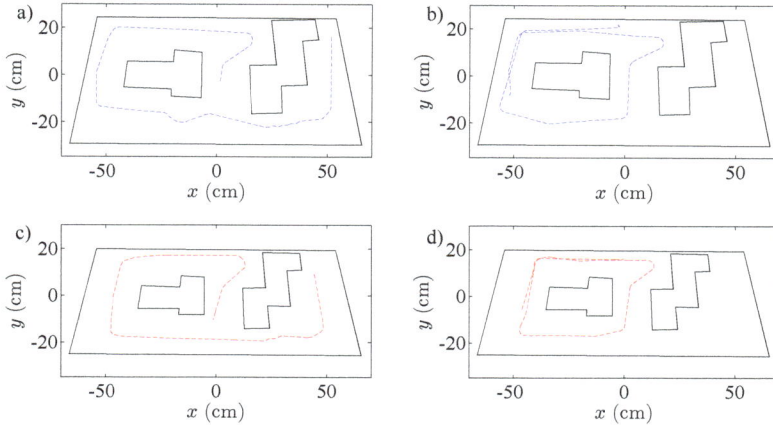

Figure 17. Trajectories of the robots controlled by the obstacle avoidance algorithm. The blue dashed line represents the discrete-time solution. (**a**) Robot 1; (**b**) Robot 2. The red dashed line is the event-based strategy. (**c**) Robot 1; (**d**) Robot 2.

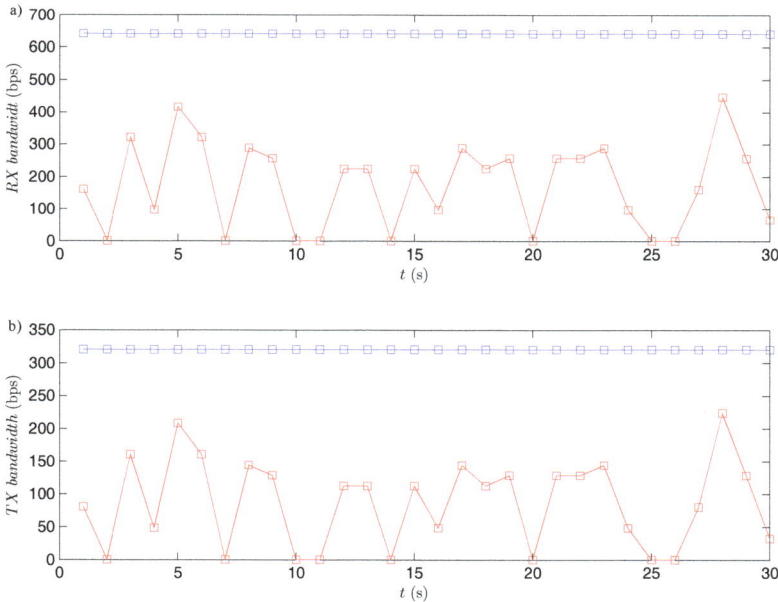

Figure 18. RF channel utilization. The blue square shows the discrete-time strategy; the red square represents the event-based strategy. (**a**) Bandwidth used in reception (from the robots to the controller); (**b**) bandwidth used in transmission (from the controller to the robots).

In Table 2, the number of accumulated transmissions of these experimented is presented.

The results in Table 2 show that the presented event-based control strategy has a higher efficiency in the communication resource usage than the discrete-time solution. In the analyzed experiments, the proposed strategy obtains an improvement of 73%.

Table 2. RF channel utilization. Number of accumulated transmissions.

t (s)	Discrete-Time	Event-Based
5	100	31
10	200	58
15	300	79
20	400	106
25	500	134
30	600	163

7. Conclusions and Future Work

The proposed event-based control strategy can be implemented in mobile robot architectures in a simple way. In this work, these strategies are applied to the mOway mobile robots platform, but this solution can be adapted to other kinds of robots following the same methodology. The control structures defined in the mobile robot only need to set one parameter, the event threshold $\bar{e}$, which defines the accuracy of the system and the performance in the radio interface. In the navigation experiments presented in this work, the proposed system has resolved the navigation problem in the same way as the discrete-time solution. The event-based solution obtains higher efficiencies in communication resource usage than the classical discrete-time solution. In the wall following algorithm, an efficiency of 15% is obtained, and for the obstacle avoidance algorithm, it is 73%.

As future work, the system is now being checked with a hybrid system, which contains several navigation algorithms (go to goal, wall following in its two versions and obstacle avoidance) working together using the architectures presented in this paper. Furthermore, the stability of the proposed solution is going to be analyzed in a theoretical way.

Acknowledgments: This work has been funded by the National Plan Project DPI2012-31303 of the Spanish Ministry of Economy and Competitiveness and FEDER funds.

Author Contributions: Rafael Socas designed the system, performed the experiments, analysed the results and prepared the first draft of the manuscript. Sebastián Dormido performed the theoretical conception and planning of the system. The manuscript was revised and modified by Raquel Dormido. Ernesto Fabregas assisted in the data acquisition and communications requirements.

Conflicts of Interest: The authors declare no conflict of interest.

References

1. Mansano, R.K.; Godoy, E.P.; Porto, A.J. The Benefits of Soft Sensor and Multi-Rate Control for the Implementation of Wireless Networked Control Systems. *Sensors* **2014**, *14*, 24441–24461.
2. Roohi, M.H.; Ghaisari, J.; Izadi, I.; Saidi, H. Discrete-time event-triggered control for wireless networks: Design and network calculus analysis. In Proceedings of the International Conference on Event-based Control, Communication, and Signal Processing (EBCCSP), Krakow, Poland, 17–19 June 2015; pp. 1–8.
3. Cloosterman, M.B.; Hetel, L.; van de Wouw, N.; Heemels, W.; Daafouz, J.; Nijmeijer, H. Controller synthesis for networked control systems. *Automatica* **2010**, *46*, 1584–1594.
4. Montestruque, L.; Antsaklis, P.J. State and output feedback control in model-based networked control systems. In Proceedings of the 41st IEEE Conference on Decision and Control, Las Vegas, NV, USA, 10–13 December 2002; Volume 2, pp. 1620–1625.
5. Montestruque, L.A.; Antsaklis, P.J. On the model-based control of networked systems. *Automatica* **2003**, *39*, 1837–1843.
6. Sun, Y.; El-Farra, N.H. Quasi-decentralized model-based networked control of process systems. *Comput. Chem. Eng.* **2008**, *32*, 2016–2029.
7. Tabuada, P. Event-triggered real-time scheduling of stabilizing control tasks. *IEEE Trans. Autom. Control* **2007**, *52*, 1680–1685.
8. Wang, X.; Lemmon, M.D. Event-triggering in distributed networked control systems. *IEEE Trans. Autom. Control* **2011**, *56*, 586–601.

9. Garcia, E.; Antsaklis, P.J. Decentralized model-based event-triggered control of networked systems. In Proceedings of the American Control Conference (ACC), Montreal, QC, Canada, 27–29 June 2012; pp. 6485–6490.

10. Guinaldo, M.; Dimarogonas, D.V.; Johansson, K.H.; Sánchez, J.; Dormido, S. Distributed event-based control for interconnected linear systems. In Proceedings of the 50th IEEE Conference on Decision and Control and European Control Conference (CDC-ECC), Orlando, FL, USA, 12–15 December 2011; pp. 2553–2558.

11. Anta, A.; Tabuada, P. To sample or not to sample: Self-triggered control for nonlinear systems. *IEEE Trans. Autom. Control* **2010**, *55*, 2030–2042.

12. Wang, X.; Lemmon, M.D. Event-Triggering in Distributed Networked Systems with Data Dropouts and Delays. In *Hybrid Systems: Computation and Control*; Springer: Berlin, Germany, 2009; pp. 366–380.

13. Mazo, M.; Tabuada, P. Decentralized event-triggered control over wireless sensor/actuator networks. *IEEE Trans. Autom. Control* **2011**, *56*, 2456–2461.

14. Dimarogonas, D.V.; Johansson, K.H. Event-triggered control for multi-agent systems. In Proceedings of the 48th IEEE Conference on Decision and Control, 2009 Held Jointly with the 2009 28th Chinese Control Conference, Shanghai, China, 15–18 December 2009; pp. 7131–7136.

15. Sánchez, J.; Visioli, A.; Dormido, S. Event-based PID control. In *PID Control in the Third Millennium*; Springer: Berlin, Germany, 2012; pp. 495–526.

16. Aström, K.J. Event Based Control. In *Analysis and Design of Nonlinear Control Systems*; Springer: Berlin, Germany, 2008; pp. 127–147.

17. Lunze, J.; Lehmann, D. A state-feedback approach to event-based control. *Automatica* **2010**, *46*, 211–215.

18. Nguyen, V.H.; Suh, Y.S. Networked estimation with an area-triggered transmission method. *Sensors* **2008**, *8*, 897–909.

19. Pawlowski, A.; Guzman, J.L.; Rodríguez, F.; Berenguel, M.; Sánchez, J.; Dormido, S. Simulation of greenhouse climate monitoring and control with wireless sensor network and event-based control. *Sensors* **2009**, *9*, 232–252.

20. Miskowicz, M. Efficiency of event-based sampling according to error energy criterion. *Sensors* **2010**, *10*, 2242–2261.

21. Li, S.; Sauter, D.; Xu, B. Fault isolation filter for networked control system with event-triggered sampling scheme. *Sensors* **2011**, *11*, 557–572.

22. Lehmann, D.; Lunze, J. Extension and experimental evaluation of an event-based state-feedback approach. *Control Eng. Pract.* **2011**, *19*, 101–112.

23. Dormido, S.; Sánchez, J.; Kofman, E. Muestreo, control y comunicación basados en eventos. *Rev. Iberoam. Autom. Inform. Ind. RIAI* **2008**, *5*, 5–26.

24. Socas, R.; Dormido, S.; Dormido, R. Event-based controller for noisy environments. In Proceedings of the Second World Conference on Complex Systems (WCCS), Agadir, Morocco, 10–12 November 2014; pp. 280–285.

25. Romero, J.A.; Pascual, N.J.; Peñarrocha, I.; Sanchis, R. Event-Based PI controller with adaptive thresholds. In Proceedings of the 4th International Congress on Ultra Modern Telecommunications and Control Systems and Workshops (ICUMT), St. Petersburg, Russia, 3–5 October 2012; pp. 219–226.

26. Siegwart, R.; Nourbakhsh, I.R.; Scaramuzza, D. *Introduction to Autonomous Mobile Robots*; MIT Press: Cambridge, MA, USA, 2011.

27. Nehmzow, U. *Robot Behaviour: Design, Description, Analysis and Modelling*; Springer Science & Business Media: Berlin, Germany, 2008.

28. Nehmzow, U. *Mobile Robotics*; Springer Science & Business Media: Berlin, Germany, 2003.

sensors

Article

Estimation of Human Arm Joints Using Two Wireless Sensors in Robotic Rehabilitation Tasks

Arturo Bertomeu-Motos *, Luis D. Lledó, Jorge A. Díez, Jose M. Catalan, Santiago Ezquerro, Francisco J. Badesa and Nicolas Garcia-Aracil

Neuro-Bioengineering Research Group, Miguel Hernandez University, Avda. de la Universidad W/N, 03202 Elche, Spain; llledo@umh.es (L.D.L.); jdiez@umh.es (J.A.D.); jose.catalan@goumh.umh.es (J.M.C.); sezquerro@umh.es (S.E.); fbadesa@umh.es (F.J.B.); nicolas.garcia@umh.es (N.G.-A.)
* Correspondence: abertomeu@umh.es; Tel.: +34-965-222-505; Fax: +34-966-658-979

Academic Editor: Gonzalo Pajares Martinsanz
Received: 27 October 2015; Accepted: 2 December 2015; Published: 4 December 2015

Abstract: This paper presents a novel kinematic reconstruction of the human arm chain with five degrees of freedom and the estimation of the shoulder location during rehabilitation therapy assisted by end-effector robotic devices. This algorithm is based on the pseudoinverse of the Jacobian through the acceleration of the upper arm, measured using an accelerometer, and the orientation of the shoulder, estimated with a magnetic angular rate and gravity (MARG) device. The results show a high accuracy in terms of arm joints and shoulder movement with respect to the real arm measured through an optoelectronic system. Furthermore, the range of motion (ROM) of 50 healthy subjects is studied from two different trials, one trying to avoid shoulder movements and the second one forcing them. Moreover, the shoulder movement in the second trial is also estimated accurately. Besides the fact that the posture of the patient can be corrected during the exercise, the therapist could use the presented algorithm as an objective assessment tool. In conclusion, the joints' estimation enables a better adjustment of the therapy, taking into account the needs of the patient, and consequently, the arm motion improves faster.

Keywords: kinematic reconstruction; neuro-rehabilitation; end-effector robots; upper limbs; MARG

1. Introduction

Robot-aided neuro-rehabilitation therapies have become an interesting field in the robotics area. There are several devices, such as exoskeletons, prosthesis or end-effector configuration robots, developed for this purpose [1,2]. They are able to help and assist the shortcomings of human beings. Post-stroke patients usually lose limb mobility due to the impairment in motor activity. Rehabilitation in this field takes an important role when it comes to improving the motor and proprioceptive activity [3,4]. In terms of the activities of daily living (ADL), the total or partial recovery of the upper limbs is the most important part in early rehabilitation. End-effector configuration robots are the most common devices used in these therapies. They are easily adapted to and easy to use by patients with different diseases.

These robots provide objective information about the trajectory followed by the end effector and the improvement in the motor recovery. However, they are not able to measure and control the arm movements. The progress in the arm joints, *i.e.*, the range of motion (ROM), is an important parameter in these kinds of therapies. This estimation requires non-invasive wearable sensors, which must be easy to place onto the patient's arm and must be extended to a clinical environment. Visual feedback of the arm configuration is studied in some rehabilitation therapies, though the arm joints cannot be measured [5,6]. This estimation can be accurately performed with optoelectronic systems based on motion tracking, even though they cannot be adapted to a rehabilitation environment [7,8]. In 2006,

Mihelj developed a method to estimate the arm joints through two accelerometers placed onto the upper arm [9]. Then, Papaleo *et al.* improved this method using a numerical integration through the augmented Jacobian in order to estimate the arm configuration with only one accelerometer [10,11]. This algorithm performs a kinematic reconstruction of the simplified human arm model with seven degrees of freedom (DoFs) assuming that the shoulder is fixed during the therapy. Due to the loss of motor function, shoulder movements cannot be avoided by the patient, and therefore, this assumption cannot be always accomplished. Thus, it is necessary to measure shoulder movements in order to correct the position of the patient during the activity. This compensation can be detected and categorized through the fusion of a depth camera with skeleton tracking algorithms [12]. However, to compute the kinematic reconstruction, the position and orientation of the shoulder with respect to the robot are necessary.

This paper presents a kinematic reconstruction algorithm of human arm joints assuming a simplified model with five DoFs. Furthermore, this method is able to estimate the shoulder movement, *i.e.*, its position and orientation. It is based on the inverse kinematics through the pseudo-inverse of the Jacobian [13]. The end-effector planar robot, called "PUPArm", with three DoFs (see Figure 1), designed and built by Neuro-Bioengineering Research Group (nBio), Miguel Hernández University of Elche, Spain, is used [14]. The accuracy of the estimated joints with respect to the real arm joints, measured through a tracking camera, is studied. In addition, the ROMs on 50 healthy subjects performing a therapy activity are evaluated in two different cases: trying not to move the shoulder during the exercise and following the movement with the trunk to reach the goal.

Figure 1. PUPArm robot.

2. Algorithm Description

2.1. Human Arm Kinematic Chain

The human arm is a complex kinematic chain that can be defined as the contribution of several robotic joints. The arm was defined as a chain of nine rotational joints by Lenarčič and Umek [15]. Only seven DoFs take part in this experiment: a spherical joint in the shoulder; an elbow joint; and a spherical joint in the wrist; as is shown in Figure 2a. On the other hand, the PUPArm robot fixes two kinds of movements: the ulnar-radial deviation and the flexion-extension of the hand; thus, abduction-adduction (q_1), flexion-extension (q_2) and internal-external rotation (q_3) of the shoulder, flexion-extension (q_4) of the elbow and pronation-supination (q_5) of the forearm comprise the kinematic chain linked through two segments: the upper arm (l_u) and the forearm (l_f). The Denavit–Hartenberg (DH) parameters of the arm are shown in Table 1, and their reference systems are shown in Figure 2b.

Table 1. DH parameters of the kinematic arm chain.

i	θ_i	d_i	a_i	α_i
1	$\pi/2 + q_1$	0	0	$\pi/2$
2	$3\pi/2 + q_2$	0	0	$\pi/2$
3	q_3	l_u	0	$-\pi/2$
4	$\pi/2 + q_4$	0	0	$\pi/2$
5	q_5	l_f	0	0

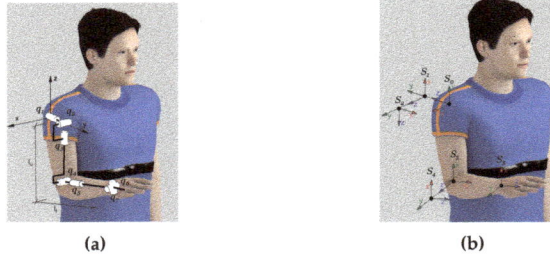

(a)

(b)

Figure 2. Human arm joints. (**a**) Simplification of human arm joints with seven DoFs; (**b**) Denavit–Hartenberg (DH) coordinate systems of the arm with five DoFs.

2.2. Integration Method

The inverse kinematics of the human arm during the exercise is based on the numerical integration through the pseudo-inverse of the Jacobian (J) [10]. The necessary devices to estimate the arm joints are: the end-effector robot; an accelerometer placed onto the upper arm and a magnetic angular rate and gravity (MARG) device placed onto the shoulder. Instantaneous joint velocities may be assessed as:

$$\dot{\vec{q}} = J^{-1}(\vec{q})\{\dot{\vec{v}}_d + K \cdot \vec{err}\} \tag{1}$$

being $\dot{\vec{v}}_d$ the Cartesian vector of the hand velocity and $\vec{err}$ the error committed due to the numerical integration. It should be noted that $\dot{\vec{v}}_d$ is the hand velocity vector with respect to the shoulder, estimated through the MARG and the accelerometer. To minimize this error, a 7×7 gain matrix K is added to this Equation [13]. Then, the current arm joints are computed as:

$$\vec{q}(t_{k+1}) = \vec{q}(t_k) + \dot{\vec{q}}(t_k)\Delta t \tag{2}$$

where $\vec{q}(t_k)$ is the previous estimated joints, $\dot{\vec{q}}(t_k)$ is the joint velocity vector obtained through Equation (1) and Δt is the sampling time. On the other hand, the initial arm joints are necessary to begin the integration method; their computation is explained in Section 2.6.

2.3. Accelerometer Orientation

If slow movements are assumed, the orientation of the accelerometer can be estimated in any position of the arm within the reachable workspace of the robot. When joints q_1 to q_5 are equal to zero, the reference position of the arm is set; a visual representation of this position is shown in Figure 2b. The acceleration acquired in the reference orientation of the accelerometer regarding the gravity, which is shown in Figure 3a, is:

$$^{acc_0}V_g = \begin{bmatrix} 0 \\ 1 \\ 0 \end{bmatrix} \tag{3}$$

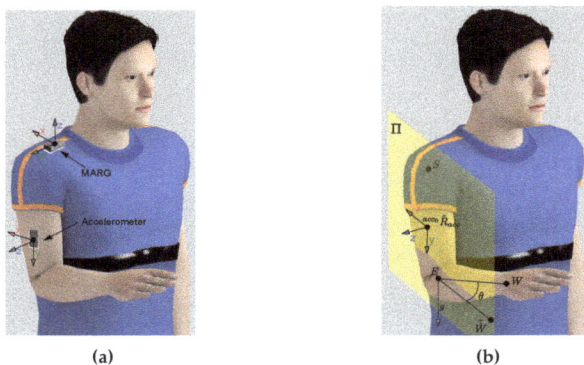

Figure 3. (a) Reference orientation of the accelerometer and the MARG. (b) Plane Π shaped by the X axis and Y axis of $^{acc_0}\tilde{R}_{acc}$.

Moreover, at any random position of the arm, $^{acc_0}V_g$ can be computed through the applied rotation to the accelerometer ($^{acc_0}R_{acc}$) as:

$$^{acc_0}V_g = {}^{acc_0}R_{acc}{}^{acc}V_g \tag{4}$$

being $^{acc}V_g$ the acceleration at this random position regarding the gravity.

Equation (4) has infinite rotation matrices over the gravity vector, though one possible solution may be computed as:

$$^{acc_0}\tilde{R}_{acc} = I + M + M^2\frac{1 - \cos(\theta)}{\sin^2(\theta)} \tag{5}$$

with:

$$M = \begin{bmatrix} 0 & -V_3 & V_2 \\ V_3 & 0 & -V_1 \\ -V_2 & V_1 & 0 \end{bmatrix}$$

$$V = {}^{acc_0}V_g \times {}^{acc}V_g$$

$$sin(\theta) = \|V\|$$

$$cos(\theta) = {}^{acc_0}V_g \cdot {}^{acc}V_g \tag{6}$$

Thereby, a plane can be shaped by the X axis and Y axis of $^{acc_0}\tilde{R}_{acc}$ (plane Π). This plane only contains the elbow point (E), but the correct orientation of the accelerometer must also contain the shoulder (S) and the wrist (W) points. Thus, the rotation angle (θ) is defined as the angle between the known wrist point and the new wrist point ($\tilde{H}$), contained in the plane Π, when it is rotated around the gravity vector (g) placed in E (see Figure 3b). Therefore, $\tilde{H}$, expressed in terms of θ, can be defined as:

$$\tilde{W} = (g \cdot \hat{W})\,g + \cos(\theta)\,(\hat{W} - (g \cdot \hat{W})\,g) - \sin(\theta)\,(g \times \hat{W}) \tag{7}$$

where $\hat{W} = (W - E)\,/\,(\|W - E\|)$ and $g = \begin{bmatrix} 0 & 0 & -1 \end{bmatrix}^T$. Then, θ can be obtained solving the following equation:

$$d\left(\tilde{W}, \Pi\right) = \frac{\left|A_\Pi \tilde{W}_x + B_\Pi \tilde{W}_y + C_\Pi \tilde{W}_z + D_\Pi\right|}{\sqrt{A_\Pi{}^2 + B_\Pi{}^2 + C_\Pi{}^2}} = 0 \tag{8}$$

having the plane Π computed as follows:

$$\tilde{P}_{acc}^x = {}^{acc_0}\tilde{R}_{acc} \begin{bmatrix} 1 & 0 & 0 \end{bmatrix}^T$$

$$\tilde{P}_{acc}^y = {}^{acc_0}\tilde{R}_{acc} \begin{bmatrix} 0 & 1 & 0 \end{bmatrix}^T$$

$$\overline{S\tilde{P}_{acc}^y} = \left(\tilde{P}_{acc}^y - S \right)$$

$$\overline{\tilde{P}_{acc}^x \tilde{P}_{acc}^y} = \left(\tilde{P}_{acc}^y - \tilde{P}_{acc}^x \right) \tag{9}$$

$$\begin{bmatrix} A_\Pi \\ B_\Pi \\ C_\Pi \end{bmatrix} = \overline{S\tilde{P}_{acc}^y} \times \overline{\tilde{P}_{acc}^x \tilde{P}_{acc}^y}$$

$$D_\Pi = \begin{bmatrix} A_\Pi & B_\Pi & C_\Pi \end{bmatrix}^T \cdot S$$

Two possible solutions are obtained through Equation (8) and, therefore, two values of ${}^{acc_0}R_{acc}$. The correct solution is one for which the Z axis is in the same direction as the cross product between the elbow-wrist segment and elbow-shoulder segment due to the reference position of the accelerometer. Finally, the rotation of the accelerometer regarding the robot is computed as:

$$^r R_{acc} = {}^r R_{acc_0} \cdot {}^{acc_0} R_{acc} \tag{10}$$

being $^r R_{acc_0}$ the reference orientation of the accelerometer concerning the robot (see Figure 3a). This orientation is required to estimate the elbow orientation and shoulder position during the exercise.

2.4. MARG Orientation

The orientation of magneto-inertial devices is usually based on Kalman filtering [16]; nevertheless, they can be quite complicated, and an extended Kalman filter is needed to linearize the problem. The orientation filter to measure the rotation of the MARG of Madgwick *et al.* is used in this algorithm [17]. The magnetic distortion that may be introduced by external sources, including metal furniture and metal structures within a building, is performed in this filter [18]. Furthermore, the orientation algorithm requires an adjustable parameter (β) that can be adjusted to the requirements of this exercise. Hence, the value of this parameter ($\beta = 5$) was established after a "trial and error" approach tested before the experiment, taking into account the features of the exercises.

This filter measures the reference quaternion of the device with respect to the Earth reference system, defined by the gravity vector and the Earth's magnetic field lines. However, the rotation of the Earth concerning the robot is unknown. If the MARG is placed in a known orientation with respect to the robot ($^R_{M_0}\hat{q}$), the acquired transformation defines the Earth frame relative to the sensor frame ($^{M_0}_E\hat{q}$), and therefore, the reference transformation between the robot and the Earth is known as:

$$^R_E\hat{q} = {}^R_{M_0}\hat{q} \otimes {}^{M_0}_E\hat{q} \tag{11}$$

Therefore, every rotation of the MARG is defined in the workspace as:

$$^R_M\hat{q} = {}^R_E\hat{q} \otimes {}^M_E\hat{q}^* \tag{12}$$

where $^M_E\hat{q}$ is the current value of the sensor. In this way, the shoulder orientation is estimated during the exercise.

2.5. Elbow and Shoulder Location

The hand, as was said before, is tightly attached to the end effector of the robot, and the ulnar-radial deviation and flexion-extension of the hand remain constant. Hence, the transformation matrix between the hand and the end effector ($^r T_w$) is known, and therefore, the elbow position may be obtained as:

$$^r P_e =^r T_w * \begin{bmatrix} 0 & 0 & -l_f & 1 \end{bmatrix}^T \tag{13}$$

The orientation of the elbow, since the rotation matrix between the elbow and the accelerometer orientation ($^{acc_0} R_e$) is known (see Figure 3a), may be calculated as:

$$^r R_e =^r R_{acc} \cdot^{acc_0} R_e \tag{14}$$

with $^r R_{acc}$ the rotation matrix computed through Equation (10). Thus, the transformation of the elbow relative to the robot remains:

$$^r T_e = \begin{bmatrix} ^r R_e & ^r P_e \\ 0 \quad 0 \quad 0 & 1 \end{bmatrix} \tag{15}$$

On the other hand, one of the most important points of this algorithm is the ability to estimate the shoulder position and orientation during the exercise. The shoulder position can be processed easily through Equation (15) as:

$$^r P_s =^r T_e * \begin{bmatrix} 0 & lu & 0 & 1 \end{bmatrix}^T \tag{16}$$

Whilst the orientation of the MARG relative to the robot is known by Equation (12), its rotation matrix $^r R_M$ is directly obtained [19]. Thus, the shoulder orientation is estimated as:

$$^r R_s =^r R_M \cdot^{M_0} R_s \tag{17}$$

where $^r R_M$ is the current rotation of the sensor with respect to the robot and $^{M_0} R_s$ the reference position of the MARG relative to the shoulder (see Figure 3a). Hence, the transformation of the shoulder relative to the robot remains:

$$^r T_s = \begin{bmatrix} ^r R_s & ^r P_s \\ 0 \quad 0 \quad 0 & 1 \end{bmatrix} \tag{18}$$

Finally, since the elbow and the shoulder location are instantaneously known, the initial conditions and the integration method can be performed.

2.6. Initial Conditions

In this algorithm, since it is based on a numerical integration, the initial conditions are required. The locations of the three main points, namely the shoulder ($^r T_s$), the elbow ($^r T_s$) and the wrist ($^r T_s$), are known. The shoulder joints (q_1, q_2 and q_3) are directly related to the matrix $^s T_e =^r T_s^{-1} \cdot^r T_e$, defined in the previous section, and they can be acquired by the spherical joint method [13]. This matrix, in terms of the corresponding joints, can be expressed by DH parameters shown in Table 1 as:

$$^{s_0} T_{s_3} =^{s_0} T_{s_1} \cdot^{s_1} T_{s_2} \cdot^{s_2} T_{s_3} = \begin{bmatrix} c_1 s_3 - c_3 s_1 s_2 & -c_2 s_1 & c_1 c_3 + s_1 s_2 s_3 & l_u c_2 s_1 \\ s_1 s_3 + c_1 c_3 s_2 & c_1 c_2 & c_3 s_1 - c_1 s_2 s_3 & -l_u c_1 c_2 \\ -c_2 c_3 & s_2 & c_2 s_3 & -l_u s_2 \\ 0 & 0 & 0 & 1 \end{bmatrix} \tag{19}$$

having $s_i = \sin(q_i)$ and $c_i = \cos(q_i)$, $i = \{1, 2, 3\}$. If the transformation matrix $^{s_0} T_{s_3}$ is defined as:

$$^{s_0} T_{s_3}(q_1, q_2, q_3) = \begin{bmatrix} n_x & n_y & n_z & p_x \\ o_x & o_y & o_z & p_y \\ a_x & a_y & a_z & p_z \\ 0 & 0 & 0 & 1 \end{bmatrix} \tag{20}$$

two possible solutions of the shoulder joints are obtained; if $q_2 \in [0 \quad \pi]$:

$$q_1 = \text{atan2}\left(-n_y, o_y\right)$$
$$q_2 = \text{atan2}\left(a_y, \sqrt{n_y^2 + o_y^2}\right) \tag{21}$$
$$q_3 = \text{atan2}\left(a_z, -a_x\right)$$

and if $q_2 \in [-\pi \quad 0]$:

$$q_1 = \text{atan2}\left(n_y, -o_y\right)$$
$$q_2 = \text{atan2}\left(a_y, -\sqrt{n_y^2 + o_y^2}\right) \tag{22}$$
$$q_3 = \text{atan2}\left(-a_z, a_x\right)$$

Thereby, the elbow joint (q_4) is directly determined with the cosine law as:

$$q_4 = \arcsin\left(\frac{l_u^2 + l_f^2 - ||H - S||^2}{2l_u l_f}\right) \tag{23}$$

and its homogeneous matrix remains:

$${}^{s_3}T_{s_4} = \begin{bmatrix} -\sin(q_4) & 0 & \cos(q_4) & 0 \\ \cos(q_4) & 0 & \sin(q_4) & 0 \\ 0 & 1 & 0 & 0 \\ 0 & 0 & 0 & 1 \end{bmatrix} \tag{24}$$

Thus, the transformation matrix between the systems s_0 and s_4 can be computed. The known matrix ${}^sT_h = {}^r T_s^{-1} \cdot {}^r T_h$ defines the transformation between the system s_0 and s_5. On the other hand, the last joint, q_5, is defined with the DH parameters as:

$${}^{s_4}T_{s_5}(q_5) = \begin{bmatrix} -\sin(q_5) & \cos(q_5) & 0 & 0 \\ \cos(q_5) & \sin(q_5) & 0 & 0 \\ 0 & 0 & 1 & 0 \\ 0 & 0 & 0 & 1 \end{bmatrix} \tag{25}$$

and therefore, q_5 is estimated as:

$$q_5 = \text{atan2}\left(-n_x, o_x\right) \tag{26}$$

Finally, two possible configurations of the arm joints are found, even though only one solution is possible. Due to the limits of the arm joints, $[-\pi/2 \quad \pi/2]$, only one solution accomplishes this restriction, and the initial position of the arm is assessed. This method can produce abrupt changes in the estimated arm joints caused by possible perturbations in the accelerometer that might lead to a non-anatomical position. Hence, since the new position depends on the latest position and the sample time, the integration method for real-time reconstruction is the best way to overcome the aforementioned drawbacks following Equations (1) and (2).

3. Results and Discussion

3.1. Experimental Exercises

With the aim of studying the arm joint estimation algorithm, with $K = diag\{1.5, 1.5, ...1.5\}$ N/ms (chosen by the "trial and error" approach tested before the experiment), two different experiments were performed. The first exercise was to compute the algorithm accuracy in terms of the arm joints and the position of the shoulder, performed by four healthy subjects. Then, a rehabilitation exercise with two different trials was performed by 50 healthy subject (aged between 20 and 72) to test the behavior of the presented algorithm. In both cases, the length of the upper arm was measured from the lateral side of the acromion to the proximal radius head, in the elbow joint. From the proximal radius head to the radial styloids, the distal part of the radius, the forearm length was measured [20]. Moreover, both experiments are performed under the same activity: 3D roulette, which may be seen in Figure 4. The activity consisted of taking a box from the perimeter and placing it in the center of the screen; hand movements are symbolized as a wrench (see Figure 4). One movement is considered when the subject goes from the center of the roulette to the perimeter and returns again to the center.

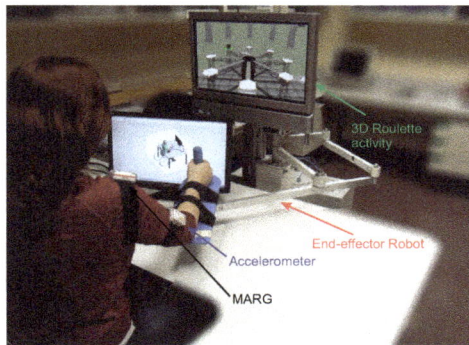

Figure 4. Subject wearing the sensors, the accelerometer and the MARG, grasping the end effector of the robot and performing the 3D roulette activity.

A magneto-inertial sensor, developed by Shimmer©, is tightly attached onto the upper arm and onto the shoulder to compute the kinematic reconstruction algorithm. The real position of the arm is computed with a six DoF optical tracking camera Optitrak V120: Trio, developed by NaturalPoint®. Specific parts attached to the hand, upper arm and forearm with retro-reflective markers were developed for this purpose. Information about the subjects who carried out the validation experiment are shown in Table 2; they performed three trials of the same exercise.

Table 2. Main subject data from the validation experiment.

ID	Age	Gender	Forearm Length (m)	Upper Arm
1	21	Male	0.23	0.32
2	51	Female	0.21	0.33
3	32	Male	0.25	0.31
4	31	Male	0.21	0.33

In the second experiment, two different trials of the same activity were performed. The first trial was intended not to move the shoulder while the exercise was being conducted, *i.e.*, without compensation with the trunk. However, the participants were asked to follow the hand movements with the shoulder in the second exercise. Each trial consisted of 24 movements.

3.2. Algorithm Validation

The mean error committed, in terms of root mean square error (RMSE) and standard deviation, is shown in Figure 5a. The mean RMSE of the joints is 0.047 rad with a standard deviation of 0.013 rad. Otherwise, the error committed on the shoulder position estimation, which may be found in Figure 5b, shows the mean RMSE committed, less than 0.87 cm, and the standard deviation, around 0.83 cm. The good results show that the error committed is small (it is hardly noticeable by the human eye), and therefore, the accuracy of the presented algorithm with respect to the real arm movements is high. A kinematic reconstruction of the arm joints and the estimation of shoulder position acquired from both methods through the presented algorithm (red dotted line) and the direct reconstruction (blue line) are pictured in Figure 6.

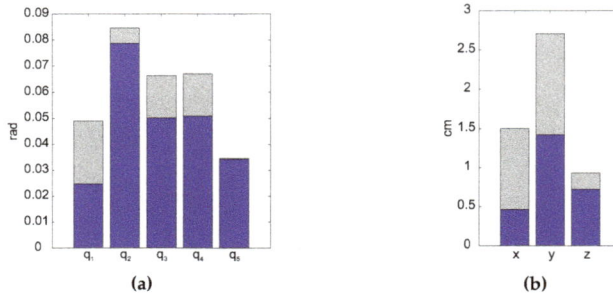

Figure 5. Error committed in the reconstruction algorithm. (**a**) Mean RMSE (blue bar) of the joints committed by the subjects and standard deviation (gray bar); (**b**) Mean RMSE (blue bar) of the shoulder position committed by the subjects and the standard deviation (gray bar): x, left/right movements; y, forward/backward movements; z, up/down movements.

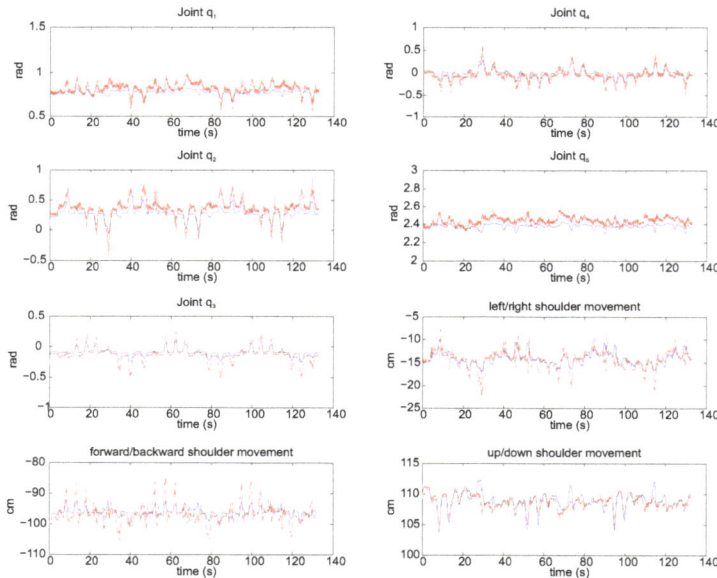

Figure 6. Joints and shoulder movements estimated through the algorithm (dotted red line) and measured through the optoelectronic system (blue line) of a subject during an exercise.

3.3. Arm Joint Range

In this experiment, the ROM between both trials, with and without compensation with the trunk, is studied. Furthermore, the shoulder movement is compared to its real position, acquired with the optoelectronical system mentioned before. To compare both groups, statistical analysis is performed through the *t*-test for paired data for each ROM. Joints 1 to 4 show significant differences ($p \leq 0.05$), but nevertheless, Joint 5, as the subject wrist is attached to the end effector of the robot, does not show significant differences ($p = 0.064$).

The estimated ROM in the exercise without compensation and with compensation is shown in Figure 7a, and the error committed might be seen in Figure 7b. It should be noted that the error committed in each joint for both exercises is smaller than six degrees. On the other hand, the ROM estimated for the trial without compensation is larger than that from the other trial. This result was expected, because the shoulder compensation affects the joint range. However, the ROMs of Joint 5 are similar, because the pronation-supination of the forearm is not affected when the compensation is performed.

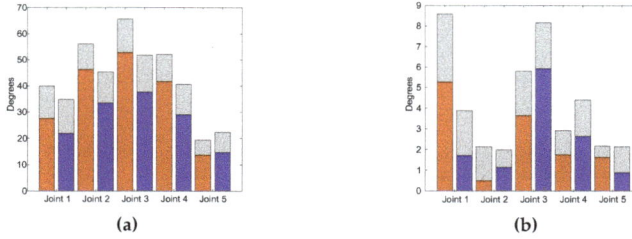

(a)   (b)

Figure 7. Representation of both trials: without compensation (orange bar) and with compensation (blue bar), and the standard deviation (gray bar) in terms of arm joints. (**a**) Estimated range of motion (ROM); (**b**) Error committed between the real ROM and the estimated ROM.

The accuracy of the shoulder position, taking into account the whole population (N = 51), is shown in Figure 8a. The estimated shoulder position with respect to the real shoulder location in a compensation trial performed by one subject can be seen in Figure 8b.

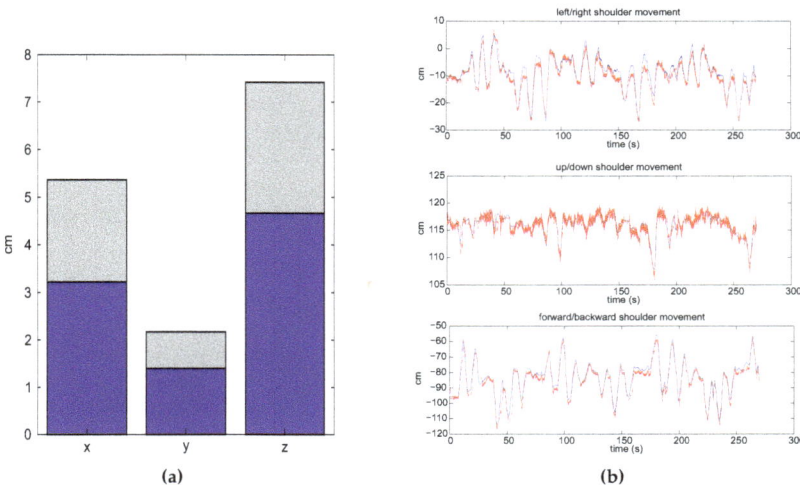

(a)   (b)

Figure 8. Shoulder movement in the compensation trial. (**a**) Mean RMSE (blue bar) committed by the population and the standard deviation (gray bar): x, left/right movements; y, forward/backward movements; z, up/down movements; (**b**) Estimated movement through the proposed algorithm (dotted red line) and the direct movement (blue line) performed by one subject.

Sensors **2015**, *15*, 120–131

4. Conclusions

In this paper, a kinematic reconstruction of the upper limbs during robot-aided rehabilitation with planar robots taking into account shoulder movements is presented. The estimated arm joints are very accurate with respect to the real position of the arm. Thus, the arm joint improvements of the patient can be measured objectively, and a better adaptation of the therapy to the patient needs can be also performed.

The measurement of the shoulder movement can be also computed accurately. To the best of our knowledge, this feature is not included in the previous algorithms where the shoulder is assumed to be fixed, even when little movements cannot be avoided during the exercise. This feature helps the therapist to correct the patient's posture during exercise for faster improvement in terms of arm mobility.

In summary, the arm joints' improvement may be included as a new objective assessment parameter in addition to the motor and proprioceptive activity and assessments scales, which are, by definition, subjective, as the Fugl–Meyer assessment [21].

Acknowledgments: This work was supported by the European Commission under FP7-ICT Contract 231143 (ECHORD (European Clearing House for Open Robotics Development)).

Author Contributions: N.G.-A., F.J.B and L.D.L. conceived of and designed the experiments. A.B.-M. and S.E. performed the experiments. A.B.-M. drafted the paper. A.B.-M., J.M.C. and J.A.D. analyzed the data. All authors read and approved the manuscript.

Conflicts of Interest: The authors declare no conflict of interest.

References

1. Nef, T.; Mihelj, M.; Colombo, G.; Riener, R. ARMin-robot for rehabilitation of the upper extremities. In Proceedings of the 2006 IEEE International Conference on Robotics and Automation, Orlando, FL, USA, 15–19 May 2006.
2. Tang, Z.; Zhang, K.; Sun, S.; Gao, Z.; Zhang, L.; Yang, Z. An upper-limb power-assist exoskeleton using proportional myoelectric control. *Sensors* **2014**, doi:10.3390/s140406677.
3. Lum, P.S.; Burgar, C.G.; Shor, P.C.; Majmundar, M.; der Loos, M.V. Robot-assisted movement training compared with conventional therapy techniques for the rehabilitation of upper-limb motor function after stroke. *Arch. Phys. Med. Rehabil.* **2002**, *83*, 952 – 959.
4. Badesa, F.; Morales, R.; Garcia-Aracil, N.; Alfaro, A.; Bernabeu, A.; Fernandez, E.; Sabater, J. Robot-assisted rehabilitation treatment of a 65-year old woman with alien hand syndrome. *Biomed. Robot. Biomech.* **2002**, doi:10.1109/BIOROB.2014.6913809.
5. Cameirao, M.; Badia, S.; Oller, E.; Verschure, P. Neurorehabilitation using the virtual reality based Rehabilitation Gaming System: methodology, design, psychometrics, usability and validation. *J. NeuroEng. Rehabil.* **2010**, doi:10.1186/1743-0003-7-48.
6. Wittmann, F.; Lambercy, O.; Gonzenbach, R.R.; van Raai, M.A.; Hover, R.; Held, J.; Starkey, M.L.; Curt, A.; Luft, A.; Gassert, R. Assessment-driven arm therapy at home using an IMU-based virtual reality system. *Rehabil. Robot.* **2015**, doi:10.1109/ICORR.2015.7281284.
7. Klopčar, N.; Lenarčič, J. Kinematic model for determination of human arm reachable workspace. *Meccanica* **2005**, *40*, 203–219.
8. Rab, G.; Petuskey, K.; Bagley, A. A method for determination of upper extremity kinematics. *Gait Posture* **2002**, doi:10.1016/S0966-6362(01)00155-2.
9. Mihelj, M. Human arm kinematics for robot based rehabilitation. *Robotica* **2006**, *24*, 377–383.
10. Papaleo, E.; Zollo, L.; Garcia-Aracil, N.; Badesa, F.; Morales, R.; Mazzoleni, S.; Sterzi, S.; Guglielmelli, E. Upper-limb kinematic reconstruction during stroke robot-aided therapy. *Med. Biol. Eng. Comput.* **2015**, *53*, 815–828.
11. Kreutz-Delgado, K.; Long, M.; Seraji, H. Kinematic analysis of 7 DOF anthropomorphic arms. *Robot. Autom.* **1990**, doi:10.1109/ROBOT.1990.126090.

12. Taati, B.; Wang, R.; Huq, R.; Snoek, J.; Mihailidis, A. Vision-based posture assessment to detect and categorize compensation during robotic rehabilitation therapy. *Biomed. Robot. Biomech.* **2012**, doi:10.1109/BioRob.2012.6290668.

13. Siciliano, B.; Sciavicco, L.; Villani, L.; Oriolo, G. *Robotics: Modelling, Planning and Control*; Springer-Verlag London: London, UK, 2009.

14. Badesa, F.J.; Llinares, A.; Morales, R.; Garcia-Aracil, N.; Sabater, J.M.; Perez-Vidal, C. Pneumatic planar rehabilitation robot for post-stroke patientes. *Biomed. Eng. Appl. Basis Commun.* **2014**, doi:10.4015/S1016237214500252.

15. Lenarčič, J.; Umek, A. Simple model of human arm reachable workspace. *IEEE Trans. Syst. Man. Cybern.* **1994**, *24*, 1239–1246.

16. Kalman, R.E. A new approach to linear filtering and prediction problems. *J. Basic Eng.* **1960**, *82*, 35–45.

17. Madgwick, S.; Harrison, A.; Vaidyanathan, R. Estimation of IMU and MARG orientation using a gradient descent algorithm. *Rehabil. Robot.* **2011**, doi:10.1109/ICORR.2011.5975346.

18. Bachmann, E.; Yun, X.; Peterson, C. An investigation of the effects of magnetic variations on inertial/magnetic orientation sensors. *Robot. Autom.* **2004**, doi:10.1109/ROBOT.2004.1307974.

19. Kuipers, J.B. *Quaternions and Rotation Sequences*; Princeton University Press: Princeton, NJ, USA, 1999.

20. Mazza, J.C. Mediciones antropométricas. Estandarización de las Técnicas de medición, Actualizada según Parámetros Internacionales. Available online: http://g-se.com/es/journals/publice-standard/articulos/mediciones-antropometricas.-estandariza-cion-de-las-tecnicas-de-medicion-actualizada-segun-parametros-internacionales-197 (accessed on 9 October 2015).

21. McCrea, P.H.; Eng, J.J.; Hodgson, A.J. Biomechanics of reaching: Clinical implications for individuals with acquired brain injury. *Disabil. Rehabil.* **2002**, *24*, 534–541.

sensors

MDPI

Article

Comparison of Three Non-Imaging Angle-Diversity Receivers as Input Sensors of Nodes for Indoor Infrared Wireless Sensor Networks: Theory and Simulation

Beatriz R. Mendoza [1], **Silvestre Rodríguez** [2,*], **Rafael Pérez-Jiménez** [2], **Alejandro Ayala** [1] and **Oswaldo González** [1]

[1] Departamento de Ingeniería Industrial, Universidad de La Laguna (ULL), 38203 La Laguna (Tenerife), Spain; bmendoza@ull.es (B.R.M.); aayala@ull.edu.es (A.A.); oghdez@ull.edu.es (O.G.)

[2] Departamento de Señales y Comunicaciones, Universidad de Las Palmas de Gran Canaria (ULPGC), 35017 Las Palmas (Gran Canaria), Spain; rperez@dsc.ulpgc.es

* Correspondence: srdguezp@ull.edu.es; Tel.: +34-922-845242

Academic Editor: Gonzalo Pajares Martinsanz

Received: 1 April 2016; Accepted: 7 July 2016; Published: 14 July 2016

Abstract: In general, the use of angle-diversity receivers makes it possible to reduce the impact of ambient light noise, path loss and multipath distortion, in part by exploiting the fact that they often receive the desired signal from different directions. Angle-diversity detection can be performed using a composite receiver with multiple detector elements looking in different directions. These are called non-imaging angle-diversity receivers. In this paper, a comparison of three non-imaging angle-diversity receivers as input sensors of nodes for an indoor infrared (IR) wireless sensor network is presented. The receivers considered are the conventional angle-diversity receiver (CDR), the sectored angle-diversity receiver (SDR), and the self-orienting receiver (SOR), which have been proposed or studied by research groups in Spain. To this end, the effective signal-collection area of the three receivers is modelled and a Monte-Carlo-based ray-tracing algorithm is implemented which allows us to investigate the effect on the signal to noise ratio and main IR channel parameters, such as path loss and rms delay spread, of using the three receivers in conjunction with different combination techniques in IR links operating at low bit rates. Based on the results of the simulations, we show that the use of a conventional angle-diversity receiver in conjunction with the equal-gain combining technique provides the solution with the best signal to noise ratio, the lowest computational capacity and the lowest transmitted power requirements, which comprise the main limitations for sensor nodes in an indoor infrared wireless sensor network.

Keywords: angle-diversity; sensor network; infrared channel; simulation; signal to noise ratio

1. Introduction

Although most wireless sensor networks used are currently based on radio frequency (RF) systems [1,2], wireless optical communications are becoming an alternative to RF technology in some well-defined indoor application scenarios, such as environments where RF emissions are forbidden or restricted (health care, nuclear and chemical plants, etc.), video/audio transmission for in-home applications, secure network access or sensor networking [3]. Optical systems do not interfere with RF systems, thus avoiding electromagnetic compatibility restrictions. Moreover, there are no current legal restrictions involving bandwidth allocation and, since radiation is confined by walls, they produce intrinsically cellular networks, which are more secure against deliberate attempts to gain unauthorized access than radio systems. Despite the recent development in the field of visible light communications (VLC), the non-directed non-line-of-sight (non-LOS) infrared (IR) radiation has also been considered as a very attractive alternative to RF waves for indoor wireless local area networks, and therefore for indoor wireless sensor networking.

Sensor networks represent a significant improvement over traditional sensors. A sensor network can range from a few sensor nodes to a few hundred nodes capable of collecting data and routing them back to the sink. Data are routed back to the sink by a multihop infrastructureless architecture through the nodes. The sink node may communicate with the task manager, for example, via Internet. A sensor node consists of four basic components: a sensing unit, a processing unit, a transceiver unit, and a power unit, although they may also have additional application-dependent components such as a location finding system, power generator and mobilizer. In a sensor network, the sensor nodes are limited in power, computational capacities, and memory. Furthermore, the nodes should communicate untethered over short distances and mainly use broadcast communication. In a multihop sensor network, communicating nodes are linked by a wireless medium. As previously mentioned, most sensor networks are based on RF technology; however, another possible method for internode communication is by infrared, which enables the use of transceivers that are cheaper and easier to build. Current infrared-based sensor networks require a line-of-sight (LOS) between transmitter and receiver [1–3]. There are two basic classification schemes for wireless IR links [4]. In the first approach, the link can be directed by employing a narrow-beam transmitter and a narrow field of view (FOV) receiver, or non-directed, with a broad-beam transmitter and a wide FOV receiver. The second scheme classifies the links according to whether or not they rely on a line-of-sight (LOS) between the transmitter and the receiver (LOS and non-LOS configurations). The directed LOS IR method is the most efficient in terms of power consumption and can achieve very high bit rates. Its drawbacks are tight alignment requirement, immobility of the receiver, and interruptions in transmission caused by shadowing. These disadvantages are overcome in non-directed non-LOS methods (referred to as diffuse links), which utilize diffuse reflections from the ceiling and walls. These methods, however, suffer from ineffective power use and multipath dispersion, which tend to greatly limit the transmission rate. In an indoor wireless optical sensor network, it is not necessary to achieve high transmission rates, but it is necessary to limit power consumption, avoid blockage and shadowing, and reduce the impact of ambient light noise. In general, the use of angle-diversity detection makes it possible to reduce the impact of ambient light noise, path loss, and multipath distortion, in part by exploiting the fact that they often receive the desired signal received from different directions [4–13]. Angle-diversity detection can be implemented in two main ways: using a composite receiver with several branches looking in different directions (non-imaging angle-diversity receiver), or using an imaging receiver consisting of imaging optics and an array of photodetectors (imaging angle-diversity receiver).

The propagation characteristics of the indoor IR channel are fully described by the channel's impulse response, which depends on multiple factors such as the room geometry, the reflection pattern of surfaces, the emitter and receiver characteristics and their relative locations. Indoor optical channel simulation can significantly enhance the study of optical wireless links. For this reason, in order to estimate the impulse response in IR wireless indoor channels, several simulation methods have been put forth [14,15], but all of them share the same problem, namely, the intensive computational effort. However, we make use of a Monte Carlo ray-tracing algorithm [16–19], which offers a lower computational cost than previous methods, especially when a high temporal resolution and a large number of reflections are required. In this paper, we study by simulation those indoor IR links that are characterised by the use of three non-imaging angle-diversity receivers as input sensor of nodes for an indoor IR wireless sensor network. The receivers, which have been proposed or studied by research groups in Spain, are the conventional angle-diversity receiver (CDR) [7,8,13], the sectored angle-diversity receiver (SDR) [11,12], and the self-orienting receiver (SOR) [20,21]. A conventional angle-diversity receiver uses multiple receiving elements or branches that are oriented in different directions, where each element employs its own filter and non-imaging concentrator, such as a compound parabolic concentrator (CPC) or hemispheric lens; a sectored receiver is a hemisphere where a set of parallels and meridians defines the photodetector boundaries; and finally, a self-oriented receiver is composed of a single-element detector that makes use of an optical front-end based on a lens, which is oriented towards the direction from which the highest signal to noise ratio (SNR) is received. The advantages achieved by angle-diversity reception depend on how the signals received in the different elements are processed and detected. When multipath distortion is significant, the optimum reception technique is maximum-likelihood combining (MLC), but its complexity is too high for many applications, and a number of simpler approaches are possible: maximal-ratio combining (MRC), selection best (SB), and equal-gain combining (EGC). When multipath distortion is negligible or it is not a significant parameter in the application scenario, such as in sensor networks, the optimum MLC reduces to MRC.

As noted earlier, current infrared-based sensor networks require a LOS link between transmitter and receiver. In this paper, we compare the use of SDR, CDR, and SOR as the input sensor for the nodes in an indoor IR wireless sensor network operating in a diffusive link for broadcast communication. This comparison is based on calculating the rms delay spread, the path loss, and the SNR when MRC, EGC, and SB combination techniques are employed. Based on the results of the simulations, we are going to demonstrate that a conventional angle-diversity receiver, used in conjunction with the equal-gain combining technique allows us to meet the requirements with the best SNR, the lowest computational capacity and the lowest transmitted power requirements. The use of angle-diversity receivers can be combined with multi-beam transmitters to significantly improve the power efficiency of diffuse links [4,10,22], which is desirable due to the power constraint of sensor nodes. This is an area for consideration in future research.

The remainder of the article is organized as follows: in Section 2, the channel model of the IR link for conventional, sectored, and self-orienting receivers using angle diversity is presented; i.e., the Monte Carlo ray-tracing algorithm used to study the signal propagation in the indoor IR channel and mathematical models employed to describe the effective signal-collection area of three receivers are described. Furthermore, we present the expressions for calculating the signal to noise ratio for an angle-diversity receiver as a function of the combining technique employed. Finally, in Section 3, we present several simulation results for comparing the performance of the use of three non-imaging angle-diversity receivers in IR links operating at low bit rates.

2. Channel Model and Signal-to-Noise Ratio

In optical wireless links, the most viable method is to employ intensity modulation (IM), in which the instantaneous power of the optical carrier is modulated by the signal. The receiver makes use of direct detection (DD), where a photodetector generates a current that is proportional to the instantaneous received optical power. The channel characteristics in an indoor optical wireless channel using IM/DD can be fully characterised by the impulse response $h(t)$ of the channel [4]:

$$I(t) = R\,x(t) \otimes h(t) + n(t) \tag{1}$$

where $I(t)$ represents the received instantaneous current at the output of the photodetector, t is the time, $x(t)$ is the transmitted instantaneous optical power, $\otimes$ denotes convolution, R is the photodetector responsivity and $n(t)$ is the background noise, which is modelled as white and Gaussian, and independent of the received signal.

2.1. Channel Impulse Response

To evaluate the impulse response of the indoor IR channel, a Monte Carlo ray-tracing algorithm was implemented. In general, the impulse response of the IR channel for arbitrary emitter E and receiver R positions can be expressed as an infinite sum of the form [14]:

$$h(t; E, R) = h^{(0)}(t; E, R) + \sum_{k=1}^{\infty} h^{(k)}(t; E, R) \tag{2}$$

where $h^{(0)}(t;E,R)$ represents the LOS impulse response and $h^{(k)}(t;E,R)$ is the impulse response of the light undergoing k reflections, i.e., the multiple-bounce impulse responses.

Given an emitter E and receiver R in an environment free of reflectors (Figure 1), with a large distance d between both, the LOS impulse response is approximately:

$$h^{(0)}(t; E, R) = \frac{1}{d^2} R_E(\theta, n) A_{eff}(\psi)\delta \left(t - \frac{d}{c} \right) \tag{3}$$

where $R_E(\theta,n)$ represents the generalized Lambertian model used to approximate the radiation pattern of the emitter, c is the speed of light and $A_{eff}(\psi)$ is the effective signal-collection area of the receiver [14].

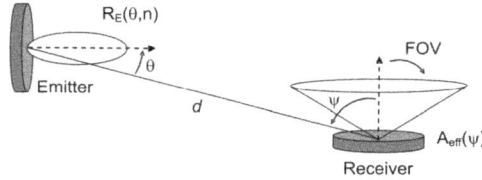

Figure 1. Emitter and receiver geometry without reflector.

In general, a bare detector, commonly called a single-element detector, achieves an effective signal-collection area of:

$$A_{eff}^{bare}(\psi) = A_R cos(\psi)\, rect \left(\frac{\psi}{FOV} \right), \;\; where \;\; rect(x) = \begin{cases} 1, & |x| \leqslant 1 \\ 0, & |x| > 1 \end{cases} \tag{4}$$

where A_R is the physical area of the receiver, and FOV is the receiver field of view (semi-angle from the surface normal).

In an environment with reflectors, however, the radiation from the emitter can reach the receiver after any number of reflections (see Figure 2). In the algorithm, to calculate the impulse response due to multiple reflections, many rays are generated at the emitter position with a probability distribution equal to its radiation pattern.

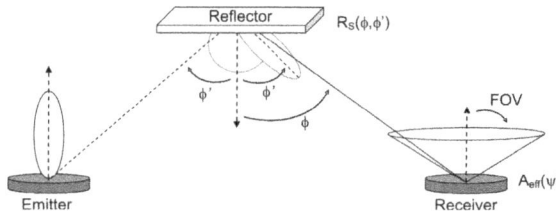

Figure 2. Emitter and receiver geometry with reflector. Reflection pattern of the surface is described by Phong's model.

The power of each generated ray is initially P_E/N, where N is the number of rays used to discretize the optical source. When a ray impinges on a surface, the reflection point becomes a new optical source, thus a new ray is generated with a probability distribution provided by that surface's reflection pattern. The process continues throughout the maximum simulation time, t_{max}. After each reflection, the power of the ray is reduced by the reflection coefficient of the surface, and the reflected power reaching the receiver ($p_{i,k}$, i-th ray, k-th time interval) is computed by:

$$p_{i,k} = \frac{1}{d^2} R_S(\varphi, \varphi') A_{eff}(\psi) \tag{5}$$

where $R_S(\varphi,\varphi')$ is the model used to describe the reflection pattern. In this work, Phong's model is used [17]. In this model, the surface characteristics are defined by three parameters: the reflection coefficient ρ, the percentage of incident signal that is reflected diffusely rd, and the directivity of the specular component of the reflection m.

Therefore, the total received power in the k-th time interval (width Δt) is calculated as the sum of the power of the N_k rays that contribute in that interval:

$$p_k = \sum_{i=1}^{N_k} p_{i,k} = \sum_{i=1}^{N_k} \frac{1}{d^2} R_S(\varphi, \varphi') A_{eff}(\psi) \tag{6}$$

Letting $M = t_{max}/\Delta t$, and assuming as the time origin the arrival of the LOS component, the impulse response after multiple reflections is given by:

$$\sum_{k=1}^{\infty} h^{(k)}(t; E, R) = \sum_{j=0}^{M-1} p_k \delta(t - j\Delta t) \tag{7}$$

Replacing Equations (3) and (7) in Equation (2), the channel impulse response can be expressed as:

$$h(t; E, R) = \frac{1}{d^2} R_E(\theta, n) A_{eff}(\psi)\delta(t) + \sum_{j=1}^{M-1} p_k \delta(t - j\Delta t) \tag{8}$$

Once the impulse response $h(t)$ is computed for a fixed emitter E and receiver R position, the main parameters that characterise the IR channel—the path loss (PL) and the rms delay spread (D)—are easily calculated as [4]:

$$PL = -10\log H(0) \qquad H(f) = \int_{-\infty}^{\infty} h(t)e^{-j2\pi tf} dt$$

$$D = \left[\frac{\int_{-\infty}^{\infty} (t-\mu)^2 h^2(t)dt}{\int_{-\infty}^{\infty} h^2(t)dt} \right]^{\frac{1}{2}} \qquad \mu = \frac{\int_{-\infty}^{\infty} t h^2(t)dt}{\int_{-\infty}^{\infty} h^2(t)dt} \tag{9}$$

2.2. Effective Signal-Collection Area Model for Non-Imaging Angle-Diversity Receivers

In general, the use of angle-diversity receivers makes it possible to reduce the impact of ambient light noise, path loss and multipath distortion, in part by exploiting the fact that they often receive the desired signal from different directions. Another advantage of angle-diversity reception is that it allows the receiver to simultaneously achieve a high optical gain and a wide FOV. In this section we describe the models for the effective signal-collection area of three non-imaging angle-diversity receivers that have been proposed or analysed by research groups in Spain. The receivers are called the conventional angle-diversity receiver, the sectored angle-diversity receiver, and the self-orienting receiver.

2.2.1. Conventional Angle-Diversity Receiver

A conventional angle-diversity receiver uses multiple receiving elements or branches that are oriented in different directions, where each element employs its own filter and non-imaging concentrator, such as a compound parabolic concentrator or hemispherical lens (see Figure 3a).

By adding a filter and concentrator to a bare detector, the effective signal–collection area of the receiver becomes:

$$A_{eff}^{c,f}(\psi) = A_R T_S(\psi)g(\psi)\cos(\psi)rect\left(\frac{\psi}{\pi/2}\right) \tag{10}$$

where $T_S(\psi)$ is the filter transmission and $g(\psi)$ the concentrator gain. Non-imaging concentrators exhibit a trade-off between gain and FOV. An idealized non-imaging concentrator [4] having an internal refractive index n achieves a constant gain expressed as:

$$g(\psi) = \frac{n^2}{sin^2\psi_c} rect\left(\frac{\psi}{\psi_c}\right) \tag{11}$$

where ψ_c is the concentrator FOV, which is usually less than or equal to $\pi/2$. In the model used, the concentrator gain is affected by the optical efficiency $\eta(\psi)$, which represents the reflection losses of the concentrator. The propagation delay introduced by the concentrator is also considered [18]:

$$g(\psi) = \frac{n^2}{sin^2\psi_c}\eta(\psi)rect\left(\frac{\psi}{\pi/2}\right), \, t(\psi) \neq 0 \tag{12}$$

Replacing $g(\psi)$ in the expression that defines the effective signal–collection area of the receiver yields:

$$A_{eff}^{c,f}(\psi) = \frac{n^2 A_R T_S(\psi)cos(\psi)}{sin^2\psi_c}\eta(\psi)rect\left(\frac{\psi}{\pi/2}\right) \tag{13}$$

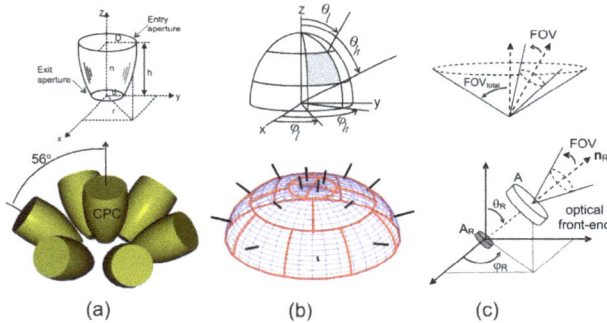

Figure 3. Non-imaging angle-diversity receiver geometries: (**a**) Conventional angle-diversity receiver; (**b**) Sectored angle-diversity receiver; (**c**) Self-orienting receiver.

In addition, in a wireless IR communications system, an optical bandpass filter can be used to limit the ambient radiation reaching the detector. A common form of bandpass filter consists of a stack of dielectric thin-film layers. By properly choosing the number of layers, their thicknesses, and their refractive indexes, it is possible to control the surface reflectance and thus the filter transmittance. The filter transmission $T_S(\psi)$ can be described fairly accurately by a simple five-parameter model [23]. In this model, for radiation of wavelength λ_0 incident at angle ψ, the filter transmission is given by:

$$T(\psi; \Delta\lambda, \psi') = \frac{T_0}{1 + \left[\frac{\lambda_0 - \lambda'(\psi;\psi')}{\Delta\lambda/2}\right]^{2m}} \tag{14}$$

where ψ' is the filter orientation, T_0 is the peak transmission at ψ', $\Delta\lambda$ is the spectral half-power bandwidth, m is the filter order and $\lambda'(\psi;\psi')$ represents the shifting to shorter wavelengths at non-normal incidences, which is described by:

$$\lambda'(\psi; \psi') = \lambda_0 \left(\frac{n_s^2 - n_1^2 sin^2\psi}{n_s^2 - n_1^2 sin^2\psi'}\right)^{\frac{1}{2}} \tag{15}$$

where n_1 is the index of the input layer and n_s is the effective index for the spacer layer. The design of the optical filter thus boils down to specifying the two parameters $\Delta\lambda$ and ψ'.

The remaining three parameters (n_s, m and T_0) should be chosen to be as large possible, while considering the constraints imposed by technology. In keeping with reference [23], we used $n_s = 2.293$, $m = 3$, and $T_0 = 0.92$. Furthermore, to provide the best utilization of the CPC and filter, the angular bandwidth $\Delta\psi$ should be equal to the concentrator FOV. Figure 3a shows a conventional angle-diversity receiver composed of seven elements, one of them oriented vertically towards the ceiling and six angled at a 56° elevation with a 60° separation in azimuth. Each element uses a bandpass optical filter and a CPC with a 50° field of view. The receiver structure resulted from a study focused on designing the conventional angle-diversity receiver that offers the best performance with respect to the path loss and the rms delay spread [13].

2.2.2. Sectored Angle-Diversity Receiver

A sectored angle-diversity receiver consists of a set of photodetectors located on a hemisphere. The space of sectored receivers enclosed between two parallels is called a crown, and a sector is the region of the crown enclosed between two equally spaced meridians. Every sector in each crown has an equal azimuth aperture, and therefore the same limiting elevation angles. In summary, the sectored receiver is defined by a set of parameters [11], Ψ, which specifies, for each crown, its number of sectors, N_S, the azimuth offset of its first sector, ε, and its limiting elevation angles, θ_l and θ_h. As an example, bottom illustration in Figure 3b shows the configuration of a sectored receiver with three crowns, which is defined by $\Psi = \{(4, 0°, 0°, 24°), (4, 20°, 24°, 53°), (8, 0°, 53°, 9°)\}$. As we can see, the first crown in the sectored receiver consists of 4 sectors ($N_S = 4$). The azimuth offset of its first sector is $\varepsilon = 0°$, and its limiting elevation angles are 0° and 24° ($\theta_l = 0°$ and $\theta_h = 24°$). Analogously to the conventional angle-diversity receiver shown in Figure 3a, the configuration of the sectored receiver was the result of a study focused on achieving the receiver structure that exhibits the minimum simultaneous rms delay spread and path loss [12].

As with a single-element detector, each sector of the sectored receiver is defined by its position, orientation, vertical and horizontal apertures, and effective signal-collection area. The region of a sector is specified by the two limiting elevation angles, θ_l and θ_h, and the two limiting azimuth angles, ϕ_l and ϕ_h, where $\theta_l < \theta_h$ and $\phi_l < \phi h$ (see top illustration in Figure 3b). Consequently, the vertical aperture is given by $(\theta_h - \theta_l)/2$, whereas the horizontal aperture is set by $(\phi_h - \phi_l)/2$. The orientation of each sector is defined by its azimuth and elevation angles (ϕ_S, θ_S), which are given by:

$$\phi_S = (\phi_l + \phi_h)/2 \quad and \quad \theta_S = (\theta_l + \theta_h)/2 \tag{16}$$

except in the case of a polar crown with a single sector where $\theta_S = 0°$ (sector aimed vertically upward, $\phi_h - \phi_l = 2\pi$, θ_h any, $\theta_l = 0$). Each sector in a sectored receiver has an effective signal-collection area given by:

$$A_{eff}(\psi) = A_R T_S(\psi)\cos(\psi)rect\left(\frac{\psi}{\pi/2}\right) \tag{17}$$

where $T_S(\psi)$ is the transmission characteristic of the filter used to limit the ambient radiation reaching each photodetector, ψ is the incidence angle of radiation with respect to the sector orientation, and A_R represents the physical area of the detector, which is described by:

$$A_R = r^2 \cdot (\phi_h - \phi_l) \cdot (\cos\theta_l - \cos\theta_h) \tag{18}$$

where r is the radius of the hemisphere. In a sectored receiver, the optical filter should be deposited or bonded onto the outer surface of the sectored receiver, i.e., onto all the sectors or photodetectors that comprise the receiver. In this sense, a bandpass or interference optical filter cannot be used because the radiation that reaches each photodetector is incident upon the filter at any value of ψ between 0° and 90°, shifting the filter transmission to shorter wavelengths and minimizing its transmission.

Therefore, the ambient light radiation and the desired signal striking the filter at wide incidence angles will not be adequately filtered. However, in order to reject ambient light, longpass absorption optical filters could be used [4], which pass light at every wavelength beyond the cutoff wavelength. They are usually made of coloured glass or plastic and their transmission characteristics are constant and substantially independent of the incidence angle, i.e., $T_S(\psi) = T_0$ for all wavelengths longer than the cutoff wavelength. Since the silicon device does not respond to wavelengths beyond about 1100 nm, the filter-photodetector combination exhibits a bandpass optical response, whose bandwidth will range from the filter cutoff wavelength to the longest wavelength at which the silicon responds. Although the sectored receiver may currently be regarded as a theoretical structure since no prototype has yet been implemented, we assume that a longpass absorption filter with a 780-nm cutoff wavelength is employed, specifically the Schott RG-780 optical filter [24]. This filter exhibits an almost constant transmission characteristic above 780 nm given by $T_0 = 0.99$, and consequently, the spectral bandwidth resulting from the filter-photodetector combination is about $\Delta\lambda = 320$ nm.

2.2.3. Self-Orienting Receiver

The third angle-diversity receiver considered is called a self-orienting receiver (see Figure 3c). The self-orienting receiver consists of a single-element detector which employs an optical front-end based on a lens [20,21]. The receiver is mounted on an electromechanical orienting system controlled by a digital signal processor (DSP). An SNR estimator and a maximum search algorithm, which is based on a modified version of Simulated Annealing (SA), are used to automatically aim the receiver at the area of the room with the highest SNR [20,21]. From the description above, we can deduce that the receiver is actually a modified version of a select-best angle diversity receiver, which allows operating with very narrow FOV. In general, the self-orienting receiver can be represented by the parameters p_R, n_R, FOV$_{total}$, A, A_R, and FOV, i.e., its position, orientation, total FOV, aperture area of the lens, physical area of photodetector and the FOV of the system formed by the detector and the optical front-end. Therefore, its effective signal-collection area can be expressed as:

$$A_{eff}(\psi) = A_R T_S(\psi) g(\psi) cos(\psi) rect \left(\frac{\psi}{\pi/2} \right) \tag{19}$$

where $T_S(\psi)$ is the filter transmission and $g(\psi) = G$ represents the optical front-end gain. Replacing the gain G by A/A_R in the expression, and assuming an ideal filter, the expression that defines the effective signal-collection area of the receiver can be expressed as:

$$A_{eff}(\psi) = A_R\, G\, cos(\psi) rect \left(\frac{\psi}{\pi/2} \right) = A\, cos(\psi) rect \left(\frac{\psi}{\pi/2} \right) \tag{20}$$

In keeping with the proposal made in [20,21], we have assumed that the self-orienting receiver uses a positive lens system with an aperture diameter of 1.5 cm as an optical concentrator, along with a circular photodetector with area $A_R = 3.53$ mm^2 and responsivity $R = 0.6$ A/W to collect the signals received. The concentrator provides a high optical gain ($G = 50$) at the expense of a narrow FOV, which is only 3°. This extremely narrow FOV allows us to use an ideal bandpass optical filter with a narrow spectral bandwidth ($\Delta\lambda = 50$ nm) for better optical noise rejection.

2.3. Signal to Noise Ratio

In general, the received signal-to-noise ratio can be expressed by:

$$SNR = (R\, P_S)^2 / \sigma^2 \tag{21}$$

where P_S is the received optical signal average power, R is the photodetector responsivity and σ^2 represents the total noise variance [4,25]. The total noise variance can be calculated by the sum of the contributions from the background light-induced shot noise and thermal noise due to the amplifier:

$$\sigma^2 = \sigma_{shot}^2 + \sigma_{thermal}^2 \tag{22}$$

The shot noise variance can be approximated by:

$$\sigma_{shot}^2 \approx 2qRI_2R_bP_{bg} \tag{23}$$

where $q = 1.6 \times 10^{-19}$ C is the electron charge, $I_2 = 0.562$ is a noise bandwidth factor, R_b is the bit rate, and P_{bg} is the incident optical power from natural and artificial ambient light.

Assuming the use of a FET-based transimpedance preamplifier [4], the thermal noise variance can be expressed as:

$$\sigma_{thermal}^2 = \frac{4kT}{R_f}I_2R_b + \frac{16\pi^2kT}{g_m}\left(\Gamma + \frac{1}{g_mR_D}\right)C_T^2I_3R_b^3 + \frac{4\pi^2KI_D^aC_T^2}{g_m^2}I_fR_b^2 \tag{24}$$

where k is Boltzmann's constant, T is absolute temperature, R_f is the feedback resistor, g_m is the transconductance, C_T is the total input capacitance of receiver, R_D is the polarisation resistance, Γ is the FET channel noise factor, K and a are the FET $1/f$ noise coefficients, I_D is the FET drain current, and I_2, I_3 and I_f are noise-bandwidth factors. Neglecting stray capacitance [10,22], the total input capacitance C_T is given by $C_T = C_d + C_g$, where C_d and C_g are the detector and FET gate capacitances, respectively. The detector capacitance C_d is proportional to the detector area A_R, i.e., $C_d = \eta\,A_R$, where η represents the fixed photodetector capacitance per unit area. Therefore, replacing C_T by $\eta\,A_R + C_g$ in Equation (24) yields:

$$\sigma_{thermal}^2 = \frac{4kT}{R_f}I_2R_b + \frac{16\pi^2kT}{g_m}\left(\Gamma + \frac{1}{g_mR_D}\right)(\eta A_R + C_g)^2 I_3R_b^3 + \frac{4\pi^2KI_D^a(\eta A_R + C_g)^2}{g_m^2}I_fR_b^2 \tag{25}$$

2.4. Signal to Noise Ratio for Angle-Diversity Receivers

Assuming no co-channel interference, a J-element angle-diversity receiver yields J receptions of the form:

$$I_j(t) = R\,x(t) \otimes h_j(t) + n_j(t), \quad j = 1, \dots, J \tag{26}$$

where $I_j(t)$ and $n_j(t)$ represent, respectively, the received instantaneous current and the total noise at the output of the j-th receiving element, t is the time, $x(t)$ is the transmitted instantaneous optical power, $h_j(t)$ is the channel between the transmitter and the j-th receiver, $\otimes$ denotes convolution, and R is the responsivity of the j-th receiver, which are considered equal for all J receiving elements.

In general, much of the performance of an angle diversity receiver depends on how the signals received in the different elements $Ij(t)$ are processed and detected. There are several possible diversity schemes that can be considered. The most common techniques include MLC, MRC, SB, and (EGC) [4]. When multipath distortion is significant, the optimum reception technique is MLC, also known as matched-filter combining (MFC). In this technique, each $I_j(t)$ is processed by a separate continuous-time matched filter $h_j(-t)$ of the j-th receiving element. The output from each matched-filter is sampled and adjusted in magnitude by a weight factor w_j, which is chosen to be inversely proportional to the power-spectral density (PSD) of the noise $n_j(t)$ at the output of the j-th receiving element. The weighted output signals are summed in order to obtain the combined signal. The main problem of MLC is its highly complex implementation, since this technique requires estimating each of the J channel impulse responses and noise PSD's separately. For this reason, MLC is not suited to many applications. There are simple alternatives to MLC that are practical to implement, such as MRC, SB and EGC, which differ depending on how the signals are weighed and combined.

Independent of the combined method and of the weight factor, and according to Equation (21), the signal to noise ratio at the output of the j-th receiving element SNR_j, and the SNR of the combined channel are defined as:

$$SNR_j = \left(R\,P_{S,j}\right)^2 / \sigma_j^2 \tag{27}$$

$$SNR_{combined} = \frac{\left(\sum\limits_{j=1}^{J} w_j R P_{S,j}\right)^2}{\sum\limits_{j=1}^{J} w_j^2 \sigma_j^2} \tag{28}$$

where $P_{S,j}$ and σ^2_j represent the received optical average power and the total noise variance of the j-th receiving element, respectively.

In MRC, the $I_j(t)$, $j = 1, \ldots, J$, are summed together with weights w_j proportional to the signal current to noise-variance ratios, thereby maximizing the SNR of the weighted sum. Simulations have shown that MRC can reduce transmitter optical power requirements by 4–6 dB in diffuse links at low bit rates [26]. Moreover, in situations where the ambient noise and the strong signal components arrive from different directions, MRC can decrease the multipath distortion in comparison to a single, wide FOV receiver. When multipath distortion is not significant, the optimum MLC reduces to MRC. MRC requires estimating the SNR's at the each of the receiving elements, thus increasing the complexity over non-diversity reception.

In order to obtain the weight factor w_j at the j-th receiving element, the J partial-derivatives of the combined SNR with respect to each of the w_j, $j = 1, \ldots, J$ are calculated and set equal to zero, so w_j becomes:

$$w_j = \frac{\sum\limits_{j=1}^{J} w_j^2 \sigma_j^2}{\sum\limits_{j=1}^{J} w_j R P_{S,j}} \cdot \frac{R P_{S,j}}{\sigma_j^2} = k_0 \cdot \frac{R P_{S,j}}{\sigma_j^2} \tag{29}$$

where k_0 is a constant independent of j, and w_j is proportional to the signal current to noise-variance ratio, as expected.

Replacing Equation (29) in Equation (28), the SNR using MRC can be calculated as the sum of each SNR of the j-th receiving elements, which is given by:

$$SNR_{MRC} = \frac{\left(\sum\limits_{j=1}^{J} \left(k_0 \cdot \frac{R P_{S,j}}{\sigma_j^2}\right) R P_{S,j}\right)^2}{\sum\limits_{j=1}^{J} \left(k_0 \cdot \frac{R P_{S,j}}{\sigma_j^2}\right)^2 \sigma_j^2} = \sum\limits_{j=1}^{J} \frac{\left(R P_{S,j}\right)^2}{\sigma_j^2} = \sum\limits_{j=1}^{J} SNR_j \tag{30}$$

The SB method chooses the branch or receiving element with the highest SNR from among all branches. This technique can often improve SNR since it separates the signal from ambient light noise, but the gains are not as large as those achieved using MRC. Simulations have shown that SB requires 1–2 dB more transmitter optical power than MRC [27]. Also, in [11], it was shown that a reduction in multipath delay spread occurs when directional narrow FOV receivers are employed in the branches, making SB a suitable technique for high bit-rate systems. On the other hand, SB is not simpler to employ than MRC, since it requires estimating the SNR at each diversity branch. The SNR using SB is given by:

$$SNR_{SB} = \max_j \left(\frac{\left(R P_{S,j}\right)^2}{\sigma_j^2}\right), \quad 1 \leqslant j \leqslant J \tag{31}$$

EGC corresponds to MRC, but without attempting to weigh the signals; that is, the weights of all the received signals $I_j(t)$ are equal to a constant. This technique increases the receiver FOV but is unable to separate the signal from ambient noise. Moreover, using EGC can result in an increase in multipath distortion, making it unsuitable for very high bit rate links. The main advantage of EGC is that it avoids the need to estimate the SNR [7]. The SNR using EGC is not dependent on the constant weight value, and its expression is given by:

$$SNR_{EGC} = \frac{\left(\sum_{j=1}^{J} wRP_{S,j}\right)^2}{\sum_{j=1}^{J} w^2\sigma_j^2} = \frac{\left(\sum_{j=1}^{J} RP_{S,j}\right)^2}{\sum_{j=1}^{J} \sigma_j^2} \tag{32}$$

3. Simulation Results and Discussion

In this section, simulations are used to compare the performance of the three non-imaging angle-diversity receivers described in the previous section and shown in Figure 3: the conventional angle-diversity receiver (CDR), the sectored angle-diversity receiver (SDR), and the self-orienting receiver (SOR). The comparison is based on calculating the rms delay spread, the path loss, and the SNR when the MRC, SB, and EGC combination techniques are employed. To this end, the simulation algorithm described in the previous section was implemented, the models for the three non-imaging angle-diversity receivers were included, and the IR signal propagation for different configurations of optical links in the room shown in Figure 4 was studied. The simulation tool implemented allows us to study the infrared signal propagation inside any simulation environment or 3D scene. The tool features two fully differentiated parts. The first is charged with defining the 3D scene, which the user can describe by means of any CAD software that is capable of generating or storing the scene in 3DS format. In our simulations, we have used the Blender graphic design program because it offers multi-platform support in a freeware product that can output a 3DS file. The second part consists of implementing the propagation model. This refers to the mathematical models that characterize the effect of each of the elements present in the simulation environment (reflecting surfaces, emitters, and receivers), and to the simulation algorithm that, aided by these models, allows the channel response to be computed. The part of the tool that implements the propagation model and into which the 3D scene is input was programmed in C++. A detailed description of the simulation tool developed was presented in [19], where the parallelization of the simulation algorithm was also discussed.

The indoor environment shown in Figure 4 was based on the channel models for a $14 \times 14 \times 3$ m^3 office, a $6 \times 6 \times 3$ m^3 living room, and an $8 \times 8 \times 3$ m^3 hospital room, which were proposed by the IEEE802.15.7 VLC work group [28]. The room selected is almost analogous to the living room, but with a rectangular instead of square shape. Furthermore, in order to make the comparison as independent as possible of the characteristics of the emitter and the furniture distribution in the room, no obstacles and a diffuse emitter were assumed.

In general, the topology of an infrared-based sensor network can vary from a simple star network to a multihop wireless mesh network. However, regardless of the network topology, the sensor nodes should be able to collect data and route them to the sink node. That is, in a wireless sensor network there is a large number of tiny transmitters collecting data and routing them to a small number of receiver nodes. For this reason, in order to compare the performance of the three angle-diversity receivers used as the input sensor of a receiver sensor node, we have proposed a simulation scenario consisting of an angle-diversity receiver located in the centre of the room and multiple transmitters uniformly distributed along the diagonal from the northwest to the southeast of the room, as shown in Figure 4b. As previously mentioned, the angle-diversity receiver is located in the centre of the room, 1 m above the floor and aimed vertically towards the ceiling. We have assumed that the CDR uses photodetectors with a physical area of $A_R = 1$ cm^2, the SOR employs a photodetector with an area of 3.53 mm^2, and the sectored receiver makes use of a hemisphere of radius $r = 1.4$ cm, meaning the largest sector has a physical area of about 1 cm^2. Specifically, the sectors in the first, second and third crown of the sectored receiver have an area of 0.27 cm^2, 0.95 cm^2 and 0.92 cm^2, respectively. Every photodetector in the three angle-diversity receivers has a responsivity of 0.6 A/W. As for the emitters, they are oriented vertically towards the ceiling, 1 m above the floor, located in thirty-seven locations uniformly distributed along the northwest-southeast diagonal of the room, and modelled as a first-order Lambertian emitter with a total emitted power of 15 mW. The remaining parameters used in the simulations match those shown in Table 1.

Figures 5 and 6 show the rms delay spread and the path loss, respectively, as a function of the distance between the emitter and receiver along a diagonal from the northwest (negative values or distances) to the southeast (positive values or distances) of the room shown in Figure 4. The results were obtained for all three angle-diversity receivers, SDR, CDR, and SOR, when MRC, EGC, and SB combination techniques are applied. In the case of SOR, which only employs a photodetector element, the combining techniques are not applicable. Furthermore, as noted earlier, the SOR receiver employs a maximum search algorithm to aim the receiver in the direction of the highest SNR. In our simulations, in order to find the reception orientation with the best SNR, first the SNR is calculated for all possible reception directions, as defined by their elevation and azimuth angles in steps of its FOV. Secondly, a higher resolution search is carried out around the best SNR obtained in the previous stage, which is bounded by the receiver FOV.

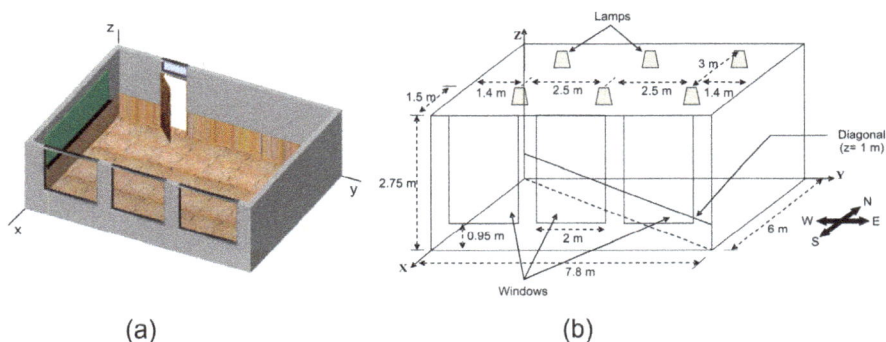

(a)   (b)

Figure 4. Graphical representation of the room: (**a**) 3D design; (**b**) Display of dimensions.

Independently of the angle-diversity receiver employed in the link, all the values obtained for the rms delay spread are below 6 ns, well above the requirements for a receiver operating in a 115 kbps link, i.e., the rms delay spread parameter is not significant when selecting the angle-diversity receiver and the combination technique that provides the best performance. As for the path loss shown in Figure 6, CDR and SOR exhibit smaller values than SDR due to the optical gain provided by the optical front-end employed for the photodetectors in both receivers. The CDR structure exhibits the smallest path loss because its photodetectors have a physical area larger than the SOR photodetector. In term of path loss, the CDR receiver offers the best power efficiency for the power transmitted by the emitter. Specifically, SOR has a path loss about 10 dBo greater than CDR when EGC is used for the emitter located in the centre of the room. The conventional angle-diversity receiver exhibits a unique behaviour when the emitter is located about 0.26 m from the centre of the room. The path loss is minimal because all the radiation is detected by the vertical element of the receiver after undergoing a single reflection. This effect also results in a minimum delay spread since all the radiation reaches the receiver at approximately the same time (see Figure 5), and in a maximum SNR because a low path loss involves a high received power (see Figure 7). In general, the sectored and conventional angle-diversity receivers exhibit the best path loss when EGC is employed, because all the power received by each photodetector element is collected or, put another way, EGC increases the receiver's total FOV. In MRC, the power received by the elements with a low SNR is attenuated, and in SB, only the power received by the element with the best SNR is taken into account. In short, EGC offers the best path loss if the noise is significant and the signal power is uniformly distributed throughout the room (broadcast communication).

Table 1. Simulation parameters.

Parameter			Value
Room:	width (x), m		6
	length (y), m		7.8
	height (z), m		2.75
Emitter:	mode (n)		1
	Power (P_E), mW		15
	position (x, y, z), m		(-, -, 1)
Receivers:	photodetectors: responsivity (R), A/W		0.6
	photodetectors: minimum power detected, W		10^{-12}
	position (x, y, z), m		(3, 3.9, 1)
CPC:	FOV		50°
	refractive index		1.8
	exit aperture, mm		5.64
Bandpass filter:	number of layers		20
	peak transmission (T_0)		0.92
	effective index (n_s)		2.293
	filter order (m)		3
	angular bandwidth ($\Delta\psi$), degrees		50
	spectral bandwidth ($\Delta\lambda$), nm		50
	λ_0, nm		810
Longpass filter:	filter transmission (T_0)		0.99
	cutoff wavelength, nm		780
	filter-photodetector combination ($\Delta\lambda$), nm		320
Tungsten lamps:	mode (n)		2
	lamp power-spectral density, W/nm		0.037
	position (x_1,y_1,z_1), m		(1.5, 1.4, 2.75)
	position (x_2,y_2,z_2), m		(4.5, 1.4, 2.75)
	position (x_3,y_3,z_3), m		(1.5, 3.9, 2.75)
	position (x_4,y_4,z_4), m		(4.5, 3.9, 2.75)
	position (x_5,y_5,z_5), m		(1.5, 6.4, 2.75)
	position (x_6,y_6,z_6), m		(4.5, 6.4, 2.75)
Window:	spectral radiant emittance, W/nm/m^2		0.2
Resolution:	Δt, ns		0.2
Bounces:	k		20
Number of rays:	N		500,000
Materials	ρ	r_d	m
Wood	0.63	0.6	3
Varnished W.	0.75	0.3	97
Cement	0.40	1.0	—
Ceramic floor	0.16	0.7	20
Glass	0.03	0.0	280

In order to obtain the SNR given by Equation (21), it is necessary to determine the total noise variance as the sum of the contributions from the background light-induced shot noise and thermal noise due to the amplifier. To this end, a bit rate of R_b = 115 kbps was considered since the study involves the application of non-imaging angle-diversity receivers to indoor IR wireless sensor networks. The shot noise can be computed using Equation (23), where the incident optical power from ambient light P_{bg} originates at the windows and six tungsten bulbs located in the ceiling of the room (see Figure 4). To compute the incident optical power from the windows, each window surface is divided into small square elements of equal area (25 cm²), and each element is modelled as a first-order Lambertian emitter with a spectral radiant emittance of 0.20 W/nm/m² [8], i.e., the noise PSD of each element is 0.5 mW/nm. The noise optical power emitted by each element can be obtained by multiplying the noise PSD by $\Delta\lambda$, which represents the spectral bandwidth of the optical filter used to limit the ambient radiation reaching the photodetector, which was set at 50 nm for the CDR and SOR structures and 320 nm for SDR. Therefore, applying the Monte Carlo ray-tracing algorithm to each element E, and using Equation (8), the incident optical power from the windows act as an ambient light (noise) source that can be expressed as a sum of the form:

$$P_{bg} = \sum_{j=1}^{N_e} \left(\sum_{j=0}^{M} h(t; E, R) \right) = \sum_{j=1}^{N_e} \left(\frac{1}{d^2} R_E(\theta, 1) A_{eff}(\psi) + \sum_{j=1}^{M-1} p_k \right) \tag{33}$$

where $M = t_{max}/\Delta t$ is the number of time intervals of width Δt, and N_e is the number of elements used to divide the window surface. Moreover, the radiant intensity from the bulbs can be modelled as Lambertian sources of second order with an optical spectral density of 0.037 W/nm. Analogously to the calculation of the noise power from ambient light, the incident optical power from the bulb can be obtained by multiplying the power spectral density by $\Delta\lambda$. As in the previous case, the optical power contribution from a bulb E can be calculated by:

$$P_{bg} = \frac{1}{d^2} R_E(\theta, 2) A_{eff}(\psi) + \sum_{j=1}^{M-1} p_k. \tag{34}$$

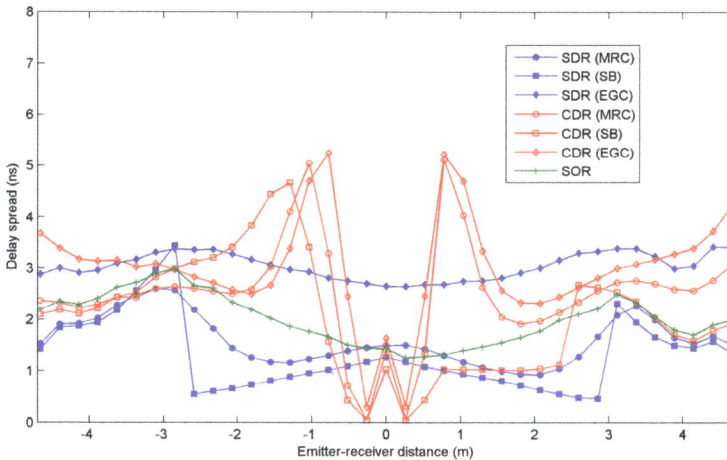

Figure 5. Delay spread for the sectored angle-diversity receiver (SDR), the conventional angle-diversity receiver (CDR) and the self-orienting receiver (SOR) as a function of the emitter-receiver distance along the diagonal from northwest (negative values) to southeast (positive values).

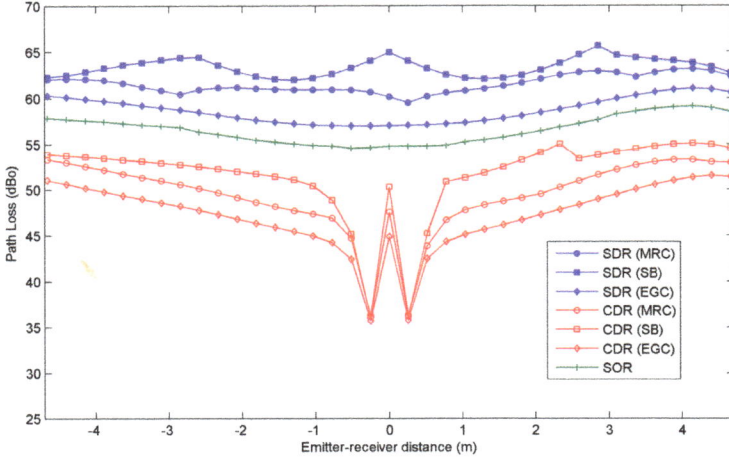

Figure 6. Path loss for the sectored angle-diversity receiver (SDR), the conventional angle-diversity receiver (CDR) and the self-orienting receiver (SOR) as a function of the emitter-receiver distance along the diagonal from northwest (negative values) to southeast (positive values).

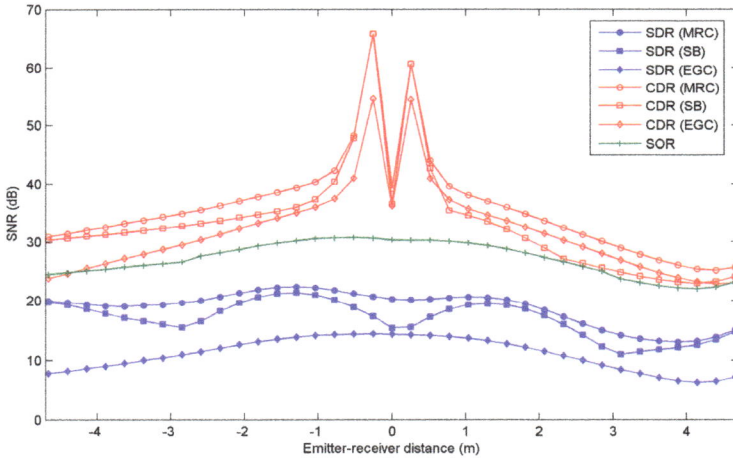

Figure 7. SNR for the sectored angle-diversity receiver (SDR), the conventional angle-diversity receiver (CDR) and the self-orienting receiver (SOR) as a function of the emitter-receiver distance along the northwest- southeast diagonal.

Assuming typical parameters for a receiver that relies on a FET-based transimpedance preamplifier, i.e., $k = 1.38 \times 10^{-23}$ J/K, $T = 295$ K, $\eta = 175$ pF/cm^2, $R_f = 10$ kΩ, $g_m = 40$ mS, $C_g = 1$ pF, $R_D = 146$ Ω, $\Gamma = 1.5$, $K = 294$ fA, $a = 1$, $I_D = 20$ mA, $I_2 = 0.562$, $I_3 = 0.0868$, and $I_f = 0.184$, Equation (25), which defines the thermal noise, can be expressed just in terms of the physical area of the photodetector. Thus, the thermal noise can be easily computed for each photodetector element of the angle-diversity receiver used in the link.

Figure 7 shows the SNR as function of the emitter-receiver distance along the northwest-southeast diagonal for the three angle-diversity receivers, when MRC, EGC, and SB are applied. In general, CDR and SDR exhibit the highest and the lowest SNR, respectively, with a difference of about 20 dB for the emitter located in the centre of the room. Independently of the receiver, the SNR degrades when the emitter is moved towards the corners of the room, similar to the path loss. Furthermore, when emitters are located along the southeast diagonal of the room, the SNRs are a little lower than along the northwest diagonal. In the southeast corner, the emitters are close to the windows, so the desired signal and the contribution from the shot noise due to the natural ambient light are received of the same direction. In short, the conventional angle-diversity receiver exhibits the best simultaneous SNR and path loss. Furthermore, as expected, for conventional and sectored angle-diversity receivers, the MRC scheme provides the best results, followed by SB and EGC. However, the MRC and SB combining techniques require estimating the SNR at each diversity branch, and the difference between the SNR obtained by MRC and EGC is only about 5 dB for the CDR structure. This is not significant when selecting the receiver and the combining technique that offer the best performance in a link operating at 115 kbps.

Assuming the emitter transmits at 115 kbps using on-off keying (OOK) with non-return-to-zero (NRZ) pulses, the channel is distortionless, the preamplifier is followed by an equalizer that converts the received pulse to one, having a raised-cosine Fourier transform with 100% excess bandwidth, and the equalizer gain is chosen so that when sampled, its output is either 0 or $2RP_S$ (A), ignoring noise. Each sample of the equalizer output contains Gaussian noise having a total variance that is the sum of contributions from shot and thermal noises. In these conditions [4], the Bit Error Rate (BER) is given by:

$$BER = Q\left(\sqrt{SNR}\right), \quad where \ Q\left(x\right) = \frac{1}{\sqrt{2\pi}} \int_x^\infty e^{-\frac{y^2}{2}} dy \tag{35}$$

According to Equation (35), to achieve a BER = 10^{-9} requires a SNR = 15.6 dB. Based on the results shown in Figure 7, the CDR and SOR receivers ensure a BER below 10^{-9} for any location along the diagonal of the room. The minimum SNR for SDR is about 10 dB, ensuring a bit error rate below 10^{-3} for a total emitted power of 15 mW. In general, regardless of the combining technique employed, CDR shows a better SNR and path loss than the SDR structure. It is also evident that in terms of both parameters, the CDR receiver provides the best performance. Although the self-orienting receiver exhibits a SNR almost similar to CDR, SOR presents a path loss worse than CDR, regardless of the combining technique applied. Furthermore, SOR needs to implement a SNR estimator, include a search algorithm that automatically aims the receiver towards the highest SNR, and incorporate an electromechanical orienting system. As was previously mentioned, the sensor nodes in a sensor network are limited in power and computational capacity. Therefore, using the SOR as the input sensor for the nodes in an indoor infrared-based sensor network requires additional power consumption and computational capacity over using a CDR in conjunction with EGC. In short, the use of a conventional angle-diversity receiver in conjunction with the EGC technique yields the best trade-off between SNR, computational capacity, and transmitted power. Finally, the use of multi-beam transmitters in conjunction with angle-diversity receivers and power efficient modulation schemes should be analyzed in future research in order to minimize the transmitted power requirements due to the communication between the sensor nodes in an infrared-based sensor network.

Acknowledgments: This work was funded by the Spanish Government (MINECO) and by the European Regional Development Fund (ERDF) programme under project TEC2013-47682-C2-2-P.

Author Contributions: R.P.-J. coordinated the research. B.R.M. and S.R. developed the models for non-imaging angle-diversity receivers, designed the simulation experiments and analysed the simulation results. The equations for the noise and signal to noise ratio were derived from the revision conducted by B.R.M., A.A., S.R. and O.G. programmed the simulation tool. The manuscript was mainly drafted by S.R. and B.R.M.

Conflicts of Interest: The authors declare no conflict of interest.

References

1. Akyildiz, I.; Su, W.; Sankarasubramaniam, Y.; VCayirci, E. A survey on sensor networks. *IEEE Commun. Mag.* **2002**, *40*, 102–114. [CrossRef]
2. Sharif, A.; Potdar, V.; Chang, E. Wireless multimedia sensor network technology: A survey. In Proceedings of the 7th IEEE International Conference on Industrial Informatics, Cardiff, Wales, UK, 23–26 June 2009; pp. 606–613.
3. Agrawal, N.; Davis, C.; Milner, S. Design and performance of a directional media access control protocol for optical wireless sensor networks. *IEEE J. Opt. Commun. Netw.* **2014**, *6*, 215–224. [CrossRef]
4. Kahn, J.M.; Barry, J.R. Wireless infrared communications. *Proc. IEEE.* **1997**, *85*, 367–379. [CrossRef]
5. Alsaadi, F.E.; Elmirghani, J.M.H. Adaptive mobile line strip multibeam MC-CDMA optical wireless system employing imaging detection in a real indoor environment. *IEEE J. Sel. Areas Commun.* **2009**, *27*, 1663–1675. [CrossRef]
6. Alsaadi, F.E.; Elmirghani, J.M.H. Adaptive mobile multicarrier code division multiple access optical wireless systems employing a beam clustering method and diversity detection. *IET Optoelectron.* **2010**, *4*, 95–112. [CrossRef]
7. Al-Ghamdi, A.G.; Elmirghani, J.M.H. Analysis of diffuse optical wireless channels employing spot-diffusing techniques, diversity receivers, and combining schemes. *IEEE Trans. Commun.* **2004**, *52*, 1622–1631. [CrossRef]
8. Carruthers, J.B.; Kahn, J.M. Angle diversity for nondirected wireless infrared communication. *IEEE Trans. Commun.* **2000**, *48*, 906–969. [CrossRef]
9. Jivkova, S.; Hristov, B.A.; Kaverhrad, M. Power-efficient multispot-diffuse multiple-input-multiple-output approach to broad-band optical wireless communications. *IEEE Trans. Veh. Technol.* **2004**, *53*, 882–888. [CrossRef]
10. Djahani, P.; Kahn, J.M. Analysis of infrared wireless links employing multibeam transmitters and imaging diversity receivers. *IEEE Trans. Commun.* **2000**, *48*, 2077–2088. [CrossRef]
11. Lomba, C.R.A.T.; Valadas, R.T.; De Oliveira Duarte, A.M. Sectored receivers to combat the multipath dispersion of the indoor optical channel. In Proceedings of the Sixth IEEE International Symposium on Personal, Indoor and Mobile Communications (PIMRC'95), Toronto, ON, Canada, 27–29 September 1995; pp. 321–325.
12. Mendoza, B.R.; Rodríguez, S.; Pérez-Jiménez, R.; González, O.; Poves, E. Considerations on the design of sectored receivers for wireless optical channels using a Monte Carlo based ray tracing algorithm. *IET Optoelectron.* **2007**, *1*, 226–232. [CrossRef]
13. Rodríguez, S.; Mendoza, B.R.; Pérez-Jiménez, R.; González, O.; García-Viera, A. Design considerations of conventional angle-diversity receivers for indoor optical wireless communications. *EURASIP J. Wirel. Commun. Netw.* **2014**, *221*. [CrossRef]
14. Barry, J.R.; Kahn, J.M.; Lee, E.A.; Messerschmitt, D.G. Simulation of multipath impulse response for indoor wireless optical channels. *IEEE J. Sel. Areas Commun.* **1993**, *11*, 367–379. [CrossRef]
15. López-Hernández, F.J.; Betancor, M.J. DUSTIN: A novel algorithm for the calculation of the impulse response on IR wireless indoor channels. *Electron. Lett.* **1997**, *33*, 1804–1805. [CrossRef]
16. López-Hernandez, F.J.; Pérez-Jiménez, R.; Santamaría, A. Ray-Tracing algorithms for fast calculation of the channel impulse response on diffuse IR-wireless indoor channels. *Opt. Eng.* **2000**, *39*, 2775–2780.
17. Rodríguez, S.; Pérez-Jiménez, R.; López-Hernádez, F.J.; González, O.; Ayala, A. Reflection model for calculation of the impulse response on IR-wireless indoor channels using ray-tracing algorithm. *Microw. Opt. Technol. Lett.* **2002**, *32*, 296–300. [CrossRef]
18. Rodríguez, S.; Pérez-Jiménez, R.; González, O.; Rabadán, J.; Mendoza, B.R. Concentrator and lens models for calculating the impulse response on IR-wireless indoor channels using a ray-tracing algorithm. *Microw. Opt. Technol. Lett.* **2003**, *36*, 262–267. [CrossRef]
19. Rodríguez, S.; Pérez-Jiménez, R.; Mendoza, B.R.; López-Hernádez, F.J.; Ayala, A. Simulation of impulse response for indoor visible light communications using 3D CAD models. *EURASIP J. Wirel. Commun. Netw.* **2013**, *7*. [CrossRef]

Sensors **2016**, *16*, 1086

20. Castillo-Vazquez, M.; Puerta-Notario, A. Self-orienting receiver for indoor wireless infrared links at high bit rates. In Proceedings of the 57th IEEE Semiannual Vehicular Technology Conference, Orlando, FL, USA, 22–25 April 2003.

21. Castillo-Vazquez, M.; García-Zambrana, A.; Puerta-Notario, A. Self-Orienting Receiver using Rate-Adaptive Transmission based on OOK Formats with Memory for Optical Wireless Communications. In Proceedings of the IEEE Global Telecommunications Conference, Dallas, TX, USA, 29 November–3 December 2004.

22. Tang, A.P.; Kahn, J.M.; Ho, K.P. Wireless infrared communications links using multi-beam transmitters and imaging receivers. In Proceedings of the IEEE International Conference on Communications, Dallas, TX, USA, 23–27 June 1996.

23. Barry, J.R.; Kahn, J.M. Link design for nondirected wireless infrared communications. *Appl. Opt.* **1995**, *34*, 3764–3776. [CrossRef] [PubMed]

24. SchottRG780 Longpass Filter Data Sheet. Available online: http://www.sydor.com/wp-content/uploads/SCHOTT-RG780-Longpass-Filter.pdf (accessed on 3 July 2016).

25. Franco, S. *Design with Operational Amplifiers and Analog Integrated Circuits*, 4th ed.; McGraw-Hill Education: Penn Plaza, NY, USA, 2015; pp. 357–360.

26. Tavares, A.M.R.; Valadas, R.J.M.T.; de Oliveira Duarte, A.M. Performance of an optical sectored receiver for indoor wireless communication systems in presence of artificial and natural noise sources. In Proceedings of the SPIE, Wireless Data Transmission, Philadelphia, PA, USA, 8 December 1995.

27. Valadas, R.T.; Duarte, A.M. Sectored receivers for indoor wireless optical communication systems. In Proceedings of the IEEE International Symposium on Personal, Indoor and Mobile Radio Communications, The Hague, The Netherlands, 18–23 September 1994.

28. LiFi Reference Channel Models: Office, Home and Hospital. Available online: https://mentor.ieee.org/802.15/dcn/15/15-15-0514-01-007a-lifi-reference (accessed on 14 June 2016).

sensors

MDPI

Article

Trust and Privacy Solutions Based on Holistic Service Requirements

José Antonio Sánchez Alcón *, Lourdes López †, José-Fernán Martínez † and Gregorio Rubio Cifuentes †

Centro de Investigación en Tecnologías Software y Sistemas Multimedia para la Sostenibilidad (CITSEM), Campus Sur Universidad Politécnica de Madrid (UPM), Ctra. de Valencia, km. 7. 28031 Madrid, Spain; lourdes.lopez@upm.es (L.L.); jf.martinez@upm.es (J.-F.M.); gregorio.rubio@upm.es (G.R.C.)
* Correspondence: jose.asanchez-alcon@upm.es; Tel.: +34-914-524-900 (ext. 20791)
† These authors contributed equally to this work.

Academic Editor: Dario Bruneo
Received: 24 September 2015; Accepted: 17 December 2015; Published: 24 December 2015

Abstract: The products and services designed for Smart Cities provide the necessary tools to improve the management of modern cities in a more efficient way. These tools need to gather citizens' information about their activity, preferences, habits, *etc.* opening up the possibility of tracking them. Thus, privacy and security policies must be developed in order to satisfy and manage the legislative heterogeneity surrounding the services provided and comply with the laws of the country where they are provided. This paper presents one of the possible solutions to manage this heterogeneity, bearing in mind these types of networks, such as Wireless Sensor Networks, have important resource limitations. A knowledge and ontology management system is proposed to facilitate the collaboration between the business, legal and technological areas. This will ease the implementation of adequate specific security and privacy policies for a given service. All these security and privacy policies are based on the information provided by the deployed platforms and by expert system processing.

Keywords: Smart Cities; Smart Grid; Internet of Things; Wireless Sensor Network; security services; privacy; personal data protection; Utility Matrix

---

1. Introduction

A smart city represents a leap forward in increasing a city's sustainable growth and strengthening city functions to provide a greater quality of life for citizens than a traditional city. It is predicted that there will be a great quantity of *"objects"* interacting continuously with citizens and which can be both collectors and distributors of information regarding their mobility, energy consumption, *etc.* As a result, cyber and real worlds are strongly linked in a smart city. Thus those *"objects"* can act as sensors and actuators to interact with the smart city [1,2]. New services based on information gathered and recorded from multiple sources can be deployed when needed. The loss of trust and privacy of citizens could be an obstacle in the interaction between smart city and citizens. Citizens with their mobile phones and other smart devices, such as wearable devices, can also act as sensors, and they can give information about their movements, habits, preferences, *etc.* One of the most significant perceived risks for citizens is the tracking of their movements and their activities through the information gathered by the objects. They also fear being included in a list of personal profiles. Analyzing these data in order to identify behaviors and habits of people, yields information that could be used in many areas, mainly in marketing.

The set of services in a smart city can be viewed as a holistic compound service comprising all single services such as urban mobility, energy consumption, critical infrastructures, public safety, health *etc.* As a result, using the appropriate Smart City technologies, sustainable management of the whole is made possible. There are many stakeholders in a Smart City, each one having their own interests. Among the major stakeholders are sponsors, services operators, and the monitored entities (some of which may be citizens). Stakeholders' interests may conflict with each other. The solution of these situations represents a challenge for legislative and regulatory entities. Therefore it is necessary to implement coherent trust and privacy policies based on legislation, since they are able to reconcile the rights and interests of all stakeholders and protect the citizens from infringements of their rights and invasions of their privacy. Nevertheless, this new range of services also requires the development of new communication architectures to minimize their vulnerability and ensure the maximum protection to citizens. Therefore it is necessary to develop and research about new mechanisms to provide safe and reliable environments. Concepts such as "*Privacy by Design (PBD)*" [3] and the mechanisms to facilitate positive or negative consent are being researched in order to build confidence and allow the users to choose. This idea also requires the participation of actors and stakeholders to protect against the possible chaos if a mass deployment of these technologies were to occur. Thus, for the *PBD* seven principles are defined [3], these deal with proactivity; prevention; privacy settings configured by default and integrated into the design, *etc.* They also deal with the visibility, transparency and designs needed to focus on the user and respect people's privacy.

This paper proposes two goals: (1) A platform to integrate the functionalities and control of the services to acquire enough capacity to generate new applications; (2) An expert system to solve the diverse legislation issues and provide options to generate a policy of trust and privacy mechanisms to apply.

Currently, several *Interconnection and Cooperation Platforms* (*ICP*) are being developed. These platforms also have to allow management of trust and privacy policies. One of them is the "*ACCUS Project* [4]". The Adaptive Cooperative Control in Urban (sub)Systems (ACCUS) platform aims to implement three innovations: (1) integration and coordination platform for urban systems; (2) new control architecture for urban subsystems and (3) general methodologies and tools for creating applications.

This paper is organized in the following manner: Section 2 shows existing related work in this research field, Section 3 shows an brief overview on ACCUS platform in the smart city, Section 4 discusses the major challenges to privacy and trust in that environment. Section 5 shows the needed elements for privacy and trust policy implementation, and in Section 6 these are applied to an example of a Smart Service in a smart city. Section 7 concludes with a summary of the major contributions of this paper and future work.

2. Related Work

The 2012 during our investigation activities about security and privacy in the "Internet of Things" field we were able to verify the huge interdependence between the selection of the mechanisms and the security and privacy countermeasures, the right legislation that must be applied to a determined IoT service and the commercial need for the cost to be as low as reasonably possible.

After a study on the state of the art about this topic reported in previous publications [5,6], three different kinds of contributions were found: (1) state of the art regarding commercial products and businesses focused on the creation of new ideas and services to be commercialized; (2) technology-based state of the art, which provides better and more efficient solutions by its natural progress; (3) legislative state of the art, which is not always homogeneous for the different markets where it is expected to be used, and with a very significant impact on the companies related to the sectors where they perform their activities. The timing that is required for each of the groups is very different and they do not always move at the same speed, thus resulting in potential risks for people, critical infrastructures, *etc.*

The study that was carried out started by analysing the selection process of the security and privacy mechanisms that were made during the specification and design process of several products and services made by two relevant companies of the sector. These companies, although unwilling to be identified in this manuscript, nevertheless provided us their support. Mechanisms were selected according to the technological solution suitable to the legal requirements that were obtained by means of the counsel requested to consultancy companies regarded as leaders of legislative cases involving Internet usage. In this process we have found a significant amount of issues. Several of them have been enumerated as follows:

1.   Consultancy costs are high, both in economic and time-to-market terms.
2.   In some cases, costs associated to security made the product or the service unviable, resulting in the cancellation of the service after significant resource expenses, or the redesign of the service, thus increasing the related costs.
3.   Concern was manifested by the companies consulted with regards to the associated cost of claims, complaints, sanctions and corporative image deterioration that suppose the impacts tied to security and privacy flaws.
4.   Both companies pointed out the convenience of having a simulator to test new ideas for products and services that could provide them with a forecast of the requirements for security and data protection in order to perform a cost evaluation prior to the start of the development phase.
5.   Heterogeneity of legislative frameworks in the different countries those companies operate in is a major issue for them.
6.   This heterogeneous legislative framework also affects the same countries or even the same service, depending on how it is used [5,6]. This also happens when one product is designed as a combination of some others, when the same terminal is used to offer several services, *etc.*
7.   Actions to be taken in a context of likely legislative changes, or when dealing with emergency level changes that may imply different features related to security and privacy.
8.   Previous knowledge of the impact that a specific legislative change will have (in terms of security and data protection on the IoT-based products and services already deployed) is prone to be helpful for the agents involved in legislation.
9.   The user tends to recklessly offer his/her trust less often. What he/she really requests is be guaranteed that their information, intimacy, security and safety will not be jeopardized by the mere fact of voluntarily using (or refusing to use) these new products and services.

After several months of study and consultations to the members of the "Internet Society" related to the legal area of knowledge, political parties, trade unions, *etc.*, the idea of having these three areas of knowledge cooperating (business, law and technology) with each other started to build up. This concept of channelling the requirements of security and privacy through a collaborative system among the areas of business, law and technology is an original contribution the authors of this manuscript (this idea has been contrasted and validated by the interlocutors previously mentioned).

This collaboration materializes itself in the collection of the entrepreneurial, legislative and technological knowledge that can be used by an expert system to provide an answer for the already mentioned main challenges.

We couldn´t find any system or integrated packet that would adapt to the automation that we were looking for. The closest were the "Legal Expert Systems", even though analysis and legislative conflicts were their focus. The expert system proposed in this paper was inspired basically on Cuadrado Gamarra´s book about expert systems in the legal field [7], and the articles of Stevens [8] and Venkateswarlu [9] as well as related references in these about "Legal Expert Systems". Within our paper the vision is a bit different; we do not try to solve legislative conflicts as is the case of the previously mentioned works [7–9], but rather what we want is to obtain the key legislative knowledge needed in a matter of security and privacy to be able to apply the needed legal imperatives, about a concrete IoT service.

On the other hand and due to the large researcher activity about it, a lot of investigation studies are available (among them the major part of the bibliography in the previous publications [5,6] mentioned before in this paper) providing mechanisms and technological countermeasures to act against the threats and attacks to the security and privacy being able to provide solutions.

Therefore, knowing the details of the IoT service that is to be developed, the legal imperatives that must be applied and a group of available technological solutions, it should be possible to manage a solution tailored for each situation. This is the purpose of the proposed expert system.

The structure of the system relies on not hindering the independent evolution of each of the spheres, each of them with its own budget capabilities, Information Technologies systems and their own route map for their own progress. The only adaptations that must be done are involving data communication, transfer and results storage; all the other actions can be performed in each of the spheres with their regular means of work.

This paper proposes an expert system to generate security and privacy policies for services in the smart city. This policy is communicated to the ACCUS platform, which is able to deploy it to the network and devices. The expert system proposed in this research has gone though various major versions since its first design. The first version was presented at the "Third Intech Conference in London, 2013" [5]. The first version was designed to decide the security level that is needed for a certain use case for a specific service. It managed the behavior for different use cases using the same WSN dedicated to health monitoring, but subjected to different legal frameworks. In that case the expert system provided its security and privacy policies to a service platform called AWARE which was then able to configure the WSN remotely.

The second version was designed in 2014, when its functionalities were expanded and the model was modified to be able to work with more than one WSN in different technologies or IoT services [6]. Thus, the expert system was improved to be able to manage the requirements of various services, taking into account the possibilities of different technologies. At the same time, the platform mentioned before was boosted to be able to communicate and configure the security mechanisms for various WSN technologies.

Lastly, this paper wishes to further expand the model of the expert system to be capable of selecting the different security and privacy levels for each one of the services in a smart city. Each newly generated services and created by combination of the existing ones must have a security level adequate, maybe can be different than the ones being used. It also offers the possibility of changing the security level in a city, depending on the possible states of alarm or emergency. In this environment, the mediating platform for the city's services is ACCUS [4]. This research paper provides a way to tackle the issues and challenges with regards to security and privacy in the Internet of Things within the framework of a smart city. These challenges have a major impact in the entrepreneurial, legislative and technological environments, and while each of them offers only one part of the solution, the final solution must come from the collaboration among all three areas. Another important issue is that flaws in security and privacy may affect people´s rights.

This paper does not propose any legislation framework and no security mechanism, but rather it describes a method to choose and apply the security services based on the collaborative environment among the business, legal and technological areas.

3. Smart Cities Applications and Urban Systems Management Using ICP

3.1. Smart City Applications

Smart city applications are grouped into several areas. One classification is proposed in [10], based on the presence of six characteristics shown in Table 1. All these areas raise new challenges in security and privacy such as transnational authentication systems for citizens and businesses, agreed frameworks for data privacy, and the sharing and collection of individual and business data, in order to make a more livable city for citizens, the performance of integrated services and urban systems (such as manager of the traffic, energy, lighting, emergency systems, or information systems) must be taken into account. This enables integrated management strengthened through mutual aid in situations that require it.

Table 1. Smart City applications.

Applications	Target
Smart Economy	Innovative spirit; Entrepreneurship; Economic image/trademarks; Productivity; Flexibility of labor market; International embeddedness.
Smart People	Level of qualification; Affinity to lifelong learning; Social and ethnic plurality; Flexibility; Creativity; Cosmopolitanism/Open-mindedness; Participation in public life.
Smart Governance	Participation in decision-making; Public and social services; Transparent governance; Political strategies/perspectives
Smart Mobility	Local accessibility; Accessibility; Availability of ICT-infrastructure; Sustainable, innovative and safe transport systems
Smart Environment	Natural conditions; Pollution; Environmental protection; Sustainable resource management
Smart Living	Cultural facilities; Health conditions; Individual safety; Housing quality; Education facilities; Social cohesion

For example, correct traffic management in emergency situations can contribute to emergency services arriving in the shortest possible time wherever they are required, and could also reinforce or restrict other resources in the same area, *etc.* To obtain this range of new applications, it is necessary to integrate both the performance and control of these autonomous systems. It should be noted that the systems providing specific services must continue to evolve independently and their integration with other systems must not impede their natural path of evolution, but must find a way that does not affect their integration with others, in a scenario of an integration of "*systems of systems*". Each system has its own internal evolution, which must not be affected by the integration process, so an integration platform that enables the possibility for each platform to maintain its functionality and control would be necessary, and to obtain with the integration the additional advantage of having enough capacity to enable the generation of new applications.

3.2. ACCUS Project

The proposal in this paper has been deployed inside the European ACCUS project, but the proposed expert system is in fact adaptable to any other platform that can control, send and receive commands and responses to/from network elements and perform the needed remote configurations.

As indicated in [4] the ACCUS project focuses on four innovations that are listed below:

- Provide an integration and coordination platform for urban systems to build new applications across urban systems.
- Provide adaptive and cooperative control architectures and the corresponding algorithms for urban subsystems in order to optimize their combined performance.
- Provide general methodologies and tools for creating real-time collaborative applications for "systems of systems".
- Seamless connectivity and semantic interoperability among all services and subsystems connected. ACCUS ICP must provide the necessary mechanisms and facilities so that present and future applications and services connected within the smart city can consult which other subsystems and services exist and what is their functionality.

Currently, the platform has two types of components: (1) core components: the components which allow the platform to provide its basic functionality such as registration, discovery, control elements, security, *etc.* and (2) city customization components: in order to enable the adaptation of the ICP platform to any city, it must have some plugins that allow this customization, e.g., event detection, location detection, data analytics, situation awareness ... A main functionality of the ICP platform is to provide the registration and discovery of the provided services by the subsystems. Figure 1 shows the basic outline of registration and discovery of the subsystems and services. In each record, the subsystems or services must be recorded semantically according to the platform semantics. In case the semantics of the service does not match, the adapter subsystem must perform the semantic conversions needed.

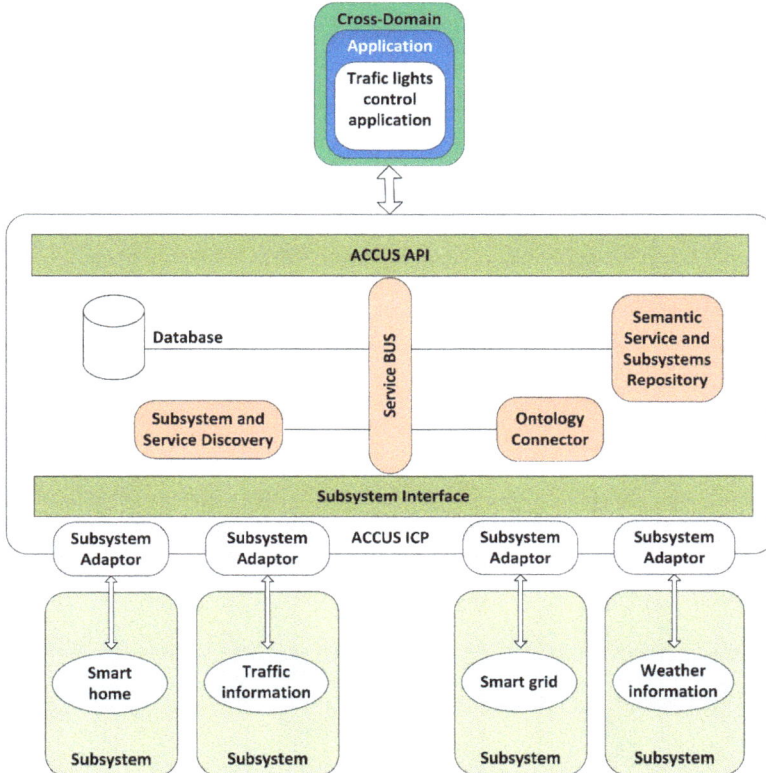

Figure 1. Registration and discovery of subsystems and services in ACCUS.

The components are connected to a service bus to exchange the messages that allow them to interoperate. The functionality of each component of ICP is:

- Service bus: provides the interconnection and cooperation of the component based on a paradigm of message exchange. It could be implemented using already existing products such as *JBOSS* [11] or *WSO2* [12].
- Subsystem and service discovery: discovers all subsystems connected to the ICP and the services provided by each subsystem. This component works in real time. It is responsible for registering subsystems in the semantic repository.
- Ontology connector: handles the translations of the data which must be added to the Semantic Repository, when these data are in a known XML format [13,14]. These translations from XMLs to RDFs are done using a previously generated mapping file, which describes, what elements/data from the source XML, must be stored as instances of classes of the ACCUS Ontology. A mapping file must be defined per each type of XML

- Semantic Service and Subsystem Repository: stores the semantic description, in a way that complies with the proper ontology of services and subsystems registered in the ICP. When a new subsystem or service is discovered their semantic description must be stored in this component.

Currently, ACCUS's architecture is in the development and demonstration phase. However, it has been evolved from an architecture developed and tested in the e-Gotham project [15] as it is shown in [16]. Thus, the architecture presented here has been partially developed and tested. A new service is registered via the following procedure as shown in Figure 2a:

(1) A request is sent to register a new service. The service previously has connected to ESB. It could be REST service, Web service or any of supported by the ESB.
(2) The request is validated against an XMLSchema so as to check whether there is any issue with the request. If the request in sot valid, will be rejected.
(3) A template can be filled with the mandatory information in XML format (semantic or non-semantic format). Of a set of templates, the most appropriate will be chosen.
(4) An XML file is sent to the Ontology Connector (Figure 2b) via OSGi interfaces.
(5) A Logical Service is created in real-time based on an Archetype. This is a key functionality since it allow to have a registrable version of the physical service.
(6) Logical service registry is acknowledged.
(7) The status of the registry can be requested via the ontology connector (Figure 2b); semantic, rdf-based responses will be obtained. To do this we use Jena API in order to build a java version of the ontology, and execute the set of parser on the ontology, let, in this way, ho manage a semantic repository in RDF.

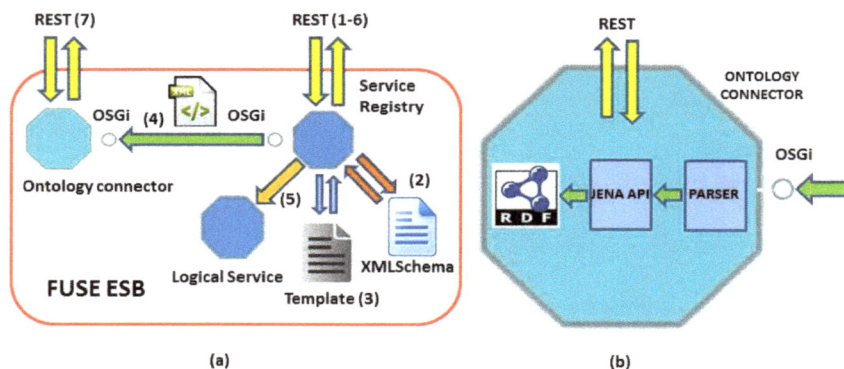

Figure 2. (a) Overall layout; (b) Ontology connector.

In Figure 3 the sequence diagram that specifies the registration process of a service is shown. When applications need to know the services to be used or included, a query must be sent to the Semantic Service and Subsystem Repository. The response to this query will be a set of XML files with the profile of the services available at a given moment.

Since the goal of the platform is to generate semantic interoperability for seamless connectivity, all the agents listed must be semantically annotated in the same way, so that a service or application can request a specific query to the Semantic Service and Subsystem Repository, and then get a reply that complies with known semantics, in this case, the ACCUS ontology. This ontology integrates the meaning of all the components of the ACCUS platform (core and city customization), the sensors and actuators, (e.g., SSN) [17], people (e.g., FOAF) [18], city model (e.g., cityGML) [19], services and subsystems in the city (for the smart grid subsystem, an *ad-hoc* ontology was developed in the e-Gotham project [16]).

The subsystems and services may be semantically annotated according to the ACCUS ontology, their own, or none whatsoever, but the ACCUS ICP must provide, as a response to the queries received, results in accordance with the ACCUS ontology. To do so, the Ontology Connector component performs the necessary transformations so that the registry in the Semantic Service and Subsystem Repository complies with the ACCUS ontology, as seen in Figure 3.

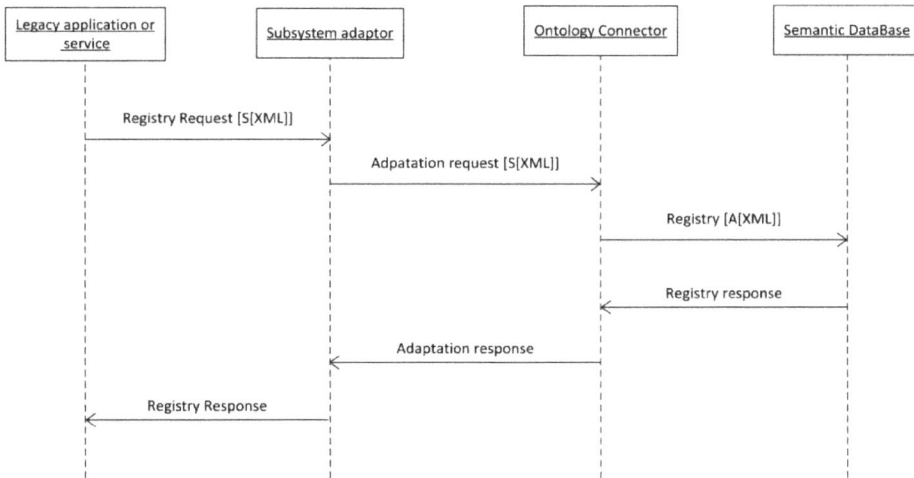

Figure 3. Registration of a service in ACCUS.

The legacy application or service does not have to know the ACCUS ontology, it will send the query with its own annotation format [S[XML]] (previously known by Ontology Connector) and with its own protocol. If necessary, a transformation protocol will be done in the Subsystem Adaptor in order to provide to the Ontology Connector the request of the legacy application or service in the field set in the communication protocol between Subsystem Adaptor and Ontology Connector. Once received the request in the original formal, it will execute the suitable transformation to ACCUS ontology syntax [A[XML]]. In this way, registration will occur in the Semantic Database. The response it will be sent to the Ontology Connector, which if necessary will transform it from ACCUS ontology syntax to Legacy Application or service syntax and Subsystem Adaptor will do the appropriate protocol transformation changes and it will be sent to the originator of the request. An example of a registration for a smart home, consisting of enhanced tele-assistance at home, can be depicted as follows:

```
<?xml version="1.0" encoding="UTF-8"?>
<service>
<profile>
        <serviceIdentification>
                EN_TEL_ASSIST_1
        </serviceIdentification>
        <functionality>
                <preconditionDescription>
                        service on
                </preconditionDescription>
```

```
            <outputDescription>
                    Celsius degrees float
            </outputDescription>
            <outputDescription>
                    CO2 level integer
            </outputDescription>
            <outputDescription>
                    Smoke presence binary
            </outputDescription>
    </functionality>
    <security>
            <policy>basic security policy</policy>
            <dataProtection>integrity</dataProtection>
            <dataProtection>autehentication</dataProtection>

    </security>
    <grounding>
            <inputMessage>
                    start
            </inputMessage>
            <outputMessage>
                    sensorID-lenghtMessage-PreviousValue-CurrentValue
            </outputMessage>
            <endPoint>
                    /icp/assist/home1
            </endPoint>
    </grounding>
</profile>
<process>
    <processID> </processID>
    <typeOfProcess>
            <atomicProcess/>
    </typeOfProcess>
    <operations>
            <operation id="read">
                    <preconditions>
                            device on service on
                    </preconditions>
                    <insANDouts>
                            <output>float</output>
                            <output>integer</output>
                            <output>binary</output>

                    </insANDouts>
            </operation>
    </operations>
</process>
<context>
    <serviceType>
            <loction> indoor</loction>
```

```
                <motion>static</motion>
        </serviceType>
        <geoCoordinates>
                <longitude> 40.33889</longitude>
                <latitude>3.628611</latitude>
        </geoCoordinates>
        <smartSpace> smart Home 1</smartSpace>
</context>
</service>
```

But, what is the real utility of ontology in this process? Since the main purpose of ontology is to represent in a standard way the meaning of contents with the goal of inferring new knowledge, semantic interoperability allows access to everything registered within the ACCUS ICP, in accordance with the ACCUS ontology. This approach enables a new service or application to send a query to the Semantic Service and Subsystem Repository in SPARQL [20], and obtain as a result information about other subsystems, services; components or devices are connected within the Smart City, their function, and form of access. Since this is a repository in RDF [21], an appropriate response will be sent back to the agent that performed the query. The discovery process of the registered services is shown in the next sequence diagram (Figure 4).

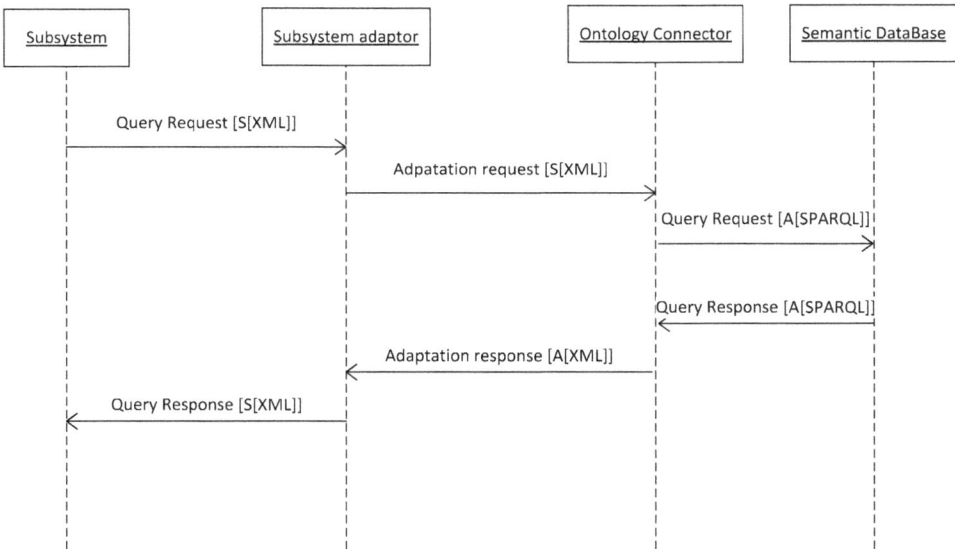

Figure 4. Discovery process in ACCUS.

If, for example, an application is willing to become aware of certain features of the services registered for selecting the more suitable one, it will generate a query that may be SPARQL-formatted or not. If necessary, it will be ported to SPARQL by the Ontology Connector module and executed in the Semantic Registry:

```
PREFIX ns:<http://www.semanticweb.org/ACCUS/1.1#>
SELECT ?ServiceIdentification ?inputDescription ?outputDescription ?policy ?data
Protection ?endPoint
WHERE {
?service ns:hasProfile ?profile.
?profile ns:hasServiceIdentification ?serviceIdentification.
?profile ns:hasFunctionality ?functionality.
?profile ns:hasSecurity ?security.
?profile ns:hasGrounding ?grounding.
?functionality ns:hasInputDescription ?inputDescription.
?functionality ns:hasOutputDescription ?outputDescription.
?security ns:hasPolicy ?ns:policy.
?security ns:hasDataProtection ?dataProtection.
?grounding ns:hasEndPoinf ?endPoint.
}
```

Response will be in an XML-formatted message according to the ontology:

```xml
<?xml version="1.0" encoding="UTF-8"?>
<services>
<service>
<profile>
        <serviceIdentification>
                EN_TEL_ASSIST_1
        </serviceIdentification>
        <functionality>
                <preconditionDescription>
                        service on
                </preconditionDescription>
                <outputDescription>
                        Celsius degrees float
                </outputDescription>
                <outputDescription>
                        CO2 level integer
                </outputDescription>
                <outputDescription>
                        Smoke Presence binary
                </outputDescription>
        </functionality>
        <security>
                <policy>basic security policy</policy>
                <dataProtection>integrity</dataProtection>
        </security>
        <grounding>
                <endPoint>
                        /icp/assist/home1
                </endPoint>
        </grounding>
```

```
</profile>
</service>
<service>.....</service>
<service>.....</service>
</services>
```

With the information therein it will be capable of inferring new knowledge that, in the context of a smart city, consists of generating new cross-domain applications and services for the city and citizens, which then must also be registered in the ACCUS ICP, producing constant feedback. Furthermore, these new applications will already use the ACCUS ontology.

3.3. Security and Privacy in a Smart City

Studies performed some years ago such as [22] recognized the importance of data privacy and personal identity among the aspects to be dealt with, not only on technical grounds, but also concerning legal frameworks:

E-Government: There are a number of technologies that will be required for the underlying infrastructure that is needed to help support this process. Fundamental technologies are key to the development of the Digital Single Market (such as authentication and privacy), and to the development of e-government in smart cities. The development of transnational authentication systems for citizens and businesses, the development of agreed frameworks for data privacy, and the sharing and collection of individual and business data, must be considered.

Health, Inclusion and Assisted Living: The key technical requirements to be addressed in this domain are: security (encryption, authentication and authorization), service discovery, scalability and survivability, persistence, interworking, community-to-community application messaging propagation, auditing and logging, location information sharing, and application service migration.

The challenge related to ICT security aspects has to be ensured by a manageable access control management system, to ensure that only authorized persons are allowed to access the data, and ensures that the data is protected to achieve confidentiality. Users should manage authorization. Dedicated authentication and logging mechanisms have to support the enforcement of access control. The challenge in this approach is that access control architecture has to enable both the decentralized storage of data, and the comprehensive access control mechanisms and enforcement that concern all parties that could have access to that data.

Intelligent Transportation Systems: the provisioning of flexible, scalable and self-optimized networks, dealing with heterogeneity, effectively exploiting location information, guaranteeing real-time exchange of data where needed, and providing security, privacy and authentication mechanisms.

Smart Grids, Energy Efficiency, and the Environment: Other challenges include: new communication and networking ICT technologies, new affordable devices that gather environment data, new intelligent algorithms for smart ubiquitous environments, new light sources, new and fair regulations that enables the mass implementation of the Intelligent Street Lighting System idea provided by different vendors; new products for global markets that enable steady economic growth; and advanced products and services based on IP to foster innovations, and economic growth based on an open innovation scheme. Recently, Sicari *et al.* presented in [23] a vision of the near future in security, privacy and trust in IoT. Finally, Weber *et al.* presented in [24] the forthcoming issues in privacy applied to IoT.

4. Challenges on Privacy and Trust

It is important to know how to classify data sources in a smart city and their relation to personal identity. These sources are the following:

- Non personal sources: data, unrelated to specific people, gathered from devices (temperature or humidity sensors, *etc.*)
- Personal sources: data, related to specific people, gathered from devices, unambiguously using user identity (social networks, *etc.*)
- Anonymous sources: data related to specific people gathered from devices, but which have been pre-processed to mask their personal identity (covering faces in video camera images, *etc.*).

It may be possible to discover user information by processing data from several sources. These situations must be considered. From a technological viewpoint the security and privacy problems can be grouped as follows: (1) Problems related to computer security and communication systems; (2) Problems related to database security, user identities and communications; (3) Problems occurring when new subsystems are added to the smart city (increasing its complexity and vulnerabilities).

4.1. Problems Related to Computer Security and Communication Systems

These are the problems such as malware (viruses, trojans, worms, backdoors, spyware, *etc.*), or bots, loggers, rootkits, DDoS attacks, lack of updates [25,26], *etc.* They are prevented by installing suitable antivirus, firewalls, honeypots, intrusion detection systems (IDS), security policies, updating and implementing system authentication measures. When the devices are localized all around the city and do not have a common control platform, updating or implementing the new policies of authentication or refusing the authorizations is difficult.

4.2. Problems Related to Database Security, User Identities and Communications

- Database security [27] The Statistical Disclosure Control (SDC) techniques consist of inserting noise or aggregations to maintain privacy, while maintaining the significant value of the data. Private Information Retrieval (PIR) techniques are based on queries asking for more than the necessary information in order to hide the specific information demanded by the user.
- To hide user identities accessing location-based services (LBS) techniques are used such as cloaking and using pseudonyms.
- Privacy in communications, advanced cryptography and access control can be used to prevent eavesdropping on the data and prevent unauthorized connection nodes to the networks with distributed devices in access public places [28].

4.3. Problems Occurring When New Subsystems Are Added to the Smart City

When new subsystems are added to the smart city the complexity and the number of vulnerabilities grow [28], which can be exploited by malicious people to harm the most vulnerable systems and enter into the other subsystems of the smart city.

- Increased interconnections among services increase the ways through which a virus can propagate. Hackers can move through the interconnections among the systems and take control.
- Dependencies among infrastructures. A failure in one of the nodes in the dependencies network could cause some cascade problems. Planning and management can alleviate the problem [29].
- The connection of the smart city with the other platforms and applications by middleware is a strategic element. Those connections must be secured, implementing confidentiality, integrity and authenticity and they must also be interoperable.
- The fact that having a great quantity of services and data sources facilitates creating new applications and services, but risks the availability of these services if a fault in any of them occurs, which would cause malfunctions an application and even make it unusable.
- In an open data context with a great quantity of information sources (both real-time and historical), publishing new data makes it difficult to ensure that they cannot be used to infer the identity of users (using correlation techniques,...). To minimize the time of intrusions and attacks, solutions that implement active reactions in a crisis scenario to curb the anomaly [30] are used.

If users believe that a system is insecure or threatening to their privacy, it will not be able to establish itself successfully in the market. Thus, in order to achieve user consent, trust in, and acceptance of smart cities, the integration of security and privacy-preserving mechanisms must be a key concern of future research. New challenges arise in the area of security and privacy, and they can be classified as follows:

Interconnecting systems that serve completely different purposes (traffic control and energy management for example), and thereby create a "system of systems", increase the complexity of such collaborating systems exponentially. As a result, the number of vulnerabilities in a smart city system will be significantly higher than that of each of its sub-systems. Furthermore, the pure interconnection of two systems might open new attack vectors that have not been considered before, when securing either of the individual systems. Therefore, research into ways of handling the increasing complexity of distributed systems from the security perspective is required, which includes: cost-effective and tamper resistant smart systems or device architectures (crypto and key management for platforms with limited memory and computation); evolutionary trust models for scalable and secure inter-system interaction; comprehensive security policy; self-monitoring and self-protecting systems, as well as development of methods for designing security and privacy into complex and interdependent systems.

The number of users, and the volume and quality of collected data, will also increase with the development of smart cities. When personal data is collected by smart meters, smart phones, smart vehicles, and other types of ubiquitous sensors, privacy becomes all the more important. The challenge is, on the one hand, in the area of identity and privacy management, where, for instance, pseudonymisation must be applied throughout the whole system, in order to separate the data collected about a user from the user's real identity. On the other hand, security technologies such as advanced encryption, access control, and intelligent data aggregation techniques, must be integrated into all systems in order to reduce the amount of personal data as much as possible, without limiting the quality of service. It is necessary to work towards interoperability of different identity management systems, as well as automatic consideration of user's preferences. It is necessary to develop also privacy mechanisms which allow users to express their preferences on service quality and data minimization.

4.4. The Challenges

All services in the smart city give rise to new security and privacy challenges and although it is not the main selling issue, users implicitly expect that the involved systems are secure and the privacy of users are kept. A successful attack will directly impact the life of people. Thus, if the users deem that the system is not secure or that it threatens their privacy or their rights, they will refuse to use IoT services, and the solution will not be able to be successfully placed in the market. From the user´s point of view, the requirement is to guarantee the protection of their privacy rights. In consequence, protecting the services of smart cities is a primary issue. So, in order to achieve user consent, and acceptance of smart cities, integration of security and privacy-preserving mechanisms must be a key concern of future research. The challenges can be several aspects:

- Handling of the increasing complexity of distributed systems from the security perspective such as the identity and privacy management such as pseudonymisation throughout the whole system, in order to separate the data about a user from its real identity.
- Integration of security technologies into systems such as advanced encryption and access control, and intelligent data aggregation techniques, *etc.*
- The context of smart cities relates to open data business models. It is possible because services become pervasive and ubiquitous and the opening of the databases will become more important.

The most important issue has to be transparency, so the end-user must be know how his/her information is being used, with clear options and secured environments, when providing services that use personal data.

5. Implementation of the Solution in the ACCUS Project Environment

The holistic service in a smart city comprises several components to provide a particular service (smart grid, smart traffic, *etc.*). Each service must have its own security mechanisms to protect itself and the personal data therein. Each particular service has its own security, but the whole service (the joint service) for throughout the city must be considered, since some security holes may exist caused for interactions among those. The whole service in the city from a holistic viewpoint could be handled with some additional security techniques. The joint service in the city as the superposition of individual services is shown in Figure 5. It shows some smart services with a real infrastructure (sensor nodes, communication paths, base station, *etc.*) defined as a *Real Smart Service* (*RSS*), coexisting with other services comprised of combining and processing the information available, defined as *Virtual Smart Services* (*VSS*) [31–33]. In this environment the components to protect are sensor nodes, communication paths, base station, and sensible data that flow through them in the *RSS*. Aggregation and information processing must be protected by security mechanism in both *RSS* and *VSS*.

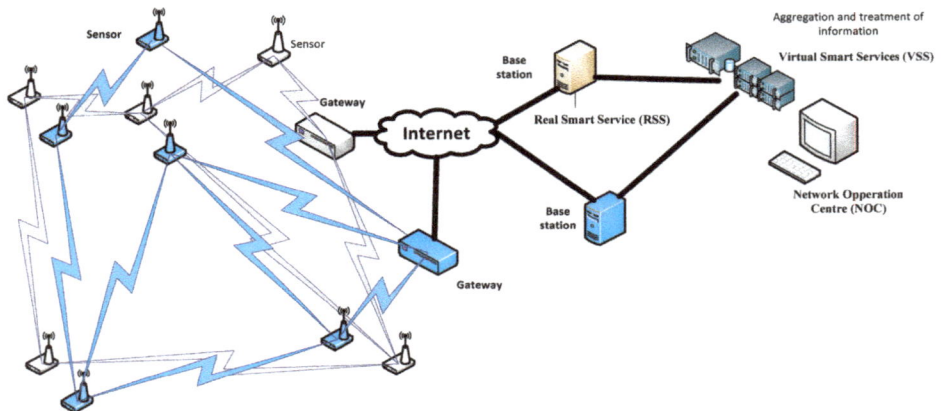

Figure 5. Joint service in the city as the superposition of individual services.

5.1. Data Protection Impact Assessment

It is important to note that each particular service is analyzed by a Data Protection Impact Assessment (DPIA) Template [34]. As a result some security holes may appear; therefore making a new DPIA Template for the holistic service could be the solution to fill these holes. A new particular service in a smart city has an effect on itself and on the holistic service, and thus the DPIA is the main tool for continuously reviewing the security mechanisms and countermeasures to satisfy the security and data protection laws. There are some services of IoT such as smart grids for which the legal analysis has already been performed in a DPIA template, but other IoT services have not got them yet. In those cases it must be made by the general method given in [34]. At the end DPIA-T must obtain the Feared events; Threat ID; Related Security & Privacy targets; Affected assets; Impact; Likelihood; Risk Level, that is, the security and privacy imperatives for the entities that must be protected.

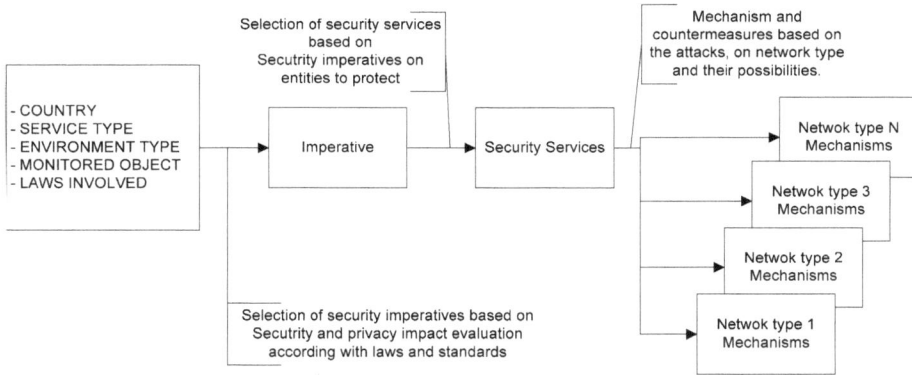

Figure 6. Selection chain of security mechanisms.

5.2. Constructing Security and Privacy Policies

As we can see, Figure 6 gives an overview of the selection chain of security mechanisms that constitute a security and privacy policy. To make it, a *DPIA template* is needed, and as a result, the security services are mapped by security imperatives based on the concept. Security services are bound by legal and regulatory frameworks. To comply with them, a *DPIA template* is very useful. Each network type has its own mechanisms and countermeasures, depending on its technology and its limitations (battery, memory, process capacity, *etc.*). A service in the same city made up of different technologies may have different mechanisms to address counterattacks on the same security service (possibly with different results). One *DPIA template* over the holistic service can give supplementary mechanisms sufficient protection. As result of this process some mechanisms and countermeasures could be modified or adjusted.

5.3. Automatic Selection System for Making Decisions over Security and Privacy Policies

To advance the state of the art, at first the relevant, available and accessible knowledge in the information resources about security and privacy is analyzed. Useful knowledge flows are singled out, such as researching reports, *etc.*, and what can be done with this knowledge is analyzed. Today there are many studies in highly targeted areas; in fact, in the technical area, there are many studies dedicated to developing new efficient security mechanisms, in order to provide specific solutions to specific cases. In the legal area, the legal implications of this new paradigm are being investigated and some ideas and projects are being developed in this regard. Moreover, companies and suppliers of equipment and networks are also devising services useful to society.

5.3.1. Overview

The basic idea of the expert system developed focuses on gathering all this knowledge generated by experts and formalizing it into knowledge bases, making it appropriate for it to be processed to obtain security policies to be applied to products and real services. This idea is developed in [6] and it is represented in Figure 7.

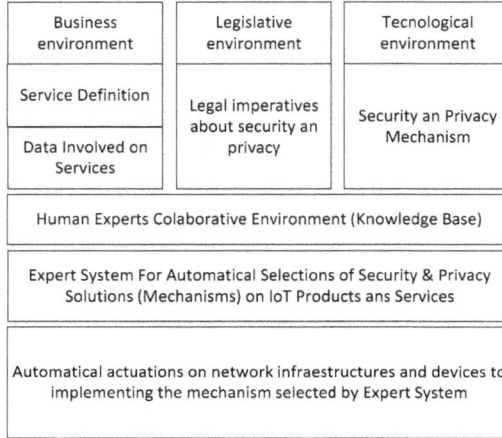

Figure 7. Involved information. Automatic Selection System for making decisions concerning security and privacy policies.

The knowledge generated by the human experts from the areas involved is stored in order to be processed. Data protection measures are selected based on this information network, so the inclusion of these knowledge areas allows for example, to certify to users and corporations that this IoT service is adequately protected.

There are three knowledge bases which are the most important part of the expert system. They contain the results of the collaborative work of the involved areas. These areas are the Business-Business Expert System (BES) about the service definition; Juridical—Legal Expert System (LES) about the Law framework; and the Technological area– Technological Selection Expert System (TSES) about attacks, security services, and mechanisms. The information flow is shown in Figure 8.

In the environment of a smart city it could be very advantageous to concentrate on a *Network Operation Centre (NOC)*, the intelligence, maintenance and deployment of actions related to security and privacy policies in the smart city. This way, all this knowledge generated by the different areas can be leveraged, and made available in an information system to act as a support of collaborative work between the areas involved, in order to find the best solutions for the protection of personal data generated in each case.

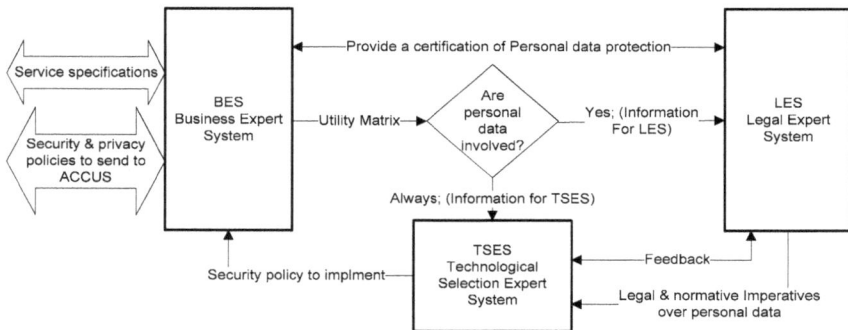

Figure 8. Expert System. Automatic Selection System for making decisions over security and privacy policies.

The system interacts with the areas that can provide the knowledge and sufficient confidence to provide quality solutions (corporations, legal and technological areas) working network environment. This interaction is represented by Figure 9.

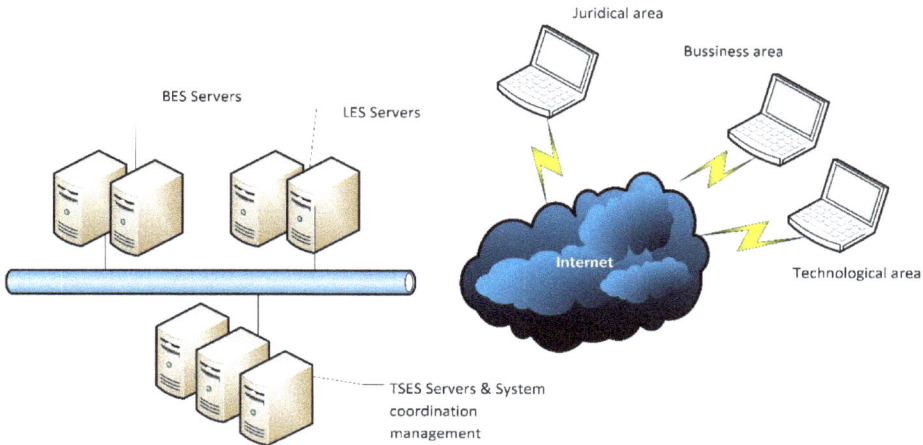

Figure 9. Working network.

This would allow the issuing of certificates to provide enough confidence for users and businesses. NOC and ICP work together. In an environment like that, companies that want to design new products and services can use the expert system to perform virtual simulations before making decisions over on the actual markets. The legal and political sectors could conduct impact assessments on society and the market about possible changes and new laws on data protection, being able to know how and to what extent existing products and services and future developments would be affected. It would also be useful for the technological sector. It could observe and assess critical aspects that need new research and innovation, or emerging issues that require technological solutions.

The current state of this development performs automatic selection which takes into account the various factors that determine the need for specific services and security mechanisms. These factors include legal and regulatory requirements for personal data protection of the product or service to be provided, its network topology, its physical characteristics, *etc.* All these elements should be considered for a robust implementation of a security system that is able to adapt to each particular case. With these elements, among others mentioned in [35], services and appropriate security mechanisms are chosen to be implemented in the design and construction of the product or service, by a decision based on certain security requirements, which must act over a set of data that must be protected by legal and regulatory requirements.

When a new service is implemented within the city, or when certain laws have been changed, it might be necessary to implement a new security and privacy policy, or to adapt the exiting one. The expert system described herein decides which policy must be applied in the smart city. Following that, every system and network element involved and distributed within the city must be reconfigured. If these actions are to be performed automatically, ACCUS must act as a mediator, that is, it must be able to translate the new security and privacy policy, received from the expert system, and it needs to generate the necessary commands and actions, adapted to the requirements of each technology of the smart city. This way, suitable mechanisms will be activated according to the security and privacy policy that each service needs to fulfill. This task of configuring and reconfiguring the security and privacy policy of the smart city is accomplished by using the chain of mediation shown in Figure 10.

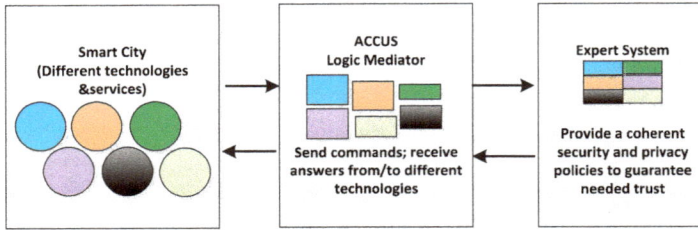

Figure 10. Mediation chain.

This figure shows the mediation chain from the expert system to provide coherent policies to the network elements in the smart city with ACCUS as mediator to manage different technologies. The different colors represent different technologies coexisting in the same environment and managed by the ACCUS platform. The expert system also must record the basic security features and privacy of each used technologies.

Figure 11 shows an outline of the interworking between the collaborative environment that generates the required knowledge and interacts with the expert system and the ACCUS platform, which understands and maintains a dialogue with the systems and elements in the city. Each of the areas involved can use the expert system. The expert system can communicate through the areas or communicate the selected policies to ACCUS to generate the necessary actuations over the smart city elements.

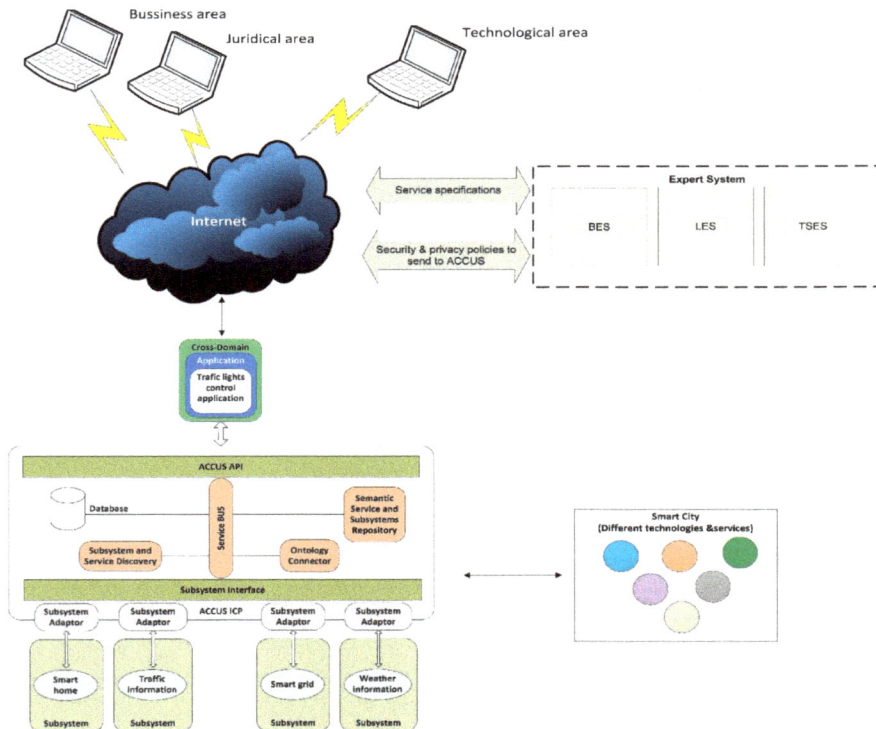

Figure 11. Overview of the system operation.

Respecting the knowledge network in the expert system, a general overview will be given in Figure 12 and is discussed in the next subsections, where the knowledge network in the system is described.

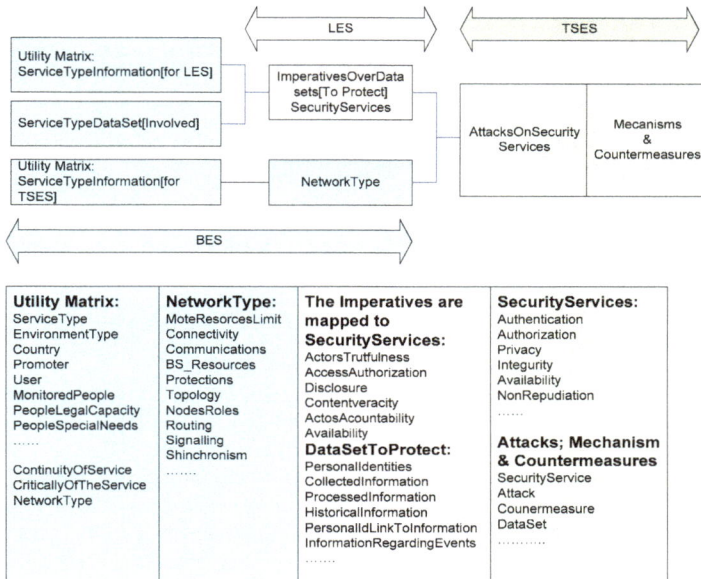

Figure 12. The most important information to select the adequate security and privacy mechanisms.

5.3.2. Business Expert System (BES)

As shown in Figure 12, the BES knowledge is composed of the service type information, the data sets that could be sensitive (BES is not sure yet if those data sets must be protected or not), and finally by the characteristics of the type of used network for this service or IoT.

The service type information is stored in the "Utility Matrix" and is composed by two parts. The first one is composed of the information of interest for the processing of legislation in the LES, that is, the necessary data to select the legislative framework for the IoT service. It is composed by information such as the type of service, the type of operating environment, and the country where it is located. It also has the Information about the promoters, users and monitored entities (if they are people, their capacity and special needs are also known). The data that is considered as sensitive are also included.

The requirements of the service, as well as the necessity of its continuity, its own criticality, and the type of network that will provide it are also included. This is precisely the part that complements the utility matrix; it is the technical information about the type of network that is relevant when selecting among the security mechanisms in the TSES. The structure of the service, its own safeguards and bug handling, the possibility of operating the nodes as standalone, signaling and synchronization, monitoring interval, network segments, transmission sharing, information aggregation and routing are also stored.

5.3.3. Legal Expert System (LES)

As shown in Figure 12, the LES knowledge is composed by the information about the legal imperatives to protect the data considered as sensitive by the legal framework. LES receives data from BES to compose the DPIA-T if does not exist yet.

As the LES processes information, data sets are put together, grouping personal and other sensitive data handled by the IoT service. A specific normative is applied to these data sets, assigning legal constraints or a necessary level of protection for a certain service type.

The service type is a function of the environment, country, promoters, users and monitored entities and their relevant legal characteristics. Actually, this information framing is coherent with the legal analysis that the NIST as well as the European Union carry out in their DPIA-T impact assessment templates.

If there is a set, $\{s_i\}$, of services that possess infrastructures, and there is a set, $\{s_j\}$, of joint services composed by subsets of $\{s_i\}$, then there will be $(i + j)$ impact assessment templates. It may also be the case that new services can be generated from elements of $\{s_j\}$, by themselves or by combining with other elements of $\{s_i\}$. In conclusion, the simplest approach is to have an impact assessment for each individual service, and one for each joint service, adapting the necessary mechanisms to each service. That way, services with an associated infrastructure are not hindered in regards to resources, achieving a tailored security. All data processing in LES is compatible with the "Data Protection Risk Assessment (DPIA)".

5.3.4. Technological Selection Expert System (TSES)

As shown in Figure 12, the TSES knowledge is composed by the security service information, and the information about the attacks, mechanism & countermeasures. The TSES processes information by mapping the imperatives in security services that need to be applied to data sets. These security services for a certain network type (resources, connectivity, communications, resources in the base station, topology, nodes, routing, signaling, and synchronism) are threatened by a list of attacks that affect the IoT service. These attacks have countermeasures (security service, network type, attacks, mechanism & countermeasure).

Luckily, nowadays there are already securities and privacy mechanisms that make it possible to use complex cryptographic mechanisms, supported by the rapidly advancing development of the hardware and operating systems of network elements. Currently, it is possible to cover the majority of cases routinely presented, and luckily there is a lot of activity focused in the creation of new solutions.

For the purpose of this paper, the security and privacy policy is considered as the set of mechanisms integrated within the services. The security policies and mechanisms' suitability analysis is selected by the degree of coverage reached while having sufficient mechanisms, and by determining the best ones for the resources of the technology to which they are applied (delay, consumption, *etc.*). If not enough mechanisms are found, a coverage alarm would be set off in the corresponding knowledge base, urging the corresponding experts to solve the issue.

Attacks can be done by both outsider and insider attackers. Insiders, however, are able to perform worst damaging actions since they used to have a higher level of permissions and system knowledge.

If an overview of insider attacks is done, it can be noticed that, for instance, SCADA systems are used in many critical infrastructure applications that have important software components, such as Human Machine Interfaces, servers with historical data, Remote Terminal Units and the communication links between them. The latter include the units used to collect information and transfer it back to the central site carrying out any necessary analysis and control, and displaying that information on an operator screen afterwards. The operators use the Human Machine Interfaces data to make supervisory decisions. Therefore, the Human Machine Interfaces, data historians, communication links, sensors threshold values and actuator normal settings [36,37] can be attacked by an outsider (the attack can be initiated from outside, by unauthorized or illegitimate users; those usually are opportunistic, deliberate, and malicious) or by insider attackers (they happen when an authorized user misuses the permissions and damages the system by sending legitimate control commands with a great impact and higher success rate; these are difficult to predict and provide protection against them). With regards to users, not only engineers (responsible for managing object libraries and user interfaces, setting grid topology, normal work condition states, setting parameters of devices, defining process set points, writing automation scripts, *etc.*) have to be considered in a secure deployment, but also operators (expected to monitor the system status in Human Machine Interfaces servers, react to alarms and some events so that the process will run correctly, execute operator commands that often prevent triggering new alarms, resolve incoming alarms, make decisions about changing topology, *etc.*). An engineer is a more powerful system user than an operator, but the transmission system is controlled by operators. Insider attacks [36,37] to be considered are:

- Unresolved alarms attacks, when alarms are not perfectly resolved (delaying or making incorrect or incomplete actuations). These situations can provoke cascading failures of major consequences if a critical security error is unnoticed. Malfunctions sometimes are a consequence of wrong, high-level management decisions, such as budget cuts, transforming the activity of human operators from one of specialized nature to a multifunctional one without enough training, *etc.*
- Misconfiguration attacks of differing nature: Overload attacks (wrong changes of topology and load transfers, which can cause overload or a power failure in a large area), Outage attacks (opening the output feeders), incorrect setting attacks (improper equipment settings which could cause equipment incorrect operation).

The incident response can be a complicated matter because one minor mistake may result in the loss of the most critical pieces of evidence and make the whole case inadmissible for a trial or other court actions. Some mechanisms against insider attacks has been mentioned and referenced in [36–41], such as:

- Detect anomalous behavior in SCADA network traffic.
- Detect anomalies based on validating protocol specifications
- A real-time anomaly detection system for unknown attacks
- Anomaly detection for insider attacks based on system logs of the SCADA system to be periodically monitored to detect anomalous behavior (it is necessary to control time periods, parameter values, content of the command orders and many more variables).
- Detecting insider attacks in SCADA by data passing through the system and include a semantic module capable of understanding user actions.
- Statistical Anomaly Detection Method (SADM) is developed in some SCADA systems, by analyzing statistical properties of alarms that will determine the normal system behavior. SADM uses statistical properties to determine whether "current behavior" deviate significantly from the "normal behavior" by using the mean and standard deviation parameters in order to set thresholds, which can be learned from observations (operator behavior).
- Reference [42] talk about any malicious behavior changes statistical properties of alarms and is identified as an anomaly (experimental scenarios have been simulated by the proposed Colored Petri Nets (CPN)-model for insider attacks).

5.4. Processing Stages

Table 2 shows the basic process of selection and the knowledge bases involved. This operation mode enables the cooperation between experts from the areas involved through knowledge generated and formalized.

Simplicity was given a priority after numerous concept tests, so that the expert system is fast and user friendly. But this is not free, the coverage and reliability of policies and security mechanisms are transferred to the performance, coverage and reliability of the information in the knowledge bases of each of the three parts, BES, LES and TSES.

Table 2. Processing stages.

Information Involved		
Input	Knowledge Base	Output
1. Services requirements	Business Knowledge	2. Utility matrix & Personal data involved
2. Utility matrix & personal data involved	Laws and standards Knowledge	3. Legal Imperatives over sensible information
3. Legal Imperatives over sensible information	Attacks, security services, mechanisms Knowledge	4. Security services & mechanisms over information pieces
4. Security services & mechanisms over information pieces.	Business Knowledge	5. Final decision
5. Final decision	Validity check	6. Legal certification

Utility matrix & personal data involved are made in a Business Expert System (BES) based on the final service requirements and the information managed. The user provides all this information through forms, in a guided way. The utility matrix is composed of two parts. The first part is comprised of the information about the type of the final service, about the country, the developers and users, the entity being monitored (persons, animals or things), the characteristics of the persons subject to monitoring (children, adults, seniors, their legal capacity, special needs, *etc.*). With this information and the data involved in the service (some of them may be personal data) the Legal Expert System (LES) is able to perform the necessary processing to obtain the legal requirements that must be implemented in order to protect the information which must be protected. The second part contains more technical information about the network type, the sensor nodes resources, the base station and the connection types, the communication used, network topology, type of routing, signaling, synchronism, if continuity of service is required or not, and the level of service criticality (critical for people, for infrastructures, *etc.*). These data are needed by the Technological Selection of security solutions Expert System (TSES) to determine the possible service vulnerabilities.

The correct selection of security services and mechanisms strongly correlates with the amount of information available about technology, topology and information extracted from the Utility Matrix [5, 6]. It is clear that not all technologies are able to support all the existing mechanisms without affecting quality of service. With the current knowledge available about the technological possibilities, it may be possible to form a synergy between the security services and mechanisms in order to obtain the minimum processing for the security and privacy required.

Laws and standards Knowledge: The LES knowledge base does not store laws; it stores the knowledge of experts in their area about the legal requirements to apply to personal data on the final service. The legal framework is obtained from the extracted information from the Utility Matrix (service type, country, environment type, if continuous monitoring is needed, or if this is a critical service or not). From this legal framework legal imperatives are extracted. Legal requirements are identified by the main concept represented:

- Actors' truthfulness is transformed into "Authenticity";
- Access authorization is transformed into "Access control";
- Disclosure or dissemination of information is transformed into "Privacy";
- Content's truthfulness is transformed into "Integrity";
- Actors' responsibility is transformed into "Non-repudiation";
- Availability and continuity of service is transformed into "Availability".

The data that may need to be protected are those that identify to the people individually and are related to their gathered data, their processed about historical data, or the complete events that identify the state or situation of the person, *etc.*

The conceptual direct relationship is established between *the LES* legal imperatives obtained and security services by Recommendation X.800 [43] in TSES. The attacks on the final service are countered by the security services which comprise countermeasures and security mechanisms. In the TSES knowledge base the security services, and the countermeasures and the security mechanisms are associated with the attacks. The obtained result is a set of mechanisms and countermeasures to achieve the security level needed. When the system is unable to find a solution for all requirements of security and privacy, a warning of insufficiency of knowledge is thrown indicating the problem encountered.

One part of the assigned work to the technological human experts of TSES is to feed the knowledge database with useful mechanisms that may be utilized. These mechanisms will be classified according to several features concerning the casuistry where they are successful when applied, their effectiveness against attacks (based both on a study made for this solution and past experience), an assessment of easiness of change and adjustment and the capacity of the each mechanism to be monitored.

The ACCUS platform directly supervises the involved security mechanisms, attack attempts and the successful attacks, and once the attack has been mitigated, the involved mechanisms will be revised, and corresponding changes are made in the platform, in the service and in the expert system.

5.5. Performance Evaluation

The expert system is responsible for controlling the level of current legislation fulfilment regarding asset protection (that is, the level of fulfilment of legal obligations). The smart city offers a collection of individual services prone to be attacked. Each of the services is subdued to some specific "legal obligations" that are implemented by means of "countermeasures and security mechanisms". "Attacks" may cause an "impact" on the assets that must be protected, so the level of fulfilment regarding current legislation is assessed on the basis of the impact that has taken place ("0" = no impact registered, "1" = impact on non-legally protected assets, "2" = impact on legally protected assets). There are three indicators:

- No legislation-based impact attacks: (Service; Attack; Impact (0 U 1)).
- Seriousness of the impacts: (Service; Attack; Impact (0, 1, 2)).
- Legislation-based impact: (Service; Attack; Impact (2)).

These indicators can be aggregated to the overall service or disaggregated in individual services from the values: $V_{n', k'}$ (Indicator value for indicator k', for the service n'). When impact is mentioned, it is referred to one value $k' \in [1, k]$ depending on the specific service $n' \in [1, n]$ it has been aggregated to. Value V is the one corresponding to the impact.

Thus, the number of attacks without legal impact, the number of attacks with legal impact, and a quantitative measure of the seriousness of the occurring impacts can be known. It is also possible to know which the most attacked services and the attacks that cause greater impact are, and therefore assess the mechanisms against the attacks.

To find the main causes that lead to unwanted impacts, auxiliary measures must be obtained to establish an improvement plan. These measures are obtained from the service platforms that can send them, or logging in periodically and processing this information from their event logs. These indicators are basically two: (1) Effectiveness of the mechanisms and countermeasures against security attacks and (2) Overhead introduced by those mechanisms in the system when these mechanisms are operating and may affect the quality of the service (traffic, delays and excessive resource consumption), *i.e.*, Overhead caused by defensive actions.

This performance control is executed by the expert system, based on an alarms system which is described below:

- Security system states:

  ○ All Seem Well (ASW), no alarm condition.
  ○ Alarm categorized as "Minor alarm", "Major alarm" or "Critical alarm".

○ An "Alarm Ceasing" condition appears when the alarm condition disappears.

- There is an alarm increasement-related category policy because of alarm accumulation and the usage of an alarm decreasement category policy if there is no alarm repetition in a certain time interval.
- When there is not any alarm the text ASW appears, and the Minor, Major and Critical alarms text with green background. When alarms appear the Minor background is blue, the Major background is yellow, and the Critical background is red. If we look at the panel and on the alarm background the number of detected alarms in each category appear highlighted. For example, supposing that in one moment there are 3 "Minor" alarms and 1 "Major" alarm, number "3" is highlighted on the Minor alarm background and number "1" on the Major alarm background. All the data of the alarms, their activity, their start and the alarm ceasing, are stored in the alarmlog. Based on the alarmlog information the statistics about the alarms are established.

This concentration of alarm information gives the possibility of generating higher-level alarms and proactive alarms. When an alarm is received, is categorized with an alarm level based on the knowledge base ($V_{n,k}$; L_{Alm}). After that, its alarm level ($A_{n,k}$) may change according to the number of repeats and time interval. Each one of the attacks on a service is associated with a mechanism, along with the impact. Noting "the time until the alarm ceases", "the associated mechanism" and the "code of completion of the action" for a mechanism, a general vision is obtained. The expert system also evaluates the features regarding the properties of the mechanisms which are:

- Flexibility: the simplicity (or complexity) degree of change and readjustment regarding security mechanisms, either due to a change in legislation or an update in the smart city alert level (pre-emptive actions, natural disasters, *etc.*) is considered too. if because of one of these reasons security levels in a smart city have to be modified, the ability of modifying them with ease (even remotely if possible) will improve the efficiency of the smart city.
- Capability of being monitored: this parameter measures the capacity of one security mechanism to be monitored. Manufacturers may provide tools to supervise the procedures that they have enabled for their equipment. If this is not the case, they must be developed by human experts and described in TSES.
- There is also knowledge coverage control over the knowledge in the BES, LES and TSES, which usually gives way to revisions and adjustments of the mechanisms.

All this information is analyzed and then enters the system improvement plan.

As a conclusion of the performance evaluation, the main intangible benefit is to provide fast answers to new legal risks about personal data protection in the IoT environment, and these answers are provided by the entity capable of doing so (juridical area) and control the results. Another benefit is to allow companies to conduct cost studies and test ideas for new products and services before beginning the development process.

The main tangible benefit is to provide a tailored security and privacy implies that use the necessaries mechanisms only. It represents as results to obtain savings in resources for sensor nodes. In some cases these savings avoid to use heterogeneous sensor network, for example. If a promoter want to provide the same service such as "health monitoring" for different user types (people, animals, or plants), each one has a different requirements on privacy. These savings can be calculated by the following expression (Equation (1)):

$$
X \cdot Y = Z; \quad
\begin{pmatrix}
b11 & . & b1k \\
. & . & . \\
. & . & . \\
bn1 & . & bnk
\end{pmatrix}
\cdot
\begin{pmatrix}
SecService1 \\
. \\
SecServicek
\end{pmatrix}
=
\begin{pmatrix}
SecServices\ UserType\ (1) \\
. \\
. \\
SecServices\ UserType\ (n)
\end{pmatrix}
\tag{1}
$$

X represents the assign matrix; Y is the matrix that represents the possibilities to provide security services for one specific technology in the final service, finally Z is the set of services assigned to one user type.

Matrix X assign the security services to user type. Each row represents the services for one user type and the each column enable "1" or disable "0", the specific security services, so b[nk] is the enable o disable value for the security service "k" for the user type "n", SecServices UserType(k) is the set of security services assigned to the user type(K).

All users have to belong to the defined user types, and each user type has a certain percentage of users inside. For example, suppose that 25% of users belong to each of the four groups; SecServices UserType(1): cows; SecServices UserType(2): horses; SecServices UserType(3): footballers; SecServices UserType(4): firefighters, according with [6], the expression is Equation (2):

$$
\begin{pmatrix} 0 & 0 & 0 & 0 \\ 1 & 1 & 0 & 0 \\ 1 & 1 & 1 & 0 \\ 1 & 1 & 1 & 1 \end{pmatrix} \cdot \begin{pmatrix} Authenticity \\ Privacy \\ Integrity \\ Availability \end{pmatrix} = \begin{pmatrix} No\ Sec \\ Aut + Priv \\ Aut + Priv + Int \\ Aut + Priv + Int + Avail \end{pmatrix} \tag{2}
$$

25% of users do not have any security services, and the others save some services. Each service can be quantified in spent of resources in terms of energy, delay, dollars, *etc.* Another tangible benefit is that by concentrating the intelligence in a NOC, not only leads to greater specialization, the manpower costs are rationalized in a coherent dimension.

6. Application Scenario for a Smart Service Providing Security, Privacy and Trust: "Living at Home Longer, Autonomously and Safely"

The goal was to develop a service for elderly people named "Living at home longer, autonomously and safely". This service is composed of one service called "Enhanced tele-assistance at home" and another one called "Safe home". The next paragraphs describe the security and privacy requirements, as well as the solutions proposed and the performance evaluation.

The first service, named "Enhanced tele-assistance at home" has been designed to specifically address the requirements of the elderly. It is composed by a gateway at home connected using land and mobile lines. It supports several protocols, such as TT21 (dual tone multi-frequency signaling or DTMF, and the sequential/single tone multi-frequency (STMF) protocol for mobile GSM/Next Generation Networks (NGN) and Telecare Home Units), TT92 (DTMF and STMF), BS8521 (DTMF), TTNEW (DTMF) so as to send/receive calls to/from the assistance center. In the home subsystem, the gateway is connected through the sink to a wearable body sensor network (ZigBee) in order to retrieve the body temperature, heart rate, and fall detector events. The system has been designed to monitor several concurrent users in the same home; all these data are sent to the assistance center through the Internet.

The second service is named "safe home" and it is addressed to a wide range of users. This service is composed by a gateway at home connected to the assistance service through the Internet. In the home subsystem, the gateway is connected through the sink to several sensors deployed throughout the house (ZigBee) providing security alarms related to the indoor temperature, CO, smoke, gas, water flood, as well as events related with open doors or windows. Those alarms can be audible and addressed to an attention center to receive assistance. When a CO, smoke or gas alarm is triggered, the actuators close the corresponding latch and open the windows.

The composite service: "Living at home longer, autonomously and safely" is basically composed of both mentioned services, with some changes in several sensors, bearing in mind that this composite service is designed for elderly people. The person's health and the living environment must be monitored. It is composed by the same gateway mentioned in the first service, with the capacity to also manage the second service using the same gateway and sink.

6.1. DPIA-T for This Service

Since there is no Data Protection Impact Assessment-Template (DPIA-T) related with this service, the expert system has to do the processing stages mentioned in Table 2. As said before, a composite service has its own security and privacy requirements, and can have substantial differences regarding the individual services. Let us see the utility matrix in order to obtain the security and privacy requirements (Tables 3 and 4).

Table 3. Utility Matrix: Service.

Utility Matrix:	Description
Service name	Living at home longer autonomously and safely.
Service Type	Health-care; Safety
Environment Type	Home
Country	Spain
Promoter	Joint venture: Health care and Home insurance companies
User	Elder people
Monitored person	People and rooms at home.
Legal capacity of person	Full legal capacity
Special needs person	Elder people with logical limitations, without special needs
Continuity of service	Push button, critical sensors for life: CO, smoke, gas, presence sensor and outside door and windows open and critical sensor for service.
Critically of the service	high
Network type	NW_Type1

The network type must be defined also.

Table 4. Utility Matrix: Network type for the service.

Network Type:	Living at Home Longer Autonomously and Safely
Network Type Name	NW_Type1
Mote resources limit	Wearable mote: Memory to store data on standalone operation
Connectivity	Radio
Communications	Wearable mote—Gateway, via radio when push button is pressed to call with assistant center. ZigBee connection between wearable node and sink for send data via internet to the assistant center.
	The home sensors—sync via ZigBee and connection via internet from gateway to service provider. In case of CO, gas or smoke alarm, is communicated to actuators to shut down the problem and open outside window and send alarm to the person.
BS Resources Limit	None, when power is down, it has batteries and connections via GSM, 3G. In home there is an emergency battery for four hours (emergency light and sensors power).
Topology	Star
Nodes Roles	The wearable node has collected basically function
	All nodes has collected basically function except window sensor node; it has an actuator function to open outside window directly when CO, gas or smoke are detected.
Routing	Routing is unicast for all sensors to Gateway.

Security imperatives in DPIA-T Format, according to the Spanish (and European) legislation on personal data protections are as follows [6]:

- Data related health must be protected or at least unlinked from the personal identity.
- Data must be fresh and true.
- The data related to the intimacy at home must be protected.

- Critical data for life safety are a priority.

The legal imperatives over sensitive information are structured in a DPIA-Template Security service-Attacks-Defences type format (Table 5) in the case of our current service. Now is the time to assign the specific protections to the data set of the composite service (Table 6). There are several changes compared with the individual services: a water flood can cause a fall, so it is considered as a critical sensor in this composite service, as well as the presence in home and presence in bed. These are important pieces of information so as to provide a good service for elderly people.

Table 5. Legal Imperatives in DPIA-T format.

DPIA-T: Living at Home Longer, Autonomously and Safely			
Security Service	**Attack**	**Target**	**Defence**
Availability	DoS	1) The physical layer is degraded and the communication among nodes is impossible (jamming).	The situation must be known to face it.
		2) A spurious node starts sending malicious data packets to the network.	
Authentication	Sybil	A node is asking for multiple IDs, and if the attack succeeds, the node is able to subvert the trust mechanism.	Restore trust mechanism by rejecting the malicious node.
	Node replication	When a node ID is copied, replicated in a new node, and then introduced in the network. From that moment on, the network accepts the node with the cloned ID as an authorized node.	Realize and revoke the malicious node.
	False node	It introduces data traffic in the network to stop legitimate nodes from communicating (injecting false data messages, requesting authorization continuously, *etc.*).	Identify the false node and discard all messages.
Integrity	Message corruption	When a message reaches the recipient with a different content than the one sent by the source. This situation is either because the message has been degraded in the transmission, or because the message has been intercepted and intentionally changed.	Ensure that messages have not been altered.
Privacy	Eavesdropping	Other devices listening in the same frequency may intercept all communications between two nodes.	Provide authentication and ciphering capabilities. Use data anonymization.
	Node subversion	When a node is captured and cryptoanalyzed the secret keys, node ID, security policies, and so forth are disclosed.	Use few data stored in each node and renew the keys.

Table 6. Data protection over data sets.

	Sensor	Reason	Tipo	Auth	Integr	Privacy	Avail	Intruders Insiders
1	Push button	Emergency	Body	-	-	-	Y	-
2	Temperature	Private information		Y	Y	Y	Y	
3	heart rate							
4	Fall detector							
5	Temperature	Auxiliary Information	Home			-		Y
6	CO					-		
7	Smoke	Vital for life				-		
8	Gas					-		
9	Water flood			Y	Y	-	Y	
10	Door	Vital for security						
11	Window					Y		
12	Presence	Vital for Service						
13	Pres in bed							

6.2. Security Services and Mechanisms

Once the data set protections of the composite service are assigned, the system must look for the appropriate mechanisms in the knowledge base to face the attacks mentioned in the DPIA-T. For these types of service and network, as shown in Table 4, the network type is a condition required to choose the mechanism to be used. In this case, the parameter "mote resources limit", has the value "wearable mote", and for these network types (in this case the mechanism must be lightweight) TSES selectw the SensoTrust proposal [44] mechanisms. All used mechanismw in this example are able to notify when an incident occurs. Each mechanism is evaluated, and assigned a value in the "Past experiences" parameter "0" non effective mechanisms; "1" effective mechanism, it can be monitored, manual reactions; "2" effective mechanism, it can be monitored, automatic reactions. This is based on the trust domains definition where each of them has a common security policy. In this case the domains are defined as follows: as we can see in the previous table, there are two major types of security and privacy required, for each one a domain is defined as shown in Table 7. "Push button" is out of domain because when the button is used the station makes a call outside the WSN.

After applying the legal imperatives (DPIA-T) to the current case, the following list of security and privacy mechanisms arises, taking into account that it can be applied as the common scheme indicated (key distributed, roles and trust policies) in SensoTrust [44,45]. Each domain has its own security and privacy policy as is shown in Tables 8 and 9.

Table 7. Trust domains defined.

	Sensor	Trust Domain	Sensor Type
1	Push button	Out of domains	
2	Temperature		Wearable
3	heart rate	Domain 1 Policy	
4	Fall detector		
5	Temperature		
6	CO		
7	Smoke	Domain 2 Policy	
8	Gas		
9	Water flood		Home
10	Door		
11	window	Domain 1 Policy	
12	Presence		
13	Presence in bed		

Table 8. Trust domain 1 policies.

Domain 1 Policy		
Security Service	**Attack**	**Countermeasure**
Availability	DoS	**Mech_DoS_1**: One alarm is triggered in the Security Manager informing about the situation
Authentication	Sybil	**Mech_Sybil_1**: In the security scheme, every node ID is preconfigured for each node and only the Security Manager (out of the WSN) has the complete list of the IDs. *In extremis*, it is possible to perform a node revocation.
	Node replication	It provides two mechanisms to avoid this attack.
		Mech_N_Repl_1: The Node ID is stored in an external entity (SM) that controls all the IDs working in the network.
		Mech_N_Repl_2: Security policy, if the SM detects that two nodes are operating with the same ID, a node revocation protocol is issued, and the node is dropped from the network.
	False node	**Mech_N_False_1**: Using the node ID, the schema is able to identify the false node and, using the domain key renewal functionality, all the messages sent by this node will be discarded.
Integrity	Message corruption	**Mech_Msg_Corrupt_1**: To avoid both issues, security schema includes the ciphering suite functionality, which allows performing a message hash (using MD5, SHA1, *etc.*).
Privacy	Eavesdropping	**Mech_Eavers_1**: To avoid data disclosure, it provides both symmetric and PKI ciphering capabilities.
		Mech_Eavers_2: Anonymization, unlinking the personal identification and his/her measure data
	Node subversion	**Mech_N_Subv_1**: To avoid it is to minimize the cryptographic and security information stored in each node. Nevertheless, all the keys in the network can be renewed.

Table 9. Trust domain 2 policies.

Domain 2 Policy		
Availability	DoS	Mech_DoS_1
Authentication	Sybil	Mech_Sybil_1
	Node replication	Mech_N_Repl_1
		Mech_N_Repl_2
	False node	Mech_N_False_1
Integrity	Message corruption	Mech_Msg_Corrupt_1

Countermeasures against outsider attacks are based on authentication, and the countermeasures against insider attacks are based on the security policies and the trust domains.

6.3. Performance Evaluation

To preserve the Quality of Service (QoS) it is necessary to know the limitations (Table 10). In the sensor nodes used for the testing purposes, the maximum power computation was limited below 20%, since it was considered that 20% of this maximum value is able to ensure proper operation. With Sybil and False node, the node load is keeping below the maximum limit defined.

Table 10. Restrictions and limitations.

	Sensor	Critical Requirement	
		Battery	Delay
1	Push button		N
2	Body Temperature	Y	
3	Heart rate		Y
4	Fall detector		
5	Home Temperature		N
6	CO		
7	Gas		Y
8	Smoke		
9	Water flood	N	
10	Outside door		
11	window		
12	Presence at home		N
13	Presence in bed		

The battery life is only important in case of Wearable devices, because gateway, sink and devices in home have battery for emergency light and sensors with enough autonomy when the electric power is fell down. The results have been obtained in laboratory for the policy more restrictive (Figures 13 and 14). Finally, it is necessary that the system is designed to provide reports about both the anomalies found (true + and −, false + and −) as their reactions.

Reporting to BES and TSES:

- No legislation-based impact attacks.
- Seriousness of the impacts.
- Legislation-based impact.
- Alarm report

Reporting to LES:

- Legislation-based impact.
- Improvement plan.

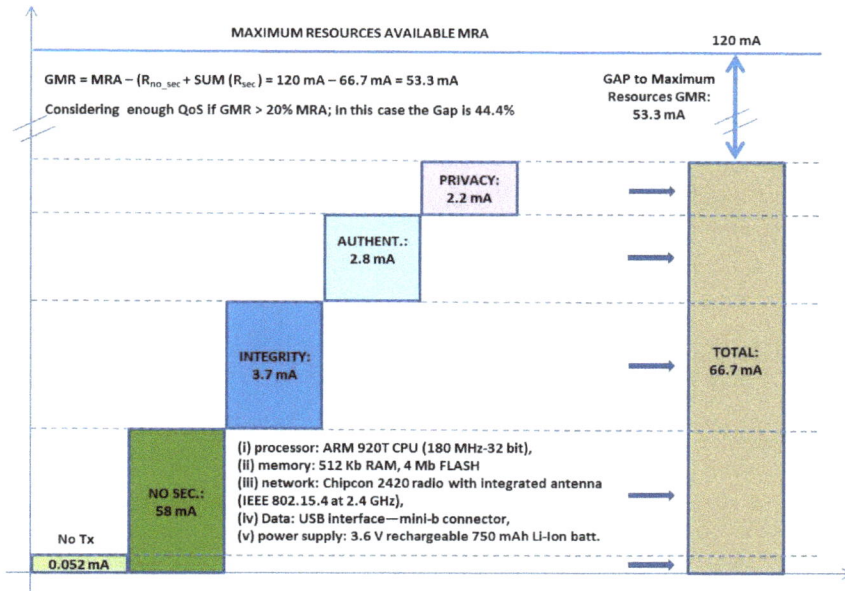

Figure 13. Energy spent *vs.* security services.

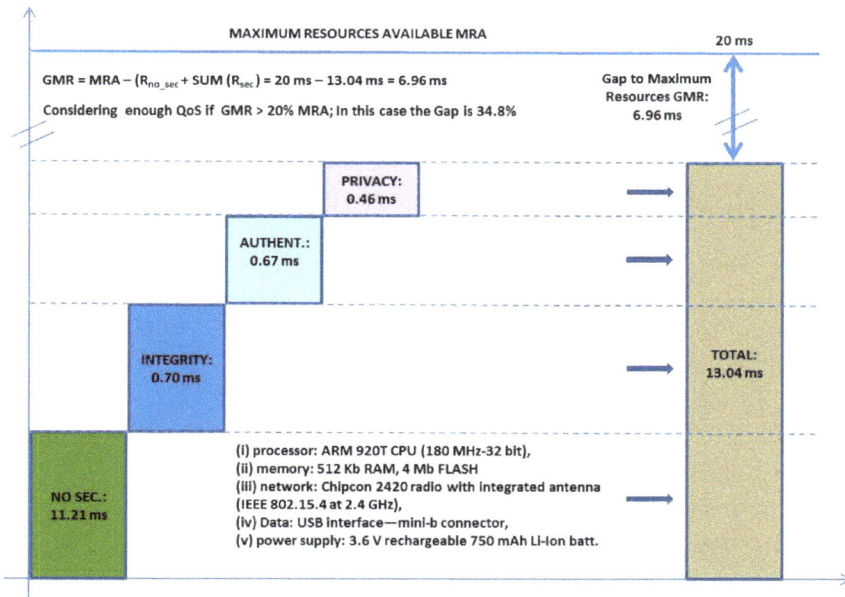

Figure 14. Delay *vs.* security services.

Respecting delays only wearable sensors, CO, Gas and smoke sensors are important while the communication with the windows actuator is connected by wire.

Some indicators can be [46]:

- Percentile 90: The time until to solve the faults.
- Considering MTBF: Mean Time Between Failures (true positives + false negatives) and MTTR: Mean Time To Repair (true positives + false negatives + false positives).
- MTTR over false positives represents the resources spent on inefficient results.
- The actuations on false negatives represent the impact when a problem is not detected on time.
- Coef (no attacks) $\geqslant$ MTBF/(MTBF + MTTR).
- Regarding the system behavior, it has obtained the following results (Tables 11 and 12).

Table 11. Citizen protection impact and reaction.

Security Service	Incidences	Impact	Resolution Time	Pending
		Impact and Reaction:		
Authentication	7	1	Manually	0
Integrity	7	1	Manually	0
Privacy	7	2	Manually	0
Other incidences	-			

These used mechanisms in the application scenario only notify about the problem, and the corrective actions are manual. Resolution time only appear for automatic actions. Respect to the operation of the entire system, the results were as follows.

Table 12. System parts behavior.

BES: Users Perspective. Forms Validation			
Validated	*Rejected*	*Validation time*	**BES-USER interaction**
10	4	70 ms	
BES: Making Utility Matrix 1st Part (Service)			
LES validations	*LES rejections*	*Process time*	**Process and BES-LES interaction**
6	0	4.2 s	
BES: Making Utility Matrix 2nd Part (Network Type for the Service)			
TSES validation	*TSES rejections*	*Process time*	**Process and BES-TSES interaction**
6	0	3.2 s	
LES: Validation Utility Matrix 1st Part			
Validated	*Rejected*	*Validation time*	**Internal Process only**
6	0	120 ms	
LES: Making DPIA-T			
TSES validations	*TSES rejections*	*Process time*	**Process and LES-TSES interaction**
6	0	3.5 s	
TSES: Validation DPIA-T			
Validated	*Rejected*	*Validation time*	**Internal Process only**
6	0	30 ms	
TSES: Validation Utility Matrix 2nd Part			
Validated	*Rejected*	*Validation time*	**Internal Process only**
6	0	50 ms	
TSES: Making Policies			
BES validation	*Rejected*	*Process time*	**Process and TSES-BES interaction**
6	0	7.8 s	
TSES: Making Policies			
ACCUS validation	*Rejected*	*Process time*	**Process and TSES-ACCUS interaction**
6	0	12.1 s	
Service Platform: Policy Validation			
Accepted	*Rejected*	*Validation time*	Internal process. is possible to do it with the information received?
6	0	340 ms	
Service Platform: Actions Generated			
Actuations completed	*Time to complete*		Generate the actuations and configure the testing nodes.
6	18.21 min		

Ten services have been perfomed for this test, four of them had bad data, and the other six were well done. Times are measured on viable services, because rejections are much faster. Interactions include processing and the information transfer between them.

6.4. Providing the Obtained Results to the ACCUS Platform

Once the mechanisms have been selected, they must be provided to ACCUS platform in order to continue the process and configure the service. The following lines present the process to communicate the policy corresponding to trusted domain number 2.

The selection criteria of security mechanisms are as follow for trust domain 2. The input and output xml can be:

```
/*Input Model*/
<?xml version="1.0" encoding="UTF-8"?>
<in:input xmlns:in="http://www.grys.org/securityServicesIN">
  <in:contextOfProtection>Smart Home  Trust Domain 2</in:contextOfProtection>
  <in:facestToProtect>
    <in:facet>  Temperature  </in:facet>
    <in:facet>  CO  </in:facet>
    <in:facet>  Smoke  </in:facet>
    <in:facet>  Gas  </in:facet>
    <in:facet>  Water Flood  </in:facet>
  </in:facestToProtect>
</in:input>
/*Output Model*/
<?xml version="1.0" encoding="UTF-8"?>
<out:recommendations xmlns:out="http://www.grys.org/securityServicesOUT">
  <out:services>
    <out:service>
      <out:name id="av">Availabilitu</out:name>
            <out:mechanism>
                   Mech_DoS_1
            </out:mechanism>
    </out:service>
    <out:service>
      <out:name id="auth">Authentication</out:name>
            <out:mechanism>
                   Mech_Sybil_1
            </out:mechanism>
          <out:mechanism>
                   Mech_N_Repl_1
                   Mech_N_Repl_2
            </out:mechanism>
            <out:mechanism>
                   Mech_N_False_1
            </out:mechanism>
    </out:service>
    <out:service>
      <out:name id="int">Integrity</out:name>
            <out:mechanism>
                   Mech_Msg_Corrupt_1
```

```
            </out:mechanism>
        </out:service>
    </out:services>

    <out:attacks>
        <out:context> Smart Home</out:context>
        <out:attack refTo="av">
            <out:description> DoS </out:description>
        </out:attack>

        <out:attack refTo="auth">
            <out:description> Sybil </out:description>
            <out:description> Node replication </out:description>
            <out:description> False Node </out:description>
        </out:attack>

        <out:attack refTo="int">
            <out:description> Message corruption </out:description>
        </out:attack>

        <out:secSerToImplement>
                        Availability Authentication Integrity
        </out:secSerToImplement>
    </out:attacks>
</out:recommendations>
```

All of this effort has secured only one smart service in the city. However, this process should be iterated for all possible service combinations.

7. Conclusions and Future Work

Collaboration between multiple areas (business, legal and technological) is critical in order to provide users with the necessary trust in the security of the service and protection of their personal data. The limitation of resources of wireless sensor networks for products and services makes it necessary to implement a tailored security schema to avoid the risk of poor quality of service when resources are exhausted. In this way, the intelligent combination of security mechanisms available could increase the service's efficiency.

In an environment where several services and technologies coexist, the intelligence of the decision-making on security policies could be integrated into the officially recognized and certified network operations centre which would provide a large capacity of management, updating the deployed security measures. Since there is no legislative uniformity, it is necessary to develop tools and methods that permit a transition period with minimal risk to people and their rights.

When citizens' quality of life greatly depends on the correct performance of smart cities, which, in turn, depend on the correct performance of systems and services (and the networks in which they are set up), it might be required to configure and reconfigure privacy and security policies. It may be due to changes in states of emergency, alarm, *etc*. Or, they may be necessary because of dynamic changes within the city. Therefore, it is important to know timely and in due form the security policies that must be implemented in each case, in a reliable way, by the expert system described. It would also be convenient to deploy and perform the actions needed on the elements of the city, fast and efficiently (decreasing human intervention to a minimum), using the mediation chain described. The key lies in the cooperative environment between the three main knowledge areas described, and in a common management, even if the systems are technically duplicated or diversified to augment security and availability.

It must be considered that, in an uncertain future, a failure in the performance of a smart city may be disastrous and could have serious consequences; far from just a mild annoyance, it could result in the endangerment of people and infrastructure. A service under attack or functioning incorrectly may activate an alarm, with no emergency triggering it. Therefore, investing in security, privacy and reliability in the performance of systems in a smart city may be worthwhile.

In the future, we will point in different directions according to the road map. On the expert system side knowledge bases should be enhanced to make possible interactions between IoT platforms and several technologies. Also is will be necessary to make life easier for the human experts providing them nice tools to manage the information. ACCUS must go on with its road map, to improve the functionalities to create new services in the smart city environment. Those new services represent good opportunities for both systems to work together to face new issues.

Acknowledgments: ACCUS (Adaptive Cooperative Control in Urban Systems) project has been partially funded by the Spanish Ministry of Industry, Energy and Tourism (Ref. ART-010000-2013-2) and by the Artemis JU (GA number 333020-1). LifeWear (Mobilized Lifestyle with Wearables) project has been funded by the Spanish Ministry of Industry, Energy and Tourism (Ref. TSI-010400-2010-100) by the Avanza Competitividad I+D+I program. The authors would like to thank CITSEM (Research Center on Software Technologies and Multimedia Systems for Sustainability, Centro de Investigación en Tecnologías Software y Sistemas Multimedia para la Sostenibilidad) from the UPM.

Author Contributions: The work presented in this paper is a collaborative effort participated by all of the authors. Jose Antonio Sánchez and Jose-Fernán Martínez have studied, defined and developed the collaborative model to solve the problem dealing with collaboration on the business, legal and technological areas. Lourdes López has defined the research theme with the purpose of solving the challenges associated to the trust and privacy problem. Gregorio Rubio Cifuentes has defined the communication between expert system the urban systems. All the authors have made written contributions to the paper, and Lourdes Lopez has reviewed the whole manuscript. All of the authors have read and approved the manuscript. The research group named Next-Generation Networks and Services (GRyS), which the authors belong to, is developing the prototypes. GRyS research group is integrated within the Technical University of Madrid (UPM) and as part of CITSEM.

Conflicts of Interest: The authors declare no conflict of interest.

References

1. Kanter, R.M.; Litow, S.S. *Informed and Interconnected: A Manifesto for Smarter Cities*; Working Paper 09-141; Harvard Business School: Boston, MA, USA, 2009.
2. Caragliu, A.; del Bo, C.; Nijkamp, P. Smart Cities in Europe. *J. Urban Technol.* **2011**, *18*, 65–82. [CrossRef]
3. Cavoukian, Ann (2011) Privacy by Design—The 7 Foundational Principles. Available online: http://www.ipc.on.ca/english/privacy/introduction-to-pbd (accessed on 24 March 2015).
4. The ACCUS (Adaptive Cooperative Control in Urban (sub)Sytems). Available online: http://projectaccus.eu/ (accessed on 24 March 2015).
5. Sánchez-Alcón, J.-A.; López, L.; Martínez-Ortega, J.-F.; Castillejo, P. Automated determination of security services to ensure personal data protection in the internet of things applications. In Proceedings of the 3rd International Conference on Innovative Computing Technology (Intech), London, UK, 29–31 August 2013; pp. 71–76. [CrossRef]
6. Sánchez-Alcón, J.-A.; López-Santidrián, L.; Martínez, J.-F. Solución para garantizar la privacidad en internet de las cosas. *Prof. Inf.* **2014**, *24*, 62–70. [CrossRef]
7. Cuadrado, N. *Aplicación de los Sistemas Expertos al Campo del Derecho*; Servicio de Publicaciones de la Universidad Complutense de Madrid: Madrid, Spain, 2004. (In Spanish)
8. Stevens, C.; Barot, V.; Carter, J. The Next Generation of Legal Expert Systems—New Dawn or False Dawn? In *Research and Development in Intelligent Systems XXVII*; Springer: London, UK, 2011; pp. 439–452. [CrossRef]
9. Venkateswarlu, N.M.; Sushant, L. Building a Legal Expert System for Legal Reasoning in Specific Domain-A Survey. *Int. J. Comput. Sci. Inf. Technol.* **2012**, *4*, 175–184. [CrossRef]

10. Giffinger, R.; Fertner, C.; Kramar, H.; Kalasek, R.; Pichler-Milanovic, N.; Meijers, E. *Smart Cities—Ranking of European Medium-Sized Cities*; Research Report; Vienna University of Technology: Vienna, Austria, 2007; Available online: http://www.smart-cities.eu/download/smart_cities_final_report.pdf (accessed on 24 March 2015).

11. JBossDeveloper 2012. Available online: http://www.jboss.org/products/fuse/overview/ (accessed on 3 December 2014).

12. Open Platform for Your Connected Business 2011. Available online: http://wso2.com/ (accessed on 3 December 2014).

13. Stoimenov, L.; Stanimirovic, A.; Djordjevic-Kajan, S. Semantic Interoperability using multiple ontologies. In Proceedings of the 8th AGILE Conference on GIScience, Estoril, Portugal, 26–28 May 2005; pp. 261–270. Available online: http://plone.itc.nl/agile_old/Conference/estoril/papers/88_Leonid%20Stoimenov.pdf (accessed on 12 January 2014).

14. Zang, M.A.; Pohl, J.G. Ontological Approaches for Semantic Interoperability. In Proceedings of the 5th Annual ONR Workshop on Collaborative Decision-Support Systems, San Luis Obispo, CA, USA, 10–11 September 2003; pp. 1–10. Available online: http://digitalcommons.calpoly.edu/cgi/viewcontent.cgi?article=1078 (accessed on 12 January 2014).

15. Sustainable Smart Grid Open System for the Aggregated Control, Monitoring and Management of Energy (e-Gotham) Project. Available online: http://www.e-gotham.eu/ (accessed on 22 May 2015).

16. De Diego, R.; Martínez, J.-F.; Rodríguez-Molina, J.; Cuerva, A. A Semantic Middleware Architecture Focused on Data and Heterogeneity Management within the Smart Grid. *Energies* **2014**, *7*, 5953–5994. [CrossRef]

17. W3C Semantic Sensor Network Incubator Group. Semantic Sensor Network Ontology. 2011. Available online: http://www.w3.org/2005/Incubator/ssn/ssnx/ssn (accessed on 21 May 2015).

18. Brickley, D.; Miller, L. FOAF Vocabulary Specification 0.99. Namespace Document 14 January 2014-Paddington Edition. Available online: http://xmlns.com/foaf/spec (accessed on 20 May 2015).

19. CityGML. Available online: http://www.opengeospatial.org/standards/citygml (acceessed on 22 May 2015).

20. Prud'hommeaux, E.; Seaborne, A. SPARQL Query Lenguaje for RDF, W3C Recommendation 15 January 2008. Available online: http://www.w3.org/TR/rdf-sparql-query (accessed on 3 April 2015).

21. Gandon, F.; Schreiber, G. RDF 1.1 XML Syntax. W3C recommendation. 25 February 2014. Available online: http://www.w3.org/TR/rdf-syntax-grammar/ (acceessed on 23 May 2015).

22. Codagnone, C.; Wimmer, M.A. *Roadmapping eGovernment Research—Visions and Measures towards Innovative Governments in 2020*; European Commission FP6 Project eGovRTD2020, Final Report; Guerinoni Marco: Clusone, Italy, 2007; Available online: https://air.unimi.it/retrieve/handle/2434/171379/176506/FinalBook.pdf (accessed on 12 January 2015).

23. Sicari, S.; Rizzardi, A.; Grieco, L.A.; Coen-Porisini, A. Security, privacy and trust in Internet of Things: The road ahead. *Comput. Netw.* **2015**, *76*, 146–164. [CrossRef]

24. Weber, R.H. Internet of things: Privacy issues revisited. *Comput. Law Secur. Rev.* **2015**, *31*, 618–627. [CrossRef]

25. Ten, C.-W.; Manimaran, G.; Liu, C.-C. Cybersecurity for Critical Infrastructures: Attack and Defense Modeling. In *Systems, Man and Cybernetics, Part A: Systems and Humans*; IEEE: Piscataway, NJ, USA, 2010; Volume 40, pp. 853–865. [CrossRef]

26. Daryabar, F.; Dehghantanha, A.; Udzir, N.I.; Sani, N.F.B.M.; Bin-Shamsuddin, S. Towards secure model for scada systems. In Proceedings of the 2012 International Conference on Cyber Security, Cyber Warfare and Digital Forensic, CyberSec 2012, Kuala Lumpur, Malaysia, 26–28 June 2012; pp. 60–64. [CrossRef]

27. Martinez-Balleste, A.; Perez-Martínez, P.A.; Solanas, A. The pursuit of citizens' privacy: A privacy-aware smart city is possible. *IEEE Commu. Mag.* **2013**, *51*, 136–141. Available online: http://ieeexplore.ieee.org/stamp/stamp.jsp?tp=&arnumber=6525606&isnumber=6525582 (accessed on 7 January 2015). [CrossRef]

28. Bartoli, A.; Hernández-Serrano, J.; Soriano, M.; Dohler, M.; Kountouris, A.; Barthel, D. Security and privacy in your smart city. In Proceedings of the Barcelona Smart Cities Congress, Barcelona, Spain, 29 November–2 December 2011; pp. 1–6. Available online: http://smartcitiescouncil.com/sites/default/files/public_resources/Smart%20city%20security.pdf (accessed on 8 February 2015).

29. Rinaldi, S.M. Modeling and simulating critical infrastructures and their interdependencies. In System Sciences, Proceedings of the 37th Annual Hawaii International Conference, Big Island, HI, USA, 5–8 January 2004; IEEE: Big Island, HI, USA, 2004; p. 8. Available online: http://ieeexplore.ieee.org/stamp/stamp.jsp?tp=&arnumber=1265180&isnumber=28293 (accessed on 12 January 2015). [CrossRef]

30. Cazorla, L.; Alcaraz, C.; Lopez, J. Towards automatic critical infrastructure protection through machine learning. In Proceedings of 8th International Conference on Critical Information Infrastructures Security, Amsterdam, The Netherlands, 16–18 September 2013; pp. 197–203. Available online: http://link.springer.com/chapter/10.1007%2F978-3-319-03964-0_18#page-1 (accessed on 12 January 2015). [CrossRef]

31. Islam, M.M.; Hassan, M.M.; Lee, G.; Huh, E. A Survey on Virtualization of Wireless Sensor Networks. *Sensors* **2012**, *12*, 2175–2207. [CrossRef] [PubMed]

32. Islam, M.M.; Huh, E. Virtualization in Wireless Sensor Network: Challenges and Opportunities. *J. Netw.* **2012**, *7*, 412–418. [CrossRef]

33. Lucas Martínez, N.; Martínez, J.F.; Hernández, V. Virtualization of Event Sources in Wireless Sensor Networks for the Internet of Things. *Sensors* **2014**, *14*, 22737–22753. [CrossRef] [PubMed]

34. Data Protection Impact Assessment Template for Smart Grid and Smart Metering systems, Smart Grid Task Force 2012-14, Expert Group 2: Regulatory Recommendations for Privacy, Data Protection and Cyber-Security in the Smart Grid Environment, 18.03.2014. Available online: https://ec.europa.eu/energy/sites/ener/files/documents/2014_dpia_smart_grids_forces.pdf (accessed on 24 March 2015).

35. Lopez, J.; Roman, R.; Alcaraz, C. Analysis of Security Threats, Requirements, Technologies and Standards in Wireless Sensor Networks. In *Foundations of Security Analysis and Design*; Springer: Berlin/Heidelberg, Germany, 2009; Volume 5705, pp. 289–338.

36. Nasr, P.M.; Varjani, A.Y. Alarm based anomaly detection of insider attacks in SCADA system. In Proceedings of the Smart Grid Conference (SGC), Tehran, Iran, 9–10 December 2014; pp. 1–6. [CrossRef]

37. Zhu, B.; Sastry, S. SCADA-Specific Intrusion Detection/Prevention Systems: A Survey and Taxonomy. In Proceedings of the 1st Workshop on Secure Control Systems (SCS'10), Stockholm, Sweden, 12 April 2010.

38. Vieira, K.; Schulter, A.; Westphall, C.B.; Westphall, C.M. Intrusion Detection for Grid and Cloud Computing. *IT Prof.* **2010**, *12*, 38–43. Available online: http://ieeexplore.ieee.org/stamp/stamp.jsp?tp=&arnumber=5232794&isnumber=5512497 (accessed on 2 September 2015). [CrossRef]

39. De Chaves, S.A.; Uriarte, R.B.; Westphall, C.B. Toward an architecture for monitoring private clouds. *IEEE Commun. Mag.* **2011**, *49*, 130–137. Available online: http://ieeexplore.ieee.org/stamp/stamp.jsp?tp=&arnumber=6094017&isnumber=6093994 (accessed on 5 September 2015). [CrossRef]

40. De Chaves, S.A.; Westphall, C.B.; Lamin, F.R. SLA Perspective in Security Management for Cloud Computing. In Proceedings of the Sixth International Conference of Networking and Services (ICNS), Cancun, Mexico, 7–13 March 2010; pp. 212–217. Available online: http://ieeexplore.ieee.org/stamp/stamp.jsp?tp=&arnumber=5460645&isnumber=5460618 (accessed on 5 September 2015). [CrossRef]

41. Farooq, A.; Petros, M. *Security for Wireless Ad Hoc Networks*; Wiley-Interscience, Cop.: Hoboken, NJ, USA, 2007; p. 247.

42. Nasr, P.M.; Varjani, A.Y. Petri net model of insider attacks in SCADA system. In Proceedings of the Conference on Information Security and Cryptology (ISCISC) (11th International ISC), Tehran, Iran, 3–4 September 2014; pp. 55–60. [CrossRef]

43. International Telecommunications Union. Recommendation X.800: Security Architecture for Open Systems Interconnection for CCITT Applications. Available online: http://www.itu.int/rec/T-REC-X.800-199103-I/en (accessed on 12 January 2015).

44. Pedro Castillejo, P.; Martínez-Ortega, J.F.; López, L.; Sánchez Alcón, J.A. SensoTrust: Trustworthy Domains in Wireless Sensor Networks. *IJDSN* **2015**. [CrossRef]

45. Rolim, C.O.; Koch, F.L.; Westphall, C.B.; Werner, J.; Fracalossi, A.; Salvador, G.S. A Cloud Computing Solution for Patient's Data Collection in Health Care Institutions. In eHealth, Telemedicine, and Social Medicine 2010, Proceedings of the Second International Conference ETELEMED '10, St. Maarten, The Netherlands, 10–16 February 2010; pp. 95–99. Available online: http://ieeexplore.ieee.org/stamp/stamp.jsp?tp=&arnumber=5432853&isnumber=5432832 (accessed on 5 September 2015). [CrossRef]

46. Herrmann, D.S. *Complete Guide to Security and Privacy Metrics: Measuring Regulatory Compliance, Operational Resilience, and ROI*; Auerbach Publications: Boca Raton, FL, USA, 2007; p. 824.

Article

Semantic Registration and Discovery System of Subsystems and Services within an Interoperable Coordination Platform in Smart Cities

Gregorio Rubio *,†, José Fernán Martínez †, David Gómez † and Xin Li †

Centro de Investigación en Tecnologías Software y Sistemas Multimedia para la Sostenibilidad (CITSEM), Universidad Politécnica de Madrid (UPM), Edificio La Arboleda, Campus Sur UPM. Ctra. Valencia, Km 7, 28031 Madrid, Spain; jf.martinez@upm.es (J.F.M.); david.gomezs@upm.es (D.G.); xin.li@upm.es (X.L.)

* Correspondence: gregorio.rubio@upm.es; Tel.: +34-913-365-509; Fax: +34-913-367-821
† These authors contributed equally to this work.

Academic Editor: Gonzalo Pajares Martinsanz
Received: 5 February 2016; Accepted: 17 June 2016; Published: 24 June 2016

Abstract: Smart subsystems like traffic, Smart Homes, the Smart Grid, outdoor lighting, etc. are built in many urban areas, each with a set of services that are offered to citizens. These subsystems are managed by self-contained embedded systems. However, coordination and cooperation between them are scarce. An integration of these systems which truly represents a "system of systems" could introduce more benefits, such as allowing the development of new applications and collective optimization. The integration should allow maximum reusability of available services provided by entities (e.g., sensors or Wireless Sensor Networks). Thus, it is of major importance to facilitate the discovery and registration of available services and subsystems in an integrated way. Therefore, an ontology-based and automatic system for subsystem and service registration and discovery is presented. Using this proposed system, heterogeneous subsystems and services could be registered and discovered in a dynamic manner with additional semantic annotations. In this way, users are able to build customized applications across different subsystems by using available services. The proposed system has been fully implemented and a case study is presented to show the usefulness of the proposed method.

Keywords: subsystem registry; subsystem discovery; service registry; service discovery; semantic interoperability; ontology; system of systems

1. Introduction

A diversity of urban subsystems, such as Intelligent Transport Management systems [1,2], Smart Buildings systems [3], Smart Gird systems [4–6], Smart Outdoor Lighting systems [7] and Smart Home systems [8] are maturely developed in urban areas. Basically, each of them is managed by self-contained embedded systems and connected with Wireless Sensor and Actuator Networks (WSANs). Different smart subsystems can work effectively providing domain-specific services to citizens, but in an isolated manner. Unfortunately, collaborations and coordination between diverse smart subsystems are missing, even though they could potentially provide more citizen-friendly services by using data/services provided by different subsystems. This implies that cities are facing an unprecedented challenge, which is integrating fragmented smart subsystems and enabling cross-domain usages of services. A paradigm shift from conventional cities to "Smart Cities" has attracted a lot of interest from the research community, governments, and industry. This is an ongoing change that has been undertaken in many cities. For instance, in Spain, there are 65 cities integrated in RECI [9] which is the Spanish Network of Smart Cities. To turn conventional cities into smart ones, a system-thinking approach is needed to facilitate interconnections between different subsystems. A new platform aiming to connect different subsystems and represent a true "system of systems" integration has been developed in the Adaptive Cooperative Control in Urban (sub)Systems (ACCUS) project [10]. This platform, called Integration and Coordination Platform (ICP), emphasizes how to integrate different subsystems and enables the development of new applications across subsystems without interfering individual updates and internal policies.

The ICP is capable of providing a variety of functionalities in order to optimize combined performance of different subsystems, thus achieving more flexible, more efficient and more robust integrated urban systems and managing different emergent behaviors. The platform is conceived as a distributed and layered architecture which offers three main groups of functions:

- Core functions: they are referred to generic platform functionalities which are offered by the Runtime Environment of the ICP. Core functions are provided by the following software components: applications servers, service broker, service repository, service bus, data broker and management, policy management, security management, workflow engine, security management and message broker and database.
- Extension functions: they are referred to ICP specific functionalities that can be used by other services and applications. Extension functions are offered by info broker, control broker, subsystem monitoring, subsystems adaptors, ontology connector, programming API and ICP ontology.
- City configurations functions: these are city specific generic functionalities available for other ICP services and applications. They are provided by event detection, location detection, data analytics, situation awareness, and reasoning.

To ensure that the ICP could fulfill the aforementioned capabilities, a key premise is to address the heterogeneity (e.g., data formats and protocols) inherent to subsystems and services provided by different subsystems and provide a unified interface of reference for available subsystems and services within the ICP. With this reference, maximum reusability of available services provided by entities (e.g., sensors, or Wireless Sensor Networks) could be enabled so as to develop new cross-domain applications. Thus, an ontology-based and automatic system for registering and discovering subsystems and services is presented in this paper. This proposed system can be embedded into the ICP in order to discover and register available heterogeneous services and subsystems in a dynamic manner. The proposed system for subsystems and services registration and discovery is able to adapt to any change that occurs in subsystems and services. Additionally, in order to abstract the heterogeneity of subsystems and services and provide a common understanding to the ICP, this proposed system employs an ontological approach to provide a formalized model, named *new Subsystem and Service Oriented Ontology (nSSOO)*, for the registry and discovery process. Thus, different subsystems and services could be semantically annotated and understood by the ICP and users as well. The proposed system for subsystem and service registration and discovery is playing an important role in ICP to enable the creation of real-time collaborative applications across subsystems by employing services available in the city. The proposed system is completely implemented and validated by using a traffic light control application use case.

This paper is organized as follows: Section 2 presents related work on existing platforms for Smart Cities and their solutions to register and discover subsystems and services within them. The proposed system for registry and discovery of subsystems and services within the ICP is shown in Section 3. Specifically, Section 3.1 highlights the main contributions that are provided by this paper. Section 3.2 shows system integration regarding Semantic Interoperability right afterwards. The newly proposed nSSOO ontology used to model the registry and discovery information base is introduced in Section 3.3. Section 3.4 provides a holistic view of the system and elaborates the specific software components that are involved. Workflow for the subsystem and service discovery and registration procedures is detailed in Section 3.5. Section 4 presents a use case about the traffic light control to validate the proposed system. Finally, in Section 5, conclusions are given and future work is pointed out as well.

2. Related Work

Technical interoperability levels have progressed in the last years with already mature solutions. However, semantic interoperability [11] remains as a key obstacle to the seamless exchange of data between services, applications or systems. However, some European projects are making significant progress in the development of platforms of interconnection, using Service Oriented Architecture (SOA) and semantic technologies to get semantic interoperability as a key factor of seamless interconnection and interoperability.

European FP7 projects that have worked on developing SOA-based platforms are as follows:

SOA4ALL [12] proposed a framework and software infrastructure that aims at integrating SOA and four complementary and evolutionary technical advances (Web, context-aware technologies, Web 2.0 and Semantic Web) into a coherent and domain-independent worldwide service delivery platform. Also, some semantic source components have been developed and the concept of "linked services" which is Semantic Web services building on the success of the Linked Open Data initiative. So, they are services that can consume Resource Description Framework (RDF) from the Web of Data and feed-back RDF to Web of Data. However, "linked services" is not the same concept as the services with semantic interoperability, especially if there are services working together that have different ontologies or no semantic representation at all.

TaToo [13] proposed a framework to allow third parties to discover environmental resources data and services on the web and add valuable information in the form of semantic annotations to these resources. This annotation allows improving the reasoning and inference power of the ontologies to create richer resource annotations. This process has the goal of optimizing discovery process of services, but not improving interoperability.

Cloud4SOA [14] proposed a solution based on the concept of using cloud computing under the paradigm of PaaS [15] to resolve the interoperability and portability issues that exist in current cloud infrastructures using the same technological providing a user-centric approach for applications that are built upon and deployed by means of Cloud resources.

SemanticHealthNet [16] proposed a set of resources to support semantic interoperability process for clinical and biomedical knowledge, but it cannot be considered as a platform.

LifeWear [17] proposed a middleware platform to interconnect wearable devices and sensors of a WSN using services semantically annotated in a compliant way to an ad-hoc ontology. The system shown in this manuscript is an important evolution of the results of this project.

Not only European projects have proposed platforms; Ryu proposed in [18] an Integrated Semantic Service Platform (ISSP) with IoT-based service support in a Smart City, addressing ontological models in various domains of a Smart City.

Hussian et al. [19] presented an integrated platform to be deployed in a Smart City. This platform could enable a unified and and people-centric access to all services provided by the Smart City. However, this proposed platform emphasized the integration of various healthcare systems within the Smart Cities. Thus, it lacks generality and it is not applicable while attempting to integrate a diversity of smart subsystems beyond healthcare systems.

A distributed platform called "Kalimucho" [20] was built to enable the design of context-aware applications based on heterogeneous devices in Smart Cities. This platform is ambitious in the sense that provides a tight collaboration between different subsystems, such as transportation and logistics, healthcare, and smart environments. The platform is conceived as "everything-as-a-service" which regards subsystems and services provided by corresponding subsystems as independent services. However, all those services are pre-registered in the platform. Dynamicity of registering and discovering new services is missing; also, semantic annotation for the different services is not considered in this proposal.

Furthermore, there have been efforts related to integration, in different ways, using semantic technics WSAN and services, as antecessors to the whole platform: Rodriguez-Molina et al. [21] proposed a semantic middleware for Wireless Sensor Networks, in order to provide integration of sensors in a body area network with other WSAN present in a smart city. Bispo et al. [22] proposed a more advanced model in Semantic Infrastructure for Wireless Sensors Networks (SITRUS) with semantic information processing to generate a semantic database focused on determining the reconfiguration of a WSAN combined with a message-oriented communication service and another one used for reconfiguration. Camarhina-Matos et al. [23] proposed using the concept of collaborative network for the integration of networks or WSAN belonging to different organizations, that involves mutual engagement of participants to solve a problem together, thus implying mutual trust and taking time, effort, and dedication. In this proposals, the network of each organization can be considered a subsystem and each application a service. Last years, with the rise of Software Defined Networks (SDNs) [24], there have been efforts bent on enhancing interoperability among the various heterogeneous wireless networks when control and information levels are separated. Kosmides [25] showed a system with a centralized network controller based in SDN applied to social networks as case of study, where the Smart City was divided in geographical zones, and each zone was considered as a subsystem. However, the concept of semantic registration and discovery of subsystems and services is not embedded in the development works implemented using SDN or Collaborative Networks.

3. Proposed System for Registration and Discovery of Subsystems and Services within ICP

In this section, a new proposal for a system made for semantic registration and discovery of subsystems and services within ICP is presented. Specifically, the main contributions that are provided by this paper are listed in Section 3.1. The integration system offering semantic interoperability is shown in Section 3.2. Section 3.3 is devoted to introducing the newly proposed nSSOO ontology. The system architecture of subsystems and service registration and discovery are described in Section 3.4. Finally, the specific procedures to facilitate the registration and discovery of subsystems and services are elaborated in Section 3.5.

3.1. Innovations

Some works shown in the previous section use registration systems developed ad-hoc or integrated in Application Servers as WildFly [26] or WSO2 [27], that they have their own registration service. However, to satisfy the registration and discovery system that ACCUS ICP needs, it is necessary to add three innovations.

The first innovation of the proposed semantic subsystem and service registration and discovery system is that it contributes to the integration and coordination of urban systems, connected to the ACCUS ICP, to build applications like monitoring, management and control that can reach beyond the borders of the individual subsystems and services. The proposed system contributes to cross-domain and cross-layer cooperation of urban subsystems and services by addressing different interoperability aspects, such as semantic interoperability.

Semantic interoperability provides means for seamlessly integrating urban subsystems, composing more complex functionalities from already existing subsystems and deploying converged scenarios. It will also enable the integration and deployment of present and future urban subsystems and processes in urban environments with little involvement from the side of either the developers or operators, in an automated way, based on common agreed ontologies and semantic artefacts.

Another aspect of interoperability addressed by the proposed system is the information and knowledge discovery. It enables every subsystem, service or application connected to the ACCUS ICP, to discover registered subsystems, services or applications, and obtain information about them, by sending a query request to the ACCUS ICP. It will respond, after giving authorization to the subsystem, with the information requested according to the defined subsystems and services ontology.

Information and knowledge discovery enables, in turn, the development of applications that combine the information about services and subsystems provided by the ACCUS ICP with the purpose of offering more complex services able to provide functionalities that subsystems and services cannot provide separately, and facilitating, thus, service composition.

The second innovation of the proposed semantic subsystem and service registry and discovery procedure, is its capability of enabling a distributed control, management and optimization infrastructure, along with the algorithms and tools required to create highly advanced urban control functions, which will be implemented through the introduction of cooperation extension over multiple urban subsystems, system layers and domains. This will improve the performance of combined urban systems at run time.

The third innovation of the proposed system is that it ensures the development and application of methodologies and tools for the implementation of real-time collaborative applications for system of systems. The methodology and tool innovation covers the entire life-cycle (i.e., from design to operation, maintenance and possibly retrofitting) of the applications developed for the integrated urban subsystems domain.

3.2. Seamless Interconnection and Semantic Interoperability

In system of systems, interoperability is the ability of two or more subsystems or components to exchange information and to use what has been interchanged [28]. According to the features shown in this paper about the ICP, the most accurate model is the Level of Conceptual Interoperability Model (LCIM) [29] because it provides a framework that divides interoperability problems into different levels; at each level, interoperability problems can be settled and a solution can be developed to solve interoperability problems that belong to that level. It contains seven levels: Level 0—No interoperability; Level 1—Technical interoperability: networks and standard communication protocols enable the interchange of data between systems; Level 2—Syntactic interoperability: adds a common structure and data format to the data interchanged in an unambiguous way; Level 3—Semantic Interoperability: adds a common interpretation of data interchanged, the meaning of information exchanged between systems is defined in an unambiguous way; Level 4—Pragmatic Interoperability: systems are aware of the specific use of exchanged data by other systems; Level 5—Dynamic Interoperability: systems are able to understand the change of states in each element depending on the decisions taken according to the use of data; Level 6—Conceptual Interoperability: global interoperability.

The system of registration and discovery proposed in this paper allows the ICP to solve technical, syntactic and semantic interoperability issues. When the meaning of data is shared among services the content of the information exchange among them is unambiguously defined, so common interpretation of the data is guaranteed. ICP uses data from several different data sources, various sensors are connected to these data sources and each sensor uses different data format, which is further processed. Various types of data are used, creating a system able to combine these data. With the data integration system it is possible to connect several heterogeneous systems and create one large system. Getting is relatively simple when the integration platform is created before the services. However, when subsystems and services are already installed and were designed without knowledge that in the future they would be integrated in an ICP merging all of them in a single platform can be a challenging task.

Another important achievement of semantic interoperability made through ICP is achieving semantic interoperability without the need to change anything regarding the mode of operation, communication, data management subsystems and services already installed in the city.

As data sources are highly heterogeneous, the ICP uses a Common Data Model as a common layer to interchange information among data sources (namely, services and subsystems).

The Common Data Model uses a shared ontology with adaptation of the information provided by subsystems and services, as each of them can provide the information in their own ontology or even in their own format, which might be non-compliant with any ontology.

Using the schemas depicted in Figures 1 and 2, all subsystems and services can be registered, discovered and used with the semantic capabilities established in the shared ICP Ontology. In this way, semantic interoperability is guaranteed, but with the advantage of not requiring any of the systems and services included in the smart city to modify their own syntax.

This shared ICP Ontology has a part specifically developed to register and discover subsystems and services in the ICP. It is specified in the next section.

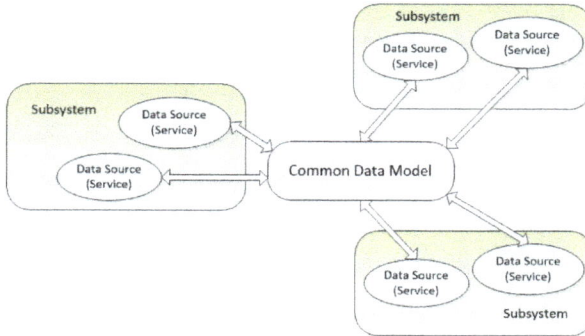

Figure 1. Common data model.

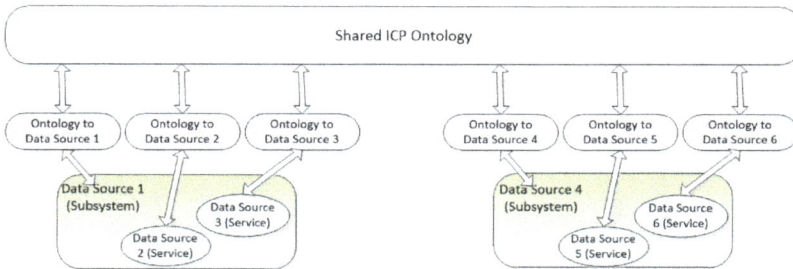

Figure 2. Integration system regarding Semantic Interoperability.

3.3. Ontology Specifications

When attempting to provide an interoperable and formalized knowledge model for subsystem, service registration and discovery processes, the new Subsystem and Service Oriented Ontology (nSSOO) is proposed. The nSSOO is conceived to comprehensively and semantically describe a variety of features about subsystems and services owned by the complex urban system. This proposed ontology makes possible the integration and registration of information provided by sensors or subsystems. Originally thought for services offered by low capability devices (sensors, PDAs, RFID tags, etc.), this ontology can also be applied to services based on normal devices or subsystems (Smart Home subsystem, Smart Traffic subsystem, etc.), as in the ACCUS project.

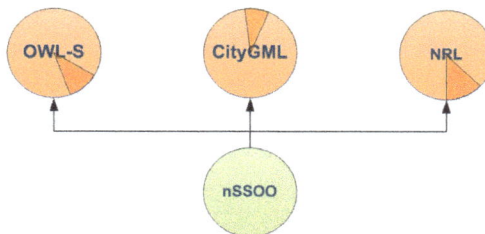

Figure 3. Proposal of nSSOO.

Figure 3 shows the hierarchical composition of the proposed nSSOO. Generally speaking, a few concepts of nSSOO are inherited from three existing and widely used ontologies which are Semantic Markup for Web Services (OWL-S) [30], City Geography Markup Language (CityGML) [31] and Security Ontology for Annotating Resources (NRL) [32]. The reusability of OWL-S, CityGML and NRL reduces the workload of developing the nSSOO and further expands its interoperability to a higher level.

The software tools used here are as follows:

- OWL-S. This ontology is used to describe semantic web services. It enables users and software agents to automatically discover, invoke and compose web resources while offering services.
- CityGML. This ontology models 3D cities taking into account multiple features, such as city geometry, topology, semantic features, and appearance characteristics. The ultimate aim of the development of CityGML is providing a common understanding for the basic entities, attributes, and relations of a 3D city model.
- NRL. It describes different types of security information including mechanisms, protocols, objectives, algorithms and credentials in various levels of detail and specificity. NRL is comprehensive, well-organized and expressive enough to describe security policies.

The most coarse-grained concepts are *Subsystem* and *Service*, which form the entire nSSOO ontology. In the following, the top-level concepts of *Subsystem* and *Service*, as shown in Figure 4, are broken down and their associated subclasses are explained in detail, along with descriptions for relationships/object properties which reflect the connections between them with the aim of providing a better understanding of the whole proposal. The primary principle of designing the nSSOO ontology is assigning different concepts with intuitive terms so that their meanings and intentions can be easily revealed. To make a clear distinction between service- and subsystem-owned ontology elements, prefixes "S_" and "SS_", as abbreviations of service and subsystem, are attached to the corresponding ontology elements.

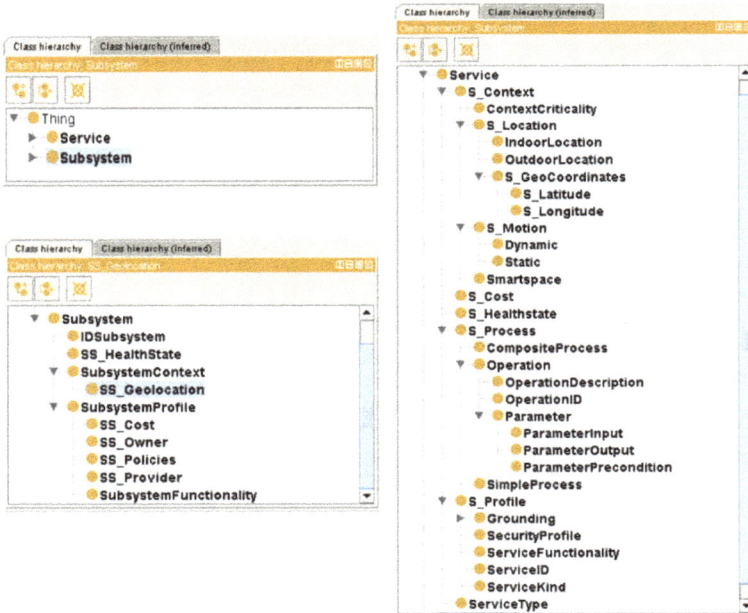

Figure 4. The visualized overall structure of nSSOO.

3.3.1. Subsystem-Related Ontology Part

The concept of *Subsystem* represents the collection of city-owned subsystems that makes measurements and provides data about specific domains (e.g., weather subsystem, smart home subsystem, intelligent transport system etc.).

Subsystem class can be unfolded into four main subclasses (see Figure 5, where the internal composition of *Subsystem* is presented with intuitive names for subclasses and relationships):

- *SubsystemContext*: the conditions in which the subsystem is provided. It is linked with *Subsystem* by an object property named *hasSSContext*.
- *SubsystemProfile*: descriptive information about the subsystem such as functionality, cost, provider, owner or usage policies. The *Subsystem* is interrelated with this class by using a *hasSSProfile* relationship. It is worth mentioning that the concepts of *SS_Geolocation* and *SS_Policies* are extracted from CityGML and NRL, respectively.
- *SS_HealthState*: information about the current health state of the subsystem. This class is connected with *Subsystem* via a *hasSSHealthState* relationship. Four different states (as potential individuals of *SS_HealthState*) are defined to describe the real status of a subsystem: *Installed* (it implies the subsystem is installed and ready to start once it receives an authorized command), *Active* (it states the subsystem is effectively running), *Suspended* (if subsystems are not required to run continuously, it is possible to make requests to pause them at any time during the active state) and *Stopped* (all the operations are stopped).
- *IDSubsystem*: a unique identification number to distinguish the subsystem. A pair of inversive (*owl:InversiveOf*) relationships (namely, *hasSSID* and *isSSIDOf*) dynamically links *Subsystem* with *IDSubsystem*.

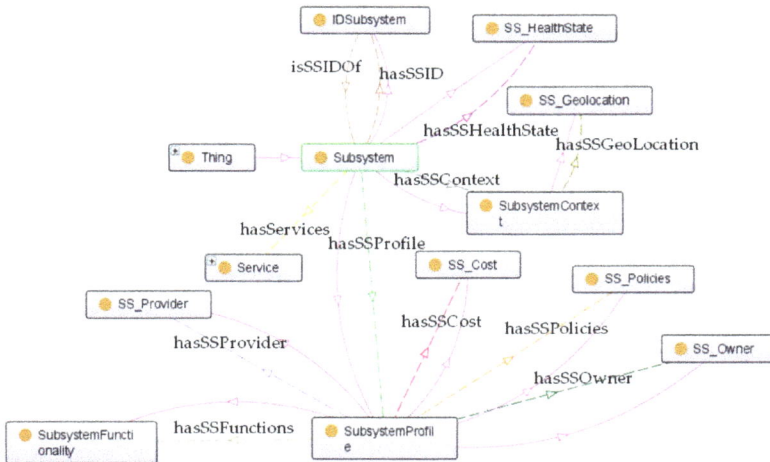

Figure 5. The internal structure of Subsystem.

3.3.2. Service-Related Ontology Part

The concept of *Service* denotes all kinds of services available within the urban system, either provided by subsystems or the ACCUS ICP platform. *Service* can be classified into six major categorizations which are *S_Cost*, *S_Context*, *S_Process*, *ServiceType*, *S_HealthState*, and *S_Profile*, as shown in Figure 6. Each subclass describes the feature of *Service* from a different point of view so that the definition of *Service* can be comprehensively represented in this model. In the following section, the breakdown of each classification will be presented.

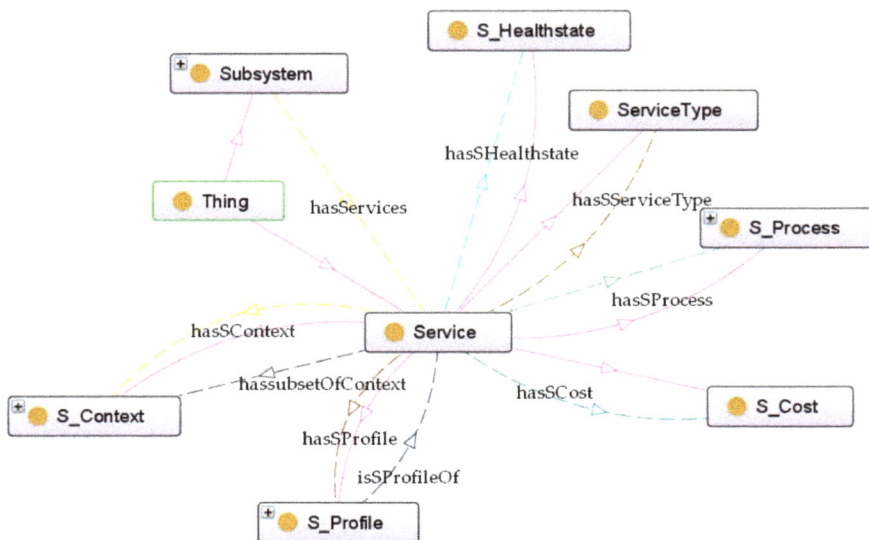

Figure 6. Top Level hierarchy of Service.

The elements that make the top level hierarchy of service are:

- *S_Cost.* It is interrelating with *Service* via a *hasSCost* relationship; this class indicates the fee to be charged to users for using the service.
- *S_Context. Service* is connected with this class by a *hasSContext* object property. It expresses the environmental conditions involved to provide the service. More details can be visualized in Figure 7. For instance, if the service is *Static*, its functionality is always provided in the same location. Otherwise, if it is *Dynamic*, such as in the case of services provided by wearable devices where the location can change, it also contains information about the *ContextCriticality* declaring whether the context is critical for the service operation or not. The *S_Location* of the service depicts whether it is provided at an indoor or an outdoor location (split into *IndoorLocation* and *OutdoorLocation* classes respectively). The *S_Geocoordinates* of the service and *Smartspace* are able to provide a unique identifier of the service context.
- *S_Process.* The element *Service* is connected with this class that provides a complete description for the logic of *Service*, via a *hasSProcess* relationship. More details can be visualized in Figure 8. The *S_Process* class is refined into atomic and aggregated/complex processes. An atomic process (*SimpleProcess*) directly takes the information generated by the environment and executes the appropriated treatment to provide the functionality. On the contrary, the aggregated process (*CompositeProcess*) provides the new functionality by composing several atomic processes. Besides, the term *Operation* makes additional descriptions for service operations. More specifically, it provides a description of the methods the service provides (*OperationDescription*), an ID for each operation (*OperationID*), and information about used parameters including input and output parameters (*ParameterInput* and *ParameterOutput*, respectively) as well as the parameter preconditions (*ParameterPrecondition*).
- *ServiceType.* This concept aims to specify the concrete type of service. This classification considers service from its source, either provided by subsystems or by the ACCUS ICP. The connection between *Service* and *ServiceType* is established by an object property named *hasSServiceType*.
- *S_HealthState.* Similar as *SS_HealthState*, the class of *S_HealthState*, linking with *Service* by a *hasSHealthsate* relationship, describes the current state of *Service*.

- *S_Profile. Service* is interrelated with the *S_Profile* class via *hasSProfile* and *isProfileOf* relationships. Different features of the service are described and attributed in *S_Profile*. As shown in Figure 9, *S_Profile* states the *ServiceID* (a unique identifier for distinguishing the service), the **ServiceKind** (a more detailed specification for the type of service which differentiates it from the ontology's point of view, being either ACCUS-compliant or non-compliant, having the service using another ontology or not), the *ServiceFunctionality* (description of what the service is capable of doing), the *SecurityProfile* (description of the security features under which the service will be provided; this concept can be further extended by NRL), and *Grounding* (particular protocols used between the service and service consumers). Regarding the *Grounding* concept, it contains a more specific description (*GroundingDescription*) of the protocol, the URI (**GroundingURI**) and the protocol (*GroundingProtocol*) of the endpoint where the application is running and also the input (*GroundingInputMessage*) and output (*GroundingOutputMessage*) messages exchanged between the service and service consumers (see Figure 9).

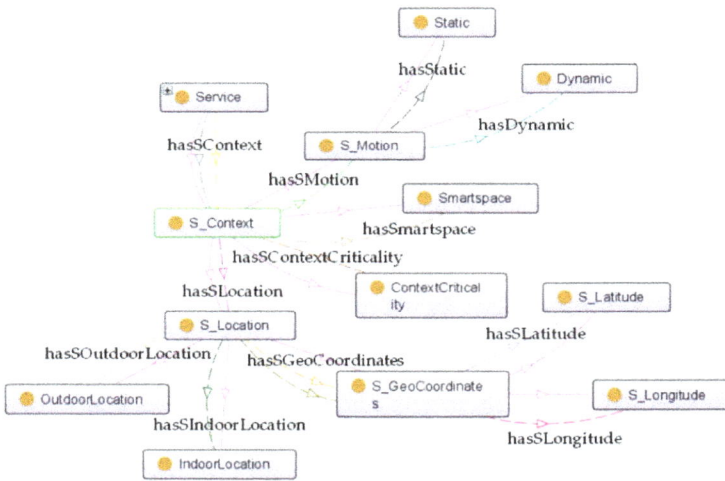

Figure 7. Internal structure of S_Context.

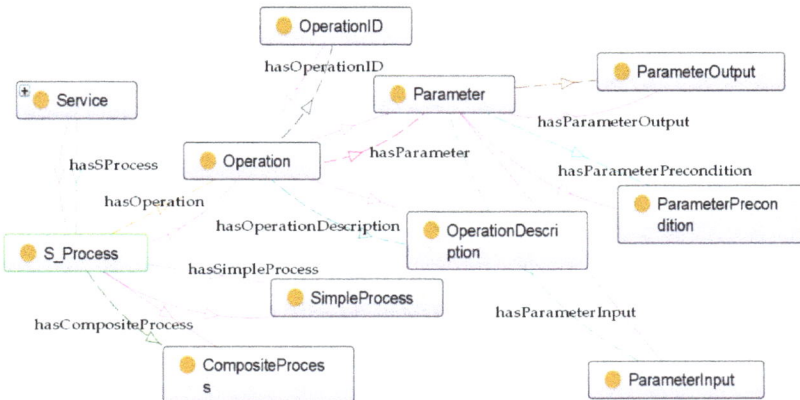

Figure 8. Internal structure of S_Process.

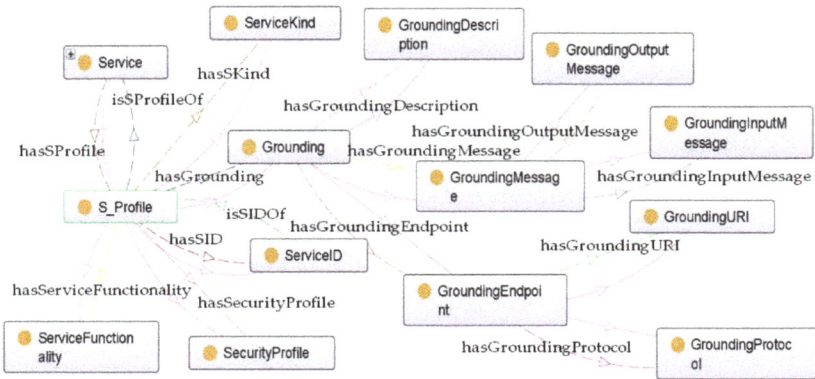

Figure 9. Internal structure of S_Profile.

3.4. Architecture of the Proposed System and Component Specifications

The ICP must provide a set of functionalities so that all subsystems belonging to a city will be properly operated. Also, it should support the development and deployment of (cross-domain) Smart City applications in any urban environment to enable the user to generate new services and applications which, in turn, will be integrated in the ICP. To achieve this, it is necessary to execute the next sequence of actions: (1) identification of ACCUS subsystems, services and applications; (2) identification and availability of external systems; (3) identification and availability of required infrastructure; (4) identification of information interchanged between ACCUS ICP and subsystems and services and (5) information provided by ACCUS ICP to cross-domain applications.

Taking into account the previously mentioned characteristics and the analysis of the information of different subsystems and services in the city, six technical features have been considered in an ICP: (1) Information and interaction, since ICP must provide services (annotated in an ICP ontology-compliant way) to enable the interaction among applications, applications and subsystems, and finally among subsystems if necessary. Two interfaces support that interaction: (a) Interface Applications—ICP, used to identify all requirements associated with the interaction among applications and ICP, considering that the information exchanged among them must be compatible with the ACCUS ontology with the idea of guaranteeing interoperability and an easy and seamless connection/disconnection of applications to the ICP. The features related with these requirements are communication with applications, an API provided to the applications and the relation with City State Database, CSDB, that stores all measurements and configuration of all WSANs and all sensors available in all subsystems deployed upon the Smart City; (b) Interface ICP-Subsystems, used to identify all requirements associated with the interaction among ACCUS ICP and subsystems; this interaction includes getting and sending information from/to the subsystem, as well as managing and controlling it, taking into account that not all subsystems are able to use the ACCUS ontology. Both interfaces must be used in the registration and discovery system, since all services and subsystems are registered and discovered by ICP while, on the other hand the cross domain application generated by users discovers the information of the services throught the semantic register and must be registered in this system, to let it be discovered by other cross domain applications; (2) Adaptive control, as the running and operating circumstances of services, subsystems applications and the own platform may vary over time, is also born in mind. The ICP has to adapt its operation to such changes; also, the control of all aspects related with the subsystems and services involved in the city is a key characteristic. Therefore, the control of the whole city depends on the ICP. What is more; (3) Security and Safety are a major concern, since the ICP is exposed to many security threats due to the security breaches likely to appear because of its dynamic and heterogeneous nature, as well as the fact that it is going to be usually operated by non-professional users in security issues; (4) Management to provide integrated management capabilities has also been conceived, as it allows both the platform as a whole and each of its software components to be managed. Applications, services and subsystems that make up the platform to suit the city are in permanent evolution too, so all the software updates and the connection/disconnection of functionalities should be done with a minimal impact on the normal operation of the platform when a new version is updated. In any case, it must be possible to return to a stable version, should significant problems appear in some updated components; (5) Development is eased by a virtual environment provided to simulate the real behaviour of the ICP. In this way, users can test the functionality of their new services or applications in a simulation environment before moving them to the operational phase; (6) Dependability [33] is employed to establish availability, reliability, safety, integrity and maintainability as its key aspects. In order to achieve this, the ICP provides self-monitoring, application state replication, plug-and-play and dynamic resource assignment.

As ICP should interconnect different city-owned subsystems where many services and applications are available but use different data formats, ICP has to create "a concept of system" which is able to combine all data formats to provide only a generic interface to applications. Therefore, the ICP must provide LCIM 3 (semantic interoperability), thus offering knowledge inference from heterogeneous data, storage of semantically enhanced data of legacy services and subsystems, integration of data and usage of data in cross-domain applications.

In order to provide semantic interoperability it is proposed that the ICP applies a global ontology, showed in Figure 10. In terms of semantic interoperability, the ontology defines the vocabulary to exchange queries and assertions among applications. Ontological commitments are agreements to use the shared vocabulary in a coherent and consistent manner [11]. Services and applications sharing a vocabulary do not need to share a knowledge base. It is not necessary to know all the characteristics of the remaining components.

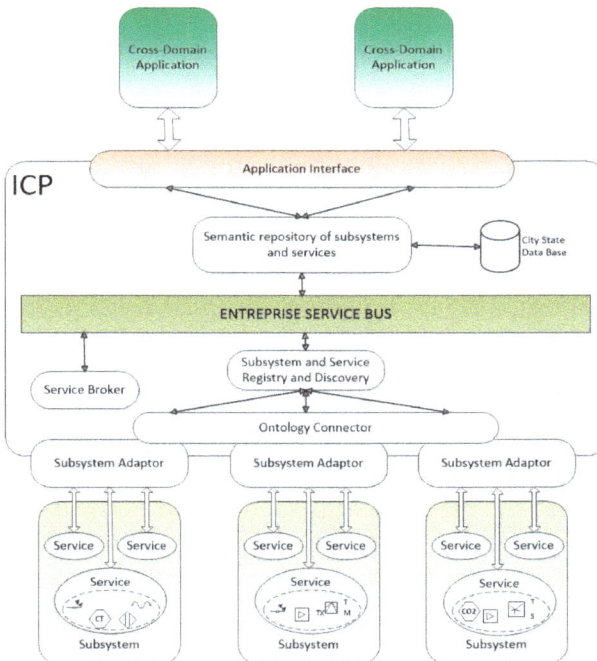

Figure 10. Proposed Integration and Coordination Platform regarding Semantic Interoperability.

If all the functions described above are to be performed, the first action to be taken is ensuring interoperability among all the Smart City systems and applications. It is strictly necessary, first of all, that the ICP itself has its own complete system of registration and discovery of subsystems and services, which it is proposed in this paper. Figure 10 shows the architecture of the whole system of ICP related to registry and discovery.

The first step to discover all subsystems connected to ICP and the services provided by each subsystem is taken by the component Subsystem and Service Registry and Discovery through Interface ICP—Subsystems. A key point is that this component works in real time. Later, each service or subsystem will provide their own description in a compliant way to nSSOO or not. Regardless of what is used, ICP provides semantic interoperability and seamless interconnection between applications and services embedded in subsystems. In order to perform semantic mapping for data interoperability enabling the transparent sharing of information among subsystems in the smart city, the mail requirement is the translation of the output of subsystems and services that are going to interchange information among them. The translations of this service are used for the registration and discovery of services and subsystems in the service repository. This fact can be achieved by parsing XML data to RDF. This way, semantic experts can define the XML to RDF mapping and non-semantic experts can work with XML files avoiding the effort of analysing ontologies for each subsystem. The service is provided by the components Subsystem adaptor and Ontology Connector. The first one facilitates the connection of subsystems and their services, and provides all necessary functions of adaptation and coordination. It is a very relevant component since it is in charge of the transformation of the information provided by services and subsystems to the ICP. The second one handles the translations of the data format which will be added to the Semantic Repository, when data are delivered in a known XML-like format [34]. These translations from XMLs to OWL or RDFs will be made using a mapping file which describes the transformation between the elements of source XML to an instance of the global ontology. A mapping file must be defined for each type of XML instance. When the description of the service or subsystem is available for the Subsystem and Service Registry and Discovery, it will establish a connection trough the Enterprise Service Bus with the component Semantic Subsystem and Service Repository to store the semantic description of services and subsystems registered in the ICP, according to the global ontology defined for the system. When a new subsystem or service is discovered by the Subsystem and Service Discovery component, its semantic description (profile) will be stored in this repository. Enterprise Service Bus provides interconnection and cooperation of the components based on a paradigm of message interchange; the ESB selected in ACCUS ICP is JBoss [35], which was preferred over two competitive alternatives: WSO2 and MULE [36]. Obviously, the component Service broker is necessary to orchestrate services and mediate between different software protocols if necessary.

Once all services and subsystems are registered, the Cross-domain Applications through the Interface Applications-ICP can execute SPARQL [37] queries and get their needed results in order to generate new applications or services, that, at the same time, will be registered.

This mode of operation allows the permanent update of all services, subsystems and applications of the Smart City.

3.5. Procedure Specifications: The Specific Workflow of Registering and Discovering Services and Subsystems

In this section, the specific process to register and discover subsystems and services is introduced.

3.5.1. Subsystems Registration

Once all subsystems and ACCUS ICP are up, subsystems must start with the registration procedure one by one. As shown in Figure 11, first, each subsystem sends a registration request to the ACCUS ICP which includes the subsystem profile in an XML document. The subsystem profile contains information to identify itself, such as functionality, geolocation, health state, provider or input and output parameters. This information is not compliant with the ACCUS ontology and therefore, it must be adapted to the ACCUS ontology for the registration subsystem and stored in the CSDB. This adaptation is done using the ontology connector service of the ACCUS ICP, which provides as output a RDF file with the information of the subsystem according to the nSSOO.

When the ACCUS ICP registers the information about the subsystem in the semantic subsystem and service repository, it automatically assigns and sends an identification number (ID) for the registered subsystem.

The information contained in *SRegistryRequest* will include information to identify the subsystem such as its geolocation, functionality, provider, owner, cost or security policies. This information can be sent in some machine-readable format such as XML or JSON.

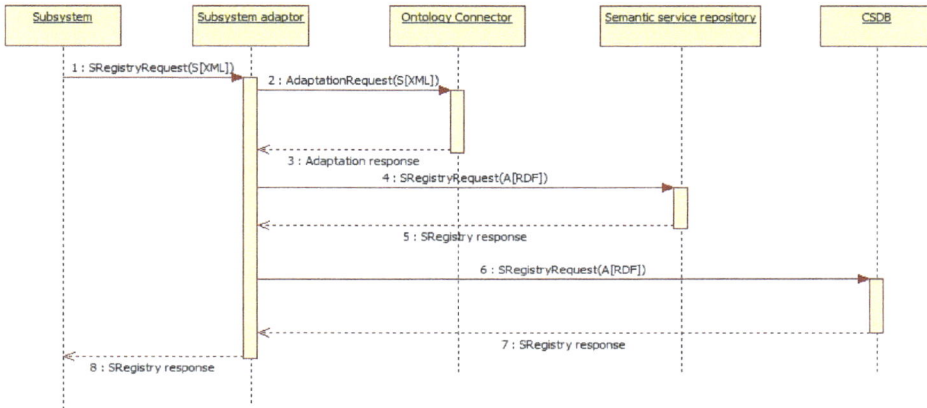

Figure 11. Subsystem registratiom sequence diagram.

3.5.2. Services Registration

When a subsystem becomes registered in the ACCUS ICP, all legacy applications and services provided by the subsystem must be registered. As shown in Figure 12, first, each legacy application or service sends a registry request to the ACCUS ICP which includes the service profile in an XML document. Service profile contains information about the service or legacy application, such as service type, functionality or information about the operations that the service can do, along with the input and output parameters involved in each operation. This information is not compliant with the ACCUS ontology, so it must be adapted to the ACCUS ontology for the service registration to have it stored in the semantic service repository as well as in the CSDB.

Once the ACCUS ICP registers the information about the service or legacy application in the semantic subsystem and service repository, it will automatically assign and send an identification number (ID) for the registered service or legacy application, as well as for each one of the operations it provides.

The information contained in *RegistryRequest* will include information to identify the service or legacy application about its profile, business logic and context. As in subsystems registry, this information can be sent in some machine-readable format such as XML or JSON.

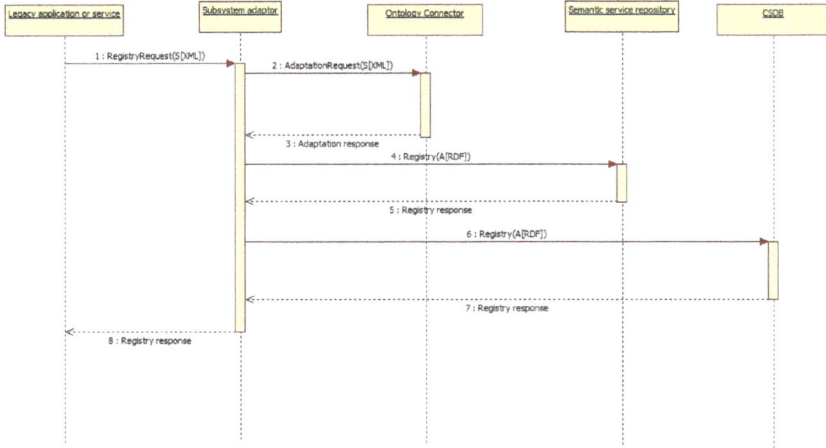

Figure 12. Service registration sequence diagram.

3.5.3. Subsystem and Service Discovery

When a subsystem, service or application wants to know about other legacy applications, services or subsystems registered in ACCUS ICP, as shown in Figure 13, it will send a SPARQL query request towards the ACCUS ICP and the ACCUS ICP will send the query request towards the Semantic Subsystem and Service Repository. Semantic Subsystem and Service Repository in the ACCUS ICP will respond with the set of results of the query in XML format. These results depend on the query executed towards the Semantic Subsystem and Service Repository. There could be different types of queries depending on the information that subsystems or applications want to obtain, e.g., *list all registered subsystems, list all registered services, list all services of a specific subsystem, retrieve all the information about a specific subsystem* or *service* or just *retrieve some specific data.*

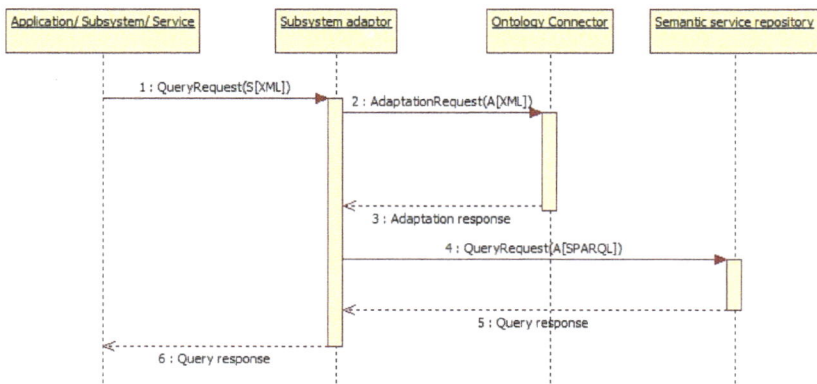

Figure 13. Discovery sequence diagram.

4. Example and Validation of Subsystem and Service Registration and Discovery

In this section, a complete example of a subsystem and a service registration and discovery is presented, based on the implementation done for the Semantic Subsystem and Service Repository of the ACCUS ICP. More specifically, using an example, the whole registration and discovery procedures will be shown, both for subsystems and services focusing on the input and output data formats, with the advantage of using specific data. Finally, validation done for evaluating the implementation will be also displayed.

4.1. Description of a Use Case Example

The proposed scenario involves a traffic control urban subsystem, which provides a traffic lights control service for managing the traffic lights cycles, along with information about the traffic density in several road intersections of a city. A smart mobility application uses the information provided by the traffic control subsystem in order to adapt traffic lights cycles to reduce and avoid traffic jams in the city.

Information provided by the traffic control subsystem is obtained by means of a Wireless Sensor Network, whose nodes will be deployed in road intersections of the city. The measurements taken by the sensors are sent via radio from the mesh network to the gateway, which comes with a special wireless node performing the base station role. This gateway is in charge of gathering data from the sensors and storing data until it is sent to the server and acts as an interface between ACCUS ICP and the Wireless Sensor Network. Then, the server, where all the data provided from the different sensors is collected and permanently stored, serves the data to the ACCUS ICP per request.

Both smart mobility application and traffic control subsystems are connected to the ACCUS ICP, which facilitates the communication between them with the services the platform provides. In order to enable this communication, traffic control subsystem must be registered in the ACCUS ICP first, so that it can be discovered and used by the smart mobility application.

Once the traffic control subsystem is connected, it sends a registration request, along with information about the subsystem, to the ACCUS ICP by means of its Subsystem Adaptor, using the corresponding method provided by the Application Interface. This method uses the Ontology Connector service to adapt the subsystem data sent in the request to an ACCUS ICP compliant format. Then, the Enterprise Service Bus locates and sends the request to the Semantic Subsystem and Service Repository, which registers the subsystem information received in a semantic repository, assigns an ID to the subsystem and returns it to the subsystem.

After subsystem registration, the traffic lights control service the subsystem provides is registered following the abovementioned procedure. When the service registration is completed, a smart mobility application can discover both the registered subsystem and service by querying Semantic Subsystem and Service Repository with SPARQL queries. In that case, information about the registered subsystem and service is returned to the smart mobility application in XML format.

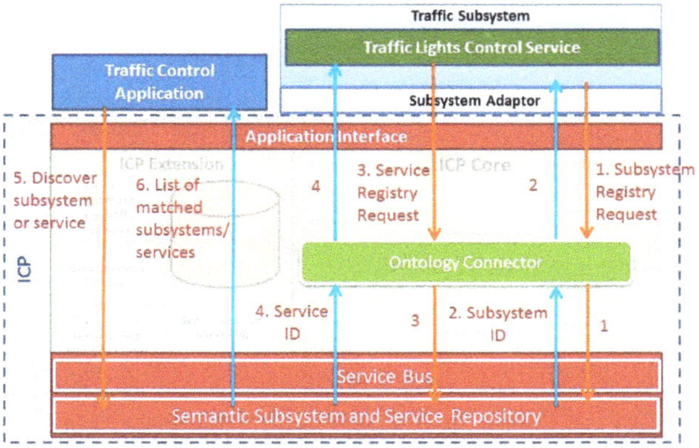

Figure 14. Example use case and registration and discovery processes.

Figure 14 shows the use case example, the whole registration and discovery processes as well as the interactions among the subsystem, the smart mobility application and the ACCUS ICP components involved in these processes.

4.2. Example of Subsystem and Service Registration and Discovery

Following the use case example presented in the previous section, and assuming that a) there is a traffic control subsystem that provides a traffic lights control service that b) has been just connected to the ACCUS ICP, two actions are carried out. As shown in Figure 15, first, the subsystem is discovered by the ICP and secondly, the subsystem sends a registration request to the ICP with information about it. Table 1 shows the information sent by the subsystem as well as the semantic annotations that the Semantic Subsystem and Service Repository will use to store that information.

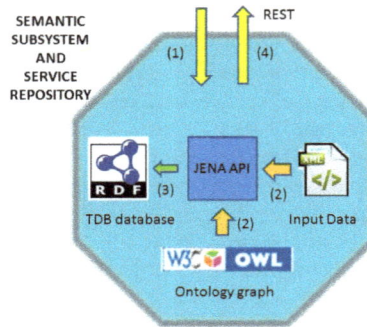

Figure 15. Semantic Subsystem and Service Repository Registration diagram.

Table 1. Subsystem data and semantic annotations.

Semantic Annotations	Subsystem Data
SS_Geolocation	Latitude: 54.3521 Longitude: 18.64637
SS_HealthState	Active
SubsystemFunctionality	Traffic Control Subsystem
SS_Provider	ACCUS
SS_Owner	Company1
SS_Cost	Free
SS_Policies	Policy1, Policy2

Again, this information can be sent in some machine readable format as, for example, in XML or JSON. ACCUS ICP receives this data via a REST interface, and Semantic Subsystem and Service Repository converts the input data into semantically annotated data formatted in RDF. To do so, Jena API methods are used in order to obtain the classes defined in the ontology graph and stored in an .owl file, as well as to instantiate them with the values of the input data. Finally, the data is stored in a triple store database provided by Jena, called Jena TDB, and a unique ID is assigned to the subsystem and returned to it through the REST interface. This interface is also used whenever a subsystem, service or application connected to the ACCUS ICP wants to discover another subsystem, service or application registered in the ICP.

Once the traffic control subsystem has been registered, the traffic lights control service it provides is registered and, for that purpose, it sends a registration request to the ACCUS ICP with the information about it. Table 2 shows the information sent by the service as well as the semantic annotations that the Semantic Subsystem and Service Repository will use to store that information.

Table 2. Subsystem data and semantic annotations.

Semantic Annotations	Service Data	Semantic Annotations	Service Data
ServiceType	Subsystem service	OperationDescription	changeState: provides the change and duration of the new state
S_HealthState	Active	ParameterPrecondition	initialState
S_Cost	Free	ParameterInput	dataTimeInterval
ServiceKind	ACCUS compliant	ParameterOutput	lightValue
ServiceFunctionality	Traffic lights control	Static	Static
SecurityProfile	securityProfile1	Dynamic	Non dynamic
GroundingDescription	change state	IndoorLocation	Non indoor
GroundingInputMessage	none	OutdoorLocation	Outdoor
GroundingOutputMessage	changes done	ContextCriticality	Critical
GroundingURI	ACCUS/trafficLightsControl	Smartspace	SS2
GroundingProtocol	REST	S_Latitude	54.3521
SimpleProcess	Simple	S_Longitude	18.64637

Then, when the ACCUS ICP receives the data, Semantic Subsystem and Service Repository registers the service, following the same procedure as for the subsystem, and finally an ID is assigned to the subsystem's service.

As soon as the subsystem and the service it provides have been registered, they can be discovered by any subsystem, service or application connected to the ACCUS ICP, thus obtaining information about them. To do so, they can query the Semantic Subsystem and Service Repository, using one of the methods that this service API provides. The results will be output in a document XML.

For example, if the smart mobility application wants to discover the traffic control subsystem previously registered, it can call listAllSubsystems method, which returns a list of all registered subsystems, or also it can call subsystemInfo method, which returns the information about the subsystem whose ID matches the one passed as an input parameter.

Therefore, calling, for example, the latter method using the traffic control subsystem ID, the following SPARQL query is executed:

```
PREFIX ns: <http://www.semanticweb.org/ACCUS/1.1#>
SELECT ?subsystemID ?subsystemFunctionality ?subsystemHealthState ?subsystemGeolocation
?subsystemProvider ?subsystemOwner ?subsystemCost ?subsystemPolicies
WHERE {ns:@id@ ns:isSSIDOf ?subsystem.
?subsystem ns:hasSSID ?subsystemID.
?subsystem ns:hasSSHealthState ?subsystemHealthState.
?subsystem ns:hasSSContext ?subsystemContext.
?subsystemContext ns:hasSSGeoLocation ?subsystemGeolocation.
?subsystem ns:hasSSProfile ?subsystemProfile.
?subsystemProfile ns:hasSSFunctions ?subsystemFunctionality.
?subsystemProfile ns:hasSSCost ?subsystemCost.
?subsystemProfile ns:hasSSPolicies ?subsystemPolicies.
?subsystemProfile ns:hasSSOwner ?subsystemOwner.
?subsystemProfile ns:hasSSProvider ?subsystemProvider.}
```

And the following output will be returned:

```
<?xml version="1.0"?>
<sparql xmlns="http://www.w3.org/2005/sparql-results#">
  <head>
  <variable name="subsystemID"/>
  <variable name="subsystemFunctionality"/>
  <variable name="subsystemHealthState"/>
  <variable name="subsystemGeolocation"/>
  <variable name="subsystemProvider"/>
  <variable name="subsystemOwner"/>
  <variable name="subsystemCost"/>
  <variable name="subsystemPolicies"/>
  </head>
  <results>
  <result>
  <binding name="subsystemID">
  <uri>http://www.semanticweb.org/ACCUS/1.1#3309</uri>
  </binding>
  <binding name="subsystemFunctionality">
  <uri>
  http://www.semanticweb.org/ACCUS/1.1#Traffic Control Subsystem
  </uri>
  </binding>
  <binding name="subsystemHealthState">
  <uri>http://www.semanticweb.org/ACCUS/1.1#Active</uri>
  </binding>
  <binding name="subsystemGeolocation">
  <uri>
  http://www.semanticweb.org/ACCUS/1.1#Latitude: 54.3521 Longitude:
  18.64637
  </uri>
  </binding>
  <binding name="subsystemProvider">
  <uri>http://www.semanticweb.org/ACCUS/1.1#ACCUS</uri>
  </binding>
  <binding name="subsystemOwner">
  <uri>http://www.semanticweb.org/ACCUS/1.1#Company1</uri>
  </binding>
  <binding name="subsystemCost">
  <uri>http://www.semanticweb.org/ACCUS/1.1#Free</uri>
  </binding>
  <binding name="subsystemPolicies">
  <uri>http://www.semanticweb.org/ACCUS/1.1#Policy1, Policy2</uri>
  </binding>
  </result>
  </results>
</sparql>
```

Similarly, if an application, subsystem or service wants to discover the traffic lights control service previously registered, it can call, the listAllServices method which returns a list of all registered services, or also the serviceInfo method, which returns information about the service whose ID matches the one passed as an input parameter. In the first case, the SPARQL query executed for listing all the registered services is the following:

```
PREFIX ns: <http://www.semanticweb.org/ACCUS/1.1#>
SELECT ?serviceID ?serviceFunctionality ?serviceType ?serviceHealthState ?serviceKind ?
serviceCost ?securityProfile
WHERE {?Resource ns:hasSServiceType ?serviceType.
?Resource ns:hasSCost ?serviceCost.
?Resource ns:hasSHealthstate ?serviceHealthState.
?Resource ns:hasSProfile ?serviceProfile.
?serviceProfile ns:hasSID ?serviceID.
?serviceProfile ns:hasServiceFunctionality ?serviceFunctionality.
?serviceProfile ns:hasSKind ?serviceKind.
?serviceProfile ns:hasSecurityProfile ?securityProfile.}
```

and the information returned will be the following:

```
<?xml version="1.0"?>
<sparql xmlns="http://www.w3.org/2005/sparql-results#">
<head>
<variable name="serviceID"/>
<variable name="serviceFunctionality"/>
<variable name="serviceType"/>
<variable name="serviceHealthState"/>
<variable name="serviceKind"/>
<variable name="serviceCost"/>
<variable name="securityProfile"/>
</head>
<results>
<result>
<binding name="serviceID">
<uri>http://www.semanticweb.org/ACCUS/1.1#4713</uri>
</binding>
<binding name="serviceFunctionality">
<uri>http://www.semanticweb.org/ACCUS/1.1#Subsystem and Service
                Repository: an ontology translator for ACCUS ICP data treatment,
                whenever the nSSOO ontology is required, that will be storing semantic
                information related with subsystems and services connected to the ICP.
</uri>
</binding>
<binding name="serviceType">
<uri>http://www.semanticweb.org/ACCUS/1.1#ACCUS ICP service</uri>
</binding>
<binding name="serviceHealthState">
<uri>http://www.semanticweb.org/ACCUS/1.1#Active</uri>
</binding>
<binding name="serviceKind">
<uri>http://www.semanticweb.org/ACCUS/1.1#ACCUS compliant</uri>
</binding>
<binding name="serviceCost">
<uri>http://www.semanticweb.org/ACCUS/1.1#Free</uri>
</binding>
```

```
<binding name="securityProfile">
<uri>http://www.semanticweb.org/ACCUS/1.1#ACCUS security profile</uri>
</binding>
</result>
<result>
<binding name="serviceID">
<uri>http://www.semanticweb.org/ACCUS/1.1#8647</uri>
</binding>
<binding name="serviceFunctionality">
<uri>http://www.semanticweb.org/ACCUS/1.1#Traffic lights control </uri>
</binding>
<binding name="serviceType">
<uri>http://www.semanticweb.org/ACCUS/1.1#Subsystem service</uri>
</binding>
<binding name="serviceHealthState">
<uri>http://www.semanticweb.org/ACCUS/1.1#Active</uri>
</binding>
<binding name="serviceKind">
<uri>http://www.semanticweb.org/ACCUS/1.1#Not ACCUS compliant</uri>
</binding>
<binding name="serviceCost">
<uri>http://www.semanticweb.org/ACCUS/1.1#Free</uri>
</binding>
<binding name="securityProfile">
<uri>http://www.semanticweb.org/ACCUS/1.1#securityProfile1</uri>
</binding>
</result>
</result>......</result>
</result>......</result>
</results>
</sparql>
```

Note that, when listing all services, information about Semantic Subsystem and Service Repository is also shown because internal ACCUS ICP active services are also registered in the semantic repository once the ICP is up.

4.3. Validation

In order to validate the practical performance of the Semantic Subsystem and Service Repository implementation, both the response time and the registration rate of the service have been tested. In order to test the response time, the timespan used for the different operations that the service provides to be executed has been measured. For each measurement, three samples have been taken in order to obtain an average value of the response time. On the other hand, the registration rate refers to the percentage of subsystems and services registered with regards to a certain number of registry requests done.

Tests have been done using an Intel(R) Core(TM)2 Quad CPU Q6600 processor @ 2.40 GHz equipped with 2.39 GHz and a RAM memory of 4 GB in a machine operating under the 64-bit Windows 7 Professional operating system.

4.3.1. Response Time

When a request is done to the Semantic Subsystem and Service Repository for the first time, this service must be initialized, so it takes more time than usual to execute the request. So, to begin with, the response time required when a request is done to the semantic repository for the first time has been measured. Results obtained are shown in Table 3.

Table 3. Response time when the first request is done to the semantic repository.

Operation	T1 (ms)	T2 (ms)	T3 (ms)	Average Time Elapsed (ms)
registerSubsystem	130	29	56	71.67
registerService	142	142	92	125.33
listAll	469	530	532	510.33
listAllSubsystems	450	447	463	453.33
listAllServices	476	490	460	475.33
getSubsystemInfo	464	498	476	479.33
getServiceInfo	815	578	564	652.33

Now, considering that the service has been initialized, the response time of the semantic repository operations has been measured in three different cases: when the repository has 100, 500 and 1000 of subsystems and services registered. The purpose is to test the normal operation of the semantic repository for different amounts of registered data. The results obtained are shown in Tables 4–6.

Table 4. Response time when there are 100 subsystems and services registered.

Operation	T1 (ms)	T2 (ms)	T3 (ms)	Average Time Elapsed (ms)
registerSubsystem	84	30	31	48.33
registerService	47	53	46	48.67
listAll	284	105	138	175.67
listAllSubsystems	67	100	74	80.33
listAllServices	20	17	18	18.33
getSubsystemInfo	8	5	5	6
getServiceInfo	99	42	27	56

Table 5. Response time when there are 500 subsystems and services registered.

Operation	T1 (ms)	T2 (ms)	T3 (ms)	Average Time Elapsed (ms)
registerSubsystem	67	66	28	53.67
registerService	76	75	41	64
listAll	603	246	158	335.67
listAllSubsystems	84	70	76	76.67
listAllServices	73	63	62	66
getSubsystemInfo	6	3	5	4.67
getServiceInfo	183	44	26	84.33

Table 6. Response time when there are 1000 subsystems and services registered.

Operation	T1 (ms)	T2 (ms)	T3 (ms)	Average Time Elapsed (ms)
registerSubsystem	78	27	35	46.67
registerService	145	42	42	76.33
listAll	398	368	771	512.33
listAllSubsystems	195	500	367	354
listAllServices	138	107	111	118.67
getSubsystemInfo	6	4	5	5
getServiceInfo	108	18	16	47.33

Analyzing the results obtained, several conclusions can be drawn. Besides the fact that the response time is higher when the first request is done due to the initialization of the semantic repository, it can also be appreciated that the response time increases with the amount of subsystems and services stored.

Comparing the time response between the different operations, it can be seen that the time response of *listAll* operation is considerably higher than in the other operations due to the higher amount of data that must be retrieved, which are all the subsystems and services stored in the semantic repository, while, for example, in *listAllSubsystems* and *listAllServices* only operation subsystems in the first case, and services in the second case are shown. Regarding the last two operations mentioned, the time response of *listAllSubsystems* is higher than the time response of *listAllServices* because in the first operation more information is shown than in the second operation. For the same reason, time response in *registerService* and *getServiceInfo* operations is higher than in *registerSubsystem* and *getSubsystemInfo*. Finally, it can be appreciated that the time response of *registerSubsystem*, *registerService*, *getSubsystemInfo*, *getServiceInfo* does not change significantly with the amount of data registered, because in these operations just one subsystem or service is registered or queried so they are not affected by the amount of data registered.

4.3.2. Registration Rate

For testing the registration rate, 5000 subsystem and service registry requests were done towards the Semantic Subsystem and Service Repository, and all the requests were successfully executed, registering the 100% of subsystems and services that requested registration.

5. Conclusions and Future Work

This paper has presented an ontology-based and automatic system for subsystem and service registry and discovery within the context of the ACCUS project. This proposed system, embedded in ACCUS ICP, is able to dynamically register and discover heterogeneous subsystems and services provided by subsystems within a Smart City so that cross-domain applications and collective optimization can be built upon the ICP by using existing services. To address the heterogeneity (e.g., data formats and protocols) of subsystems and services, a new ontology, named nSSOO, has been proposed and employed by the system to provide a formalized vocabulary for the registration and discovery processes. The nSSOO has been developed on the basis of three existing ontologies, including OWL-S, CityGML, and NRL. By complying with this ontology, heterogeneous subsystems and services provided by individual subsystems can share a same understanding which results in a formal and homogeneous appearance of the ICP. The proposed ontology, from a global point of view, is an important contribution to achieve semantic interoperability in the ICP and it has been presented with detailed explanations for inner composition. In addition to that, different software components which form the whole system have been shown with their main functionalities explained. The proposed ontology-based scheme for subsystem and service discovery and registry has been elaborated with a set of sequence diagrams that present the specific workflow of inner components involved in this scheme. Furthermore, after presenting the specific procedures to ease the subsystem and service discovery and registry, a complete example about discovering and registering a traffic control system and a traffic lights control service has been provided to show the performance of the proposed scheme. Different kinds of queries for information stored in ICP have also been introduced.

The system proposed to register and discover subsystems and services has been proven to be useful to interconnect different subsystems and services. What is more, it could abstract the heterogeneity of different subsystems and services so as to provide a homogeneous interface for applications or other services inside the ICP. Though this scheme aims to create an accurate reference framework for available information (e.g., subsystems, services, and applications) within the Smart Cities, it could be also possible to adapt it to other domains, such as underwater robotics. Developers can become aware of the services that are working in the Smart city, which are their features and how

Sensors **2016**, *16*, 955

can be accessed, to design their applications by making use of those services provided by subsystems. Also, developers are isolated of the problem of transforming data protocols.

Future work could be focused on the following aspects:

- The proposed subsystem and service discovery and registry scheme should be tested in more scenarios in a real city. In ACCUS project, the city that has been chosen to deploy the pilot is Gdansk, in Poland.

- The relationships among different services are a crucial factor for application developers when they design brand new applications. Future work should focus on examining the similarity degree of different services. For instance, services able to provide similar functionalities can be alternatives if the ideal service to be used is not available. A potential solution could be including information about relationships of different services in relevant service profiles.

- The nSSOO ontology should evolve to richly describe more features of subsystems and services. For example, it is possible to extend nSSOO with some new classes using FOAF [38] to include additional aspects of information about people, such as roles like provider or owner. Another potential extension of the ontology could be including concepts about event-driven services.

- Future emphasis can also be put on including decision-making related algorithms in the nSSOO ontology. e.g., MADISE [39] ontology can be reused and integrated with the nSSOO ontology.

Acknowledgments: This paper is a result of research made by the authors in the Adaptive and Cooperative Control in Urban Subsystems (ACCUS) project, ARTEMIS 2012-1. SP1-JT1-ARTEMIS-2012-ASP7-Embedded System supporting sustainable urban life, SP1-JT1-ARTEMIS-2012-ASP3: Embedded systems in Smart environments. The authors would like to thank CITSEM (Research Center on Software Technologies and Multimedia Systems for Sustainability, Centro de Investigación en Tecnologías Software y Sistemas Multimedia para la Sostenibilidad) from the Technical University of Madrid (UPM).

Author Contributions: The research result presented in this article is a collaborative work of all authors. Rubio has prepared the architecture of the ICP. Rubio, Gómez and Li have defined the ontology. Li has implemented the ontology. Rubio and Gómez have defined the sequence of registering and discovering subsystems and service. Gómez has prepared and implemented the SPARQL queries. Martínez has given technical and conceptual support for the entire article. All authors have participated in writing this article. All of the authors have reviewed and finally approved the manuscript.

Conflicts of Interest: The authors declare no conflict of interest.

References

1. Nellore, K.; Hancke, G. A Survey on Urban Traffic Management System Using Wireless Sensor Networks. *Sensors* **2016**, *16*, 157. [CrossRef] [PubMed]
2. Moreno, A.; Perallos, A.; López-de-Ipiña, D.; Onieva, E.; Salaberria, I.; Masegosa, A. A Novel Software Architecture for the Provision of Context-Aware Semantic Transport Information. *Sensors* **2015**, *15*, 12299–12322. [CrossRef] [PubMed]
3. Ghayvat, H.; Mukhopadhyay, S.; Gui, X.; Suryadevara, N. WSN- and IOT-Based Smart Homes and Their Extension to Smart Buildings. *Sensors* **2015**, *15*, 10350–10379. [CrossRef] [PubMed]
4. Rodríguez-Molina, J.; Martínez, J.-F.; Castillejo, P.; de Diego, R. SMArc: A Proposal for a Smart, Semantic Middleware Architecture Focused on Smart City Energy Management. *Int. J. Distrib. Sens. Netw.* **2013**, *2013*, 1–17. [CrossRef]
5. Zhou, L.; Rodrigues, J.J.P.C. Service-oriented middleware for smart grid: Principle, infrastructure, and application. *IEEE Commun. Mag.* **2013**, *51*, 84–89. [CrossRef]
6. Hernandez, L.; Baladrón, C.; Aguiar, J.M.; Calavia, L.; Carro, B.; Sanchez-Esguevillas, A.; Cook, D.J.; Chinarro, D.; Gómez, J. A Study of the Relationship between Weather Variables and Electric Power Demand inside a Smart Grid/Smart World Framework. *Sensors* **2012**, *12*, 11571–11591. [CrossRef]
7. Sung, W.-T.; Lin, J.-S. Design and Implementation of a Smart LED Lighting System Using a Self Adaptive Weighted Data Fusion Algorithm. *Sensors* **2013**, *13*, 16915–16939. [CrossRef]
8. Warriach, E.U.; Kaldeli, E.; Lazovik, A.; Aiello, M. An Interplatform Service-Oriented Middleware for the Smart Home. *Int. J. Smart Home* **2013**, *7*, 115–142.

9. RECI. Red Española de Ciudades Inteligentes. Available online: http://www.redciudadesinteligentes.es (accessed on 28 March 2016).
10. The ACCUS (Adaptive Cooperative Control in Urban (sub)Systems) Project. Available online: http://projectaccus.eu/ (accessed on 31 March 2016).
11. Zang, M.A. Ontological Approaches for Semantic Interoperability. In Proceedings of the 5th Annual ONR Workshop on Collaborative Decision-Support Systems, San Luis Obispo, CA, USA, 10–11 September 2003.
12. Services Oriented Architecture for All. Available online: http://www.soa4all.ue (accessed on 10 April 2016).
13. Tagging Tool Based on a Semantic Discovery Framework. Available online: www.tatoo-fp7.eu (accessed on 11 March 2016).
14. Cloud4SOA. Available online: http://www.cloud4soa.com (accessed on 10 March 2016).
15. Butler, B. *PaaS Primer: What Is Platform as Services and Why Does It Matter?*; Network World: Framingham, MA, USA, 2013.
16. Semantichealthnet. Available online: http://www.semantichealthnet.eu (accessed on 27 March 2016).
17. Mobilized Lifestyle with Wearables (LifeWear). Available online: https://itea3.org/project/lifeware.html (accesed 21 June 2016).
18. Ryu, M.; Kim, J.; Yun, J. Integrated semantics services platform for the internet of things: A case study of a smart office. *Sensors* **2015**, *15*, 2137–2160. [CrossRef] [PubMed]
19. Hussain, A.; Wenbi, R.; da Silva, A.L.; Nadher, M.; Mudhish, M. Health and emergency-care platform for the elderly and disabled people in the Smart City. *J. Syst. Softw.* **2015**, *110*, 253–263. [CrossRef]
20. Benhaourech, A.; Aaroud, A.; Roose, P.; Zine-Dine, K. Study and comparison of smart city dedicated platforms: Case of the Kalimucho platform. In Proceedings of the 2014 5th Workshop on Codes, Cryptography and Communication Systems (WCCCS), El Jadida, Morocco, 27–28 November 2014; pp. 161–166.
21. Rodriguez-Molina, J.; Martinez, J.F.; Castillejo, P.; López, L. Combining Wireless Sensors networks ans semantic middleware for an internet of things-based sportman/woman monitoring application. *Sensors* **2013**, *13*, 1787–1835. [CrossRef] [PubMed]
22. Bispo, K.A.; Rosa, N.S.; Cunha, P.R. SITRUS: Semantic Infraestructure for Wireless Sensor Networks. *Sensors* **2015**, *15*, 27436–27469. [CrossRef] [PubMed]
23. Camarinha-Matos, L.M.; Afsarmanesh, H.; Galeano, N.; Molina, A. Collaborative networked organizations—Concepts and practice in manufacturing entreprises. *Comput. Ind. Eng.* **2009**, *57*, 46–60. [CrossRef]
24. Li, C.-S.; Liao, W. Software Defined Networks. *IEEE Commun. Mag.* **2013**, *51*, 113–114. [CrossRef]
25. Kosmides, P.; Adamopoulo, E.; Demestichas, K.; Theologou, M.; Anagnostou, M.; Rouskas, A. Socially Aware Heterogeneous Wirelees Networks. *Sensors* **2015**, *15*, 13705–13724. [CrossRef] [PubMed]
26. WildFly. Available online: http://www.wildfly.org (accessed on 25 May 2016).
27. WSO2: Open platform for your connected Bussiness. Available online: http://wso2.com (accessed on 25 May 2016).
28. Abel, D.J.; Ooi, B.C.; Tan, K.L.; Tan, S.H. Towards integrated geographical information processing. *J. Geogr. Inf. Sci.* **1998**, *12*, 353–371. [CrossRef]
29. Turnitsa, C.D. Extending the levels of conceptual interoperability model. In Proceedings of the IEEE Summer Computer Simulation Conference, Philadelphia, PA, USA, 24–28 July 2005.
30. Martin, D.; Burstein, M.; Hobbs, J.; Lassila, O.; McDermott, D.; McIlraith, S.; Narayanan, S.; Paolucci, M.; Parsia, B.; Payne, T.; et al. OWL-S: Semantic Markup for Web Services. W3C Member Submission 22 November 2004. Available online: http://www.w3.org/Submission/OWL-S (acceesed on 15 April 2015).
31. CityGML. Available online: http://www.opengeospatial.org/standards/citygml (accessed on 25 May 2016).
32. Kim, A.; Luo, J.; Kang, M. Security Ontology for Annotating Resources. In Proceedings of the 5th International Conference on Ontology, Databases and Applications on Semantics (ODBASE; 05), Agia Napa, Cyprus, 31 October–4 November 2005; pp. 1483–1499.
33. Avizienis, A.; Laprie, J.; Randell, B.; Landwehr, C. Basic concepts and taxonomy of dependable and secure computing. *IEEE Trans. Dependable Secur. Comput.* **2004**, *1*, 11–33. [CrossRef]
34. Stoimenov, L.; Stanimirovic, A.; Dordevic-Kajan, S. Semantic interoperability using multiples ontologies. In Proceedings of the AGILE 2005 8th AGILE Conference on GIScience, Estoril, Portugal, 26–28 May 2005; pp. 261–270.

35. Jbossdeveloper. Available online: http://www.jboss.org/products/fuse/overview (accessed on 25 May 2016).

36. Mulesoft. Available online: http://www.mulesoft.com/platform/soa/mule-esb-open-source-esb (accessed on 25 May 2016).

37. Prud'hommeaux, E.; Seaborne, A. SPARQL Query Lenguaje for RDF, W3C Recommendation 15 January 2008. Available online: http://www.w3.org/TR/rdf-sparql-query (acceessed on 25 March 2016).

38. Brickley, D.; Miller, L. FOAF Vocabulary Specification 0.99. Namespace Document 14 January 2014—Paddington Edition. Available online: http://xmlns.com/foaf/spec (accessed on 25 March 2016).

39. Kornyshova, E.; Denecker, R. Decision making method family madise: Validations within the Requirements Engineering Domain. In Proceedings of the 6th International Conference on Research Challenges in Information Science RCIS, Valencia, Spain, 1–10 May 2012; pp. 1–10.

Article

On the Feasibility of Wireless Multimedia Sensor Networks over IEEE 802.15.5 Mesh Topologies

Antonio-Javier Garcia-Sanchez [1,*], Fernando Losilla [1], David Rodenas-Herraiz [2], Felipe Cruz-Martinez [1] and Felipe Garcia-Sanchez [1]

[1] Department of Information and Communication Technologies, Universidad Politécnica de Cartagena (UPCT), Campus Muralla del Mar, E-30202 Cartagena, Spain; fernando.losilla@upct.es (F.L.); felipecruz91@hotmail.es (F.C.-M.); felipe.garcia@upct.es (F.G.-S.)

[2] Computer Laboratory, University of Cambridge, William Gates Building, 15 JJ Thompson Avenue, Cambridge CB3 0FD, UK; dr424@cl.cam.ac.uk

* Correspondence: antoniojavier.garcia@upct.es; Tel.: +34-968-326-538; Fax: +34-968-325-973

Academic Editor: Gonzalo Pajares Martinsanz
Received: 1 March 2016; Accepted: 27 April 2016; Published: 5 May 2016

Abstract: Wireless Multimedia Sensor Networks (WMSNs) are a special type of Wireless Sensor Network (WSN) where large amounts of multimedia data are transmitted over networks composed of low power devices. Hierarchical routing protocols typically used in WSNs for multi-path communication tend to overload nodes located within radio communication range of the data collection unit or data sink. The battery life of these nodes is therefore reduced considerably, requiring frequent battery replacement work to extend the operational life of the WSN system. In a wireless sensor network with mesh topology, any node may act as a forwarder node, thereby enabling multiple routing paths toward any other node or collection unit. In addition, mesh topologies have proven advantages, such as data transmission reliability, network robustness against node failures, and potential reduction in energy consumption. This work studies the feasibility of implementing WMSNs in mesh topologies and their limitations by means of exhaustive computer simulation experiments. To this end, a module developed for the Synchronous Energy Saving (SES) mode of the IEEE 802.15.5 mesh standard has been integrated with multimedia tools to thoroughly test video sequences encoded using H.264 in mesh networks.

Keywords: WMSN; IEEE 802.15.5; mesh topology; multimedia; WSN

---

1. Introduction

Wireless Sensor Networks (WSNs) [1] consist of a large number of low-cost and energy-constrained devices. These devices transmit data acquired by a number of in-built sensors to one or several data collection units, named data sinks, which are in charge of sending data to remote database servers for further data processing and storage. Wireless Multimedia Sensor Networks (WMSNs) [2] are regarded as a particular type of WSN where nodes are interfaced with sensors for visual and sound data acquisition, such as Complementary Metal Oxide Semiconductor (CMOS) cameras and microphones. This technology has drawn the attention of the WSN research community, encouraged by the challenge of providing support applications such as video surveillance, environmental monitoring, target tracking and traffic management, where multimedia services are required.

Compared with low-data-sampling, low-data-transmission rate WSN applications, multimedia applications impose stringent requirements in terms of high sampling rate, data transmission reliability, data latency and data throughput. Reliable transmission of multimedia data in WMSNs is challenging due to the inherent technology limitations, such as interference between devices or limited data transmission bandwidth. Multimedia communications also require a high utilization of the transmission channel, increasing communication activity within the network and considerably decreasing the operational lifetime of battery-powered nodes. This clearly contrasts with the traditional WSN/WMSN requirements imposed by, on the one hand, the use of short-range communications over low-bandwidth links (maximum bitrate of 250 kbps at 2.4 GHz) and, on the other hand, the energy constraints of sensor devices.

Traditional routing in WSNs is also a problem in WMSNs. Many state-of-the-art multi-hop proposals for WSNs consider tree-based topological configurations where only a few number of nodes are likely to forward all the data traffic. Because of the intensive use of these nodes, the network may not fulfil the Quality of Service (QoS) required by the multimedia application and, more importantly, it will lead to the overuse and depletion of the batteries of nodes in the path. In addition, traditional routing mechanisms usually only deal with the transmission to a sink node and not to any arbitrary node of the network, which hampers the development of collaborative applications or the use of multiple sinks.

Mesh topologies, in addition to enabling multiple routing paths between two nodes outside of radio communication range with each other, have proven advantages against other topological configurations, such as tree-based or cluster-based topologies, which include data transmission reliability, network robustness against node failures, and potential reduction in energy consumption [3]. Nodes arranged in a mesh topology are generally easier to deploy and potentially easier to maintain as they are able to reconfigure themselves in order to adapt to the conditions brought by the deployment environment. Among the different mesh solutions available for low-power wireless networks, the low-rate part of the IEEE 802.15.5 (hereafter IEEE 802.15.5) [4] offers suitable functionalities for enabling multimedia applications compared with other state-of-the-art network/MAC layer standards. This standard assures that any node is able to communicate with any other node of the network through different routes, enhancing data transmission reliability and robustness against node failures. It also emphasizes simplicity, adding, for low-rate WPAN, a thin mesh sublayer on top of the IEEE 802.15.4 MAC sublayer, the most widely adopted point-to-point communication standard for low-rate wireless personal area networks (LR-WPANs). This results in an easy migration of applications using the 802.15.4 standard and in a greater scalability, enabling large-scale deployments. In addition, the standard provides two radio duty-cycle mechanisms, named Asynchronous Energy Saving (ASES) and Synchronous Energy Saving (SES). These mechanisms are intended to make efficient use of the available battery energy for, respectively, asynchronous and synchronous communications, thereby prolonging the battery life. For all the aforementioned reasons, IEEE 802.15.5 can be, a priori, a suitable and representative WMSN solution for the transmission of video in mesh topologies.

However, according to our knowledge, there are very few studies in the scientific literature combining mesh topologies with multimedia applications for low power devices. Under these circumstances, the purpose of this paper is to contribute to the development of multimedia applications in the WMSN arena, fulfilling the WSN/WMSN specific requirements, assuring the QoS of the multimedia service and achieving a reasonable performance of the network in terms of, for instance, lifetime and data transmission reliability. To achieve these goals, we have designed a novel tool, available for download, that evaluates, by means of computer simulation, the performance of video transmissions using the SES energy-saving mechanism of IEEE 802.15.5. For the evaluation of the standard, Quarter Common Intermediate Format (QCIF) video sequences are encoded using the H.264/AVC compression standard and transmitted through WMSN mesh networks of different sizes. The main results will be discussed, highlighting the most important metrics such as latency, jitter, time between packets, message delivery ratio, power consumption, lifetime and Peak Signal-to-Noise Ratio (PSNR). The rest of this paper is organized as follows: Section 2 introduces related work in the field of WMSNs. Section 3 provides the necessary technological background about video encoding with H.264 as well as WMSN operation based on the IEEE 802.15.5 standard. Section 4 describes the simulation tool and simulation environment. Section 5 shows and analyzes simulation results, which are further discussed in Section 6. Section 7 concludes this work.

2. Related Work

In spite of the proven advantages arising from the use of the IEEE 802.15.5 standard, to the knowledge of the authors, there is no other work that makes use of this technology in WMSNs. There are however a number of proposals that use mesh communication for transmitting multimedia flows [5]. However, they do not consider Wireless Sensor Networks but rather focus on less energy-constrained standards such as 802.11. In this regard, the purpose of this work is twofold: first, to carry out a first evaluation of the performance of a mesh standard, the IEEE 802.15.5, for video transmission; second, to provide simulation tools to help evaluate further improvements for multimedia transmission made to the original standard.

The performance of WSNs and WMSNs should be assessed using real WSN/WMSN hardware before actual deployment, for example, using a test bed in the laboratory. However, it is best practice to make sure that no software/hardware errors are introduced before test-bed experiments are carried out. Finding and correcting these errors is usually a tedious and time-consuming task, and simulation is a powerful tool that helps do this in a much faster and easier way than dealing directly with hardware. In this regard, Pham [6] studied the communication performance of the most popular WSN commercial platforms to date in data-intensive applications. Results showed significant delays limiting the maximum data throughput at both the sender and the receiver. The best results were obtained using MicaZ devices [7]. Pham also showed that more effort should be made on the development of new hardware platforms, but also that more realistic simulation models are needed to consider the constraints of these commercial devices. Farooq *et al.* [8] studied the feasibility of performing the evaluation of new multimedia proposals in large WSN testbeds available for research usage. They concluded that, despite the existence of powerful hardware platforms for multimedia applications, there has been little work yet to integrate them into testbeds. Consequently, simulations are the most cost-effective solution for evaluating WMSN proposals.

Some researches focused on comparing the performance of data-intensive WMSN applications using different MAC and network protocols for diverse layouts. Ammar *et al.* [9] compared the performance obtained in WMSN by using two well-known WSN MAC protocols, S-MAC and 802.15.4, and two routing protocols, Ad hoc On-Demand Distance Vector (AODV) routing and its extension Ad hoc On-demand Multipath Distance Vector (AOMDV) routing. They remarked on the importance of optimizing the network layer protocols. In addition, their simulation results showed a better performance of the IEEE 802.15.4 standard in terms of network lifetime, which is supported by the large number of WMSN proposals based on this standard. Andrea *et al.* [10] studied this standard in one-hop data-intensive networks with star topology. They simulated a video surveillance system by continuously transmitting images to a sink node. Among other outcomes, their study determined that the Packet Error Rate for this kind of traffic, where packets have the maximum size allowed by the IEEE 802.15.4 standard, presents a significant increase as the network grows in comparison with typical WSNs with scalar sensors and short packets. In the same line, Pekhteryev *et al.* [11] performed a similar research in ZigBee networks that showed the difficulties in transmitting images over multi-hop networks. Although they only compared one-hop and two-hop schemes, they observed an important decrease in the percentage of recoverable images at the receiver for the two-hop scenario (from 2.6% to 20.5% using JPEG images, while for JPEG 2000 images the tests for the two-hop scenario could not completed due to interferences from other nodes).

There are other studies that focus on optimizing wireless performance communication in WMSNs in different ways. One of them considers cross-layer approaches to improve QoS. García-Sánchez *et al.* [12] used application-level QoS parameters to tune the MAC and physical layers of the popular IEEE 802.15.4 standard. An important increase in the maximum throughput for one-hop multimedia transmissions was achieved by means of a series of optimizations which maximized the amount of data transmitted and avoided collisions due to imperfect synchronization and other issues. In addition, the proposed solution remained compliant with the original standard. Farooq *et al.* [13] proposed another cross-layer architecture for multi-hop clustered networks where, upon congestion, nodes are requested to reduce the data transmission rate to avoid congestion. The authors used a differentiated services approach that classifies data traffic according to six different classes and where nodes producing less data and low priority data are penalized more than those producing high volumes of data and high priority data. Alanazi *et al.* [14] evaluated other QoS routing protocols for real-time WMSN applications, among which the Pheromone Termite (PT) protocol [15] offered better performance. The protocol, though, relies on both ad-hoc MAC and, more importantly, physical layers, which makes its implementation on real devices very difficult.

In video transmission, the importance of each video frame can be used to select different QoS profiles. Kandris *et al.* [16] developed a video distortion model that made it possible to predict the effect on video quality of dropping each packet. In case of congestion, typically found when transmitting multimedia flows, this model is used to selectively drop less significant packets prior to transmission, therefore improving performance at the expense of additional computing load. The authors also used a hierarchical routing protocol with asymmetrical communications, where the sink is able to transmit directly to all nodes, with no intermediary hops. Zaidi *et al.* [17] developed a multipath protocol that obtained three different paths toward a sink node and, based on Bit Error Rate and delay, selected the best of them for the transmission of the most important video frames (the I frames, explained in Section 3).

In general, multipath communication has traditionally been used to increase robustness or reliability in WMSNs. For example, the MMSPEED protocol [18] uses probabilistic multipath forwarding to control the number of paths based on the required delivery probability. In this way, depending on the packet loss rate and QoS requirements, it can send multiple copies of a same packet to ensure the delivery of data. In WMSNs, multipath can also be used to increase the throughput in data-intensive flows. Teo *et al.* [19] proposed a routing protocol, I2MR, that transmits simultaneously through two paths. In order to avoid interference between them, the protocol uses the shortest path between source and destination as the primary path and marks one- and two-hop neighbors of nodes in the primary path as the interference zone. A secondary path as well as a backup path are then constructed using nodes outside the interference zone. This protocol is intended for less energy-constrained 802.11 networks, but there are similar proposals using more energy-constrained standards. Maimour [20] developed the MR2 protocol. This protocol, which is better suited for dense WSN deployments, is also interference-aware, and constructs paths incrementally. After obtaining a new path, all of the nodes neighboring to the path, which potentially may cause interference, are notified and put in a passive mode and cannot be used for new paths. Therefore, avoiding interference and saving energy. Li *et al.* [21] proposed GEAM, a Geographic Energy-Aware Multipath routing scheme that divides the network into what the authors named "districts". The distance between them is sufficiently large to avoid transmission interference. To send a packet, GEAM assigns the packet to a district and forwards it to the sink using the greedy algorithm. Consequently, it is necessary that each node is aware of its location. Bidai *et al.* [22] developed another multipath routing scheme for video transmission that combines proactive routing, which provides fast response to events of interest, with reactive routing. In this way, upon the detection of an event, the transmission of data packets to the sink starts immediately and, simultaneously, a route discovery phase begins where new paths to the sink are discovered. These paths are selected according to a metric defined by the authors that estimates their amount of interference. Simulation results demonstrate that selecting the less interfering path can increase throughput and decrease packet loss.

Table 1 shows the main contributions of the abovementioned proposals, as well as other features. As it can be observed, many of the proposals use either the IEEE 802.11 standard or the non-beacon mode of the IEEE 802.15.4 standard, which does not define how duty-cycling should be carried out. The 802.11 standard allows higher data transmission bit rates, therefore allowing for higher data throughput. However, it is not well suited for WSN because of its power consumption requirements. Similarly, 802.15.4 without a sleep/wake-up schedule, *i.e.*, duty-cycle, draws a considerably higher amount of power, albeit not as high as with 802.11. In the case of short-range transmissions, such as the typically used in WSNs, transmission power is lower than the power spent in reception mode [23], therefore the most effective measure to limit power consumption is to make devices remain as much time as possible in low power (idle) states. For this same reason, battery-powered ZigBee-based solutions, where the non-beacon enabled mode is used, are known to have shorter network lifetimes simply because it forces all nodes with routing capability to be always on, with the subsequent battery drain. Finally, other routing schemes used in the discussed proposals exhibit other concerns. Hierarchical routing tend to overload nodes in the top layers of the hierarchy, which are closer to the sink, and, a priori, have more complex route repair procedures (*i.e.*, it takes some time to adapt to node failures, provided that the protocol is designed to perform route repair). Multipath schemes may overload some nodes if the path selection criteria lead to always choosing the same paths, potentially affecting network lifetime.

Table 1. Related work overview.

Reference	Main Contributions	Routing	MAC	Energy-Saving
Garcia-Sanchez *et al.* [12]	Throughput enhancement Cross-layer optimization	One-hop	802.15.4 compliant	Duty-cycle
Farooq *et al.* [13]	Architecture for QoS provisioning, congestion control	Hierarchical (clustered)	Not applicable	Not applicable
PT [15]	Congestion control, determine link capacity	Hierarchical (cluster-tree)	OMAC	CSMA-CA improvement
Kandris *et al.* [16]	Selectively drop packets	Asymmetric, Hierarchical (clustered)	Not specified	Energy-aware routing
Zaidi *et al.* [17]	Prioritized multipath forwarding	Multipath (path selection by QoS)	Not specified	Energy-aware routing
Teo *et al.* [19]	Interference-minimized multipath	Multipath (throughput enhancement)	802.11(DCF)	Not specified
MR2 [20]	Incremental, interference-aware multipath	Multipath (throughput enhancement)	802.15.4	Turn off interfering devices
GEAM [21]	Interference-aware multipath	Multipath (geographic greedy forwarding)	802.11	Not specified
Bidai *et al.* [22]	Interference-minimized multipath	Multipath (throughput enhancement) ZigBee based	802.15.4	Not specified

As stated before, the purpose of this work is to demonstrate the feasibility of video transmission using the IEEE 802.15.5 mesh standard, which is designed to create highly robust and reliable communication links. This standard allows for a very low power operation by defining how nodes should alternate periods of activity and inactivity to save energy and how these same nodes should coordinate with one another for a reliable transmission of data. This work is also intended to provide the scientific community with simulation tools for further evaluation of new protocols and improvements for WMSNs.

3. Technological Background

The next sections describe the details of the technologies related to this paper that are more relevant to the simulations performed. First, the main parameters involved in H.264 video encoding are explained. Second, the low-rate part of the IEEE 802.15.5 standard is introduced.

3.1. H.264 Video Encoding

As already stated, the transmission of video in low-power WSNs is challenging due to their inherent constraints. The transmission of large amounts of data poses additional problems such as contention for medium access, which increases jitter and further decreases throughput. It is clear that in order to cope with these restrictions video compression is needed. As an example, a typical QCIF sequence (176 × 144 pixels) at 30 frames per second requires almost 6 Mbps to be transmitted uncompressed via streaming. However, the nominal transmission rate of IEEE 802.15.4-compliant devices, such as MICAz [7] or TelosB [24], is 250 kbps when operating in the 2.4 GHz ISM band.

H.264 [25], also known as MPEG-4 Part 10 or MPEG-4 Advanced Video Coding (AVC), is one of the most popular standards for video encoding. It is about 1.5 to 2 times more efficient than MPEG-4 (Part 2) encoding, resulting in smaller compressed files. This is crucial for WMSNs, which are severely constrained by the transmission rate and energy consumption of nodes. The main benefit arising from the use of H.264 in WMSNs is the reduction in the number of data messages necessary for transmitting a video frame. This further helps reduce the network congestion and prolong the nodes lifetime. As a drawback, H.264 encoding requires more energy at the source node. However, according to [26], the use of H.264 results in a better lifetime in networks with at least five hops between the sensor nodes and the data sink and also in smaller networks with more complex video contents or more stringent video quality requirements. A more detailed study about energy spent in video encoding can be found in [27].

Several encoding parameters used by H.264 (GOP size, Quantization Parameter and CRF mode), which will be referred to subsequently, are described next.

The Group of Pictures (GOP) structure specifies the order in which different types of frames are arranged in inter-frame encoding techniques. H.264 uses three different types of frames:

- I-frames (intra-coded frames). I-frames are encoded independently of all other frames. They are the largest frames and provide the best quality, but they are also the least efficient from a compression perspective.
- P-frames (predictive-coded frames). P-frames reference redundant information contained in previous frames (I or P frames). Therefore, they contain a lesser amount of information.
- B-frames (bidirectional predictive-coded frames). B-frames can reference both previous and future frames. From a compression perspective, they are the most efficient frames.

Encoders usually require the GOP size as a parameter, which is the number of frames in a GOP structure. Since a GOP only contains an I-frame at the beginning of the structure, this parameter sets the separation between two I-frames.

For each frame, the Quantization Parameter (QP) regulates how much spatial detail is saved or discarded. In this paper, rather than using a fixed QP for each frame, the Constant Rate Factor (CRF) mode is used instead. With this mode, the encoder tries to achieve a constant perceptual quality level for all frames by adjusting the QP according to the characteristics of each frame. That is, it uses higher compression rates for frames with high motion, as the loss of detail is harder to notice in moving objects. A CRF parameter, ranging from 0 to 51, is defined to control the desired quality of the video, where lower values result in better quality and higher values in more compressed video sequences.

In this work, the value of the CRF has been set to 24 since this offers a balance between video quality and compression. In Table 2 different metrics are compared for CRF values of 10 and 24 using the same video sequence. It can be observed that a value of 24 decreases the bitrate of the compressed video and, consequently, the transmission time over a 6 × 6 grid mesh network. The drop in quality of the video is very small, as can be observed by comparing Figure 1.

Table 2. Comparison of bitrate, latency (Latency, see Section 5.1) and PSNR (Peak Signal-to-Noise Ratio (PSNR), objective video (or signal) quality metric, analyzed in Section 5.6) according to CRF.

CRF	Bitrate (kbps)	Latency (s)	Mean PSNR (dB)
10 *(medium-low compression level)*	247	356.704	50.936
24 *(medium-high compression level)*	53	84.042	41.730

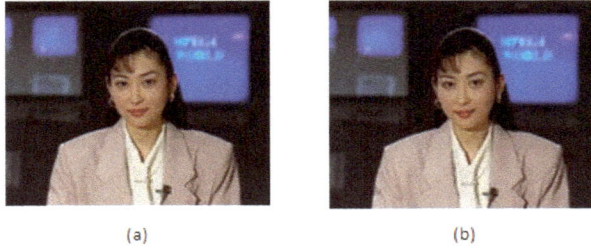

(a)   (b)

Figure 1. (a) Akiyo QCIF, CRF = 10; (b) Akiyo QCIF, CRF = 24.

3.2. IEEE 802.15.5 Low Rate

IEEE 802.15.4 standard defines Physical (PHY) and Medium Access Control (MAC) specifications for Low-Rate Wireless Personal Networks (LR-WPAN). This standard does not define the network layer and, therefore, does not offer routing capabilities by itself. Consequently, with the aim of providing an efficient multi-hop scheme, the IEEE 802.15.5 standard emerged in 2009. This new standard has a low-rate part, which will be the focus of the paper, consisting of a set of recommendations, which, according to the standard, provide an architectural framework that enables low-power, low-rate WPAN devices to promote interoperable, stable, and scalable wireless mesh topologies [28].

The objective of the standard is to support features such as unicast, multicast and reliable broadcast over multi-hop mesh links, synchronous and asynchronous communication for power saving, trace route functionality and portability of end devices [28]. In addition, being a mesh standard, it provides route redundancy, which enhances network reliability. Another key feature of the standard is that it fosters simplicity. There is no need for route discovery, reducing communication overhead, or for storing routing tables for all possible destinations. One of its main advantages over other mesh networking approaches is that it has a very similar set of service access points to the 802.15.4 MAC sublayer and, therefore, migration from non-mesh IEEE 802.15.4 networks to mesh networks is a straightforward process. In this regard, IEEE 802.15.5 aims at standardizing mesh networking over IEEE 802.15.4.

Figure 2 shows a comparison of the topologies of an IEEE 802.15.5 network with a cluster-tree IEEE 802.15.4 network. The main difference lies in the mesh links (represented by dashed lines), which connect any node with any neighbor node within communication range. The types of device defined in an IEEE 802.15.5 network are also shown in the Figure. A *mesh coordinator* is the root of the logical tree of the mesh network. It creates and manages the mesh network, and may also serve as the reference clock when network-wide synchronization is required. *Mesh devices* are responsible for intelligently relaying sensor data toward their intended destination. Finally, *end devices* are generally nodes with in-built sensors and do not include mesh routing capability.

The establishment of the network topology is key to the operation of IEEE 802.15.5. During this process, logic addresses are assigned to nodes according to their position in the network. By binding logic addresses to the network topology, the Tunneled Direct Link Setup (TDLS) routing protocol [29] suggested by the IEEE 802.15.5 standard can forward packets without performing route discovery, eliminating the associated latency and reducing communication overhead as well as avoiding the need for explicit route repair. This protocol, which measures route quality in terms of hop count, balances traffic among all the most suitable paths.

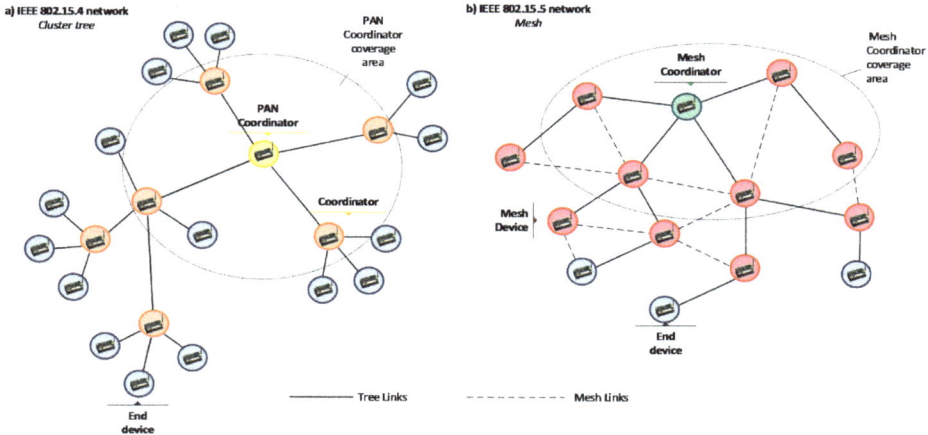

Figure 2. Topology examples. (**a**) IEEE 802.15.4 cluster tree; (**b**) IEEE 802.15.5 topology.

3.2.1. Energy-Saving Mechanisms

In other WSNs solutions such as Zigbee, nodes' radios are forced to remain on listening for any incoming transmissions. This involves a considerable power consumption and, consequently, a reduction of the lifetime of nodes. To cope with this problem, the low-rate part of the IEEE 802.15.5 standard has two energy-saving modes, called Asynchronous Energy Saving (ASES) and Synchronous Energy Saving (SES). These modes define the duty cycles of nodes, that is, the time during which their transceivers are active or inactive. ASES and SES are intended, respectively, for asynchronous and synchronous communications.

The mechanisms proposed by the IEEE 802.15.5 standard, the ASES and the SES mechanisms, have some similarities regarding the time structure in which they schedule active and inactive periods of the radio. Figure 3 shows this duty-cycle schedule for every node. The *wakeup interval* (WI) refers to the length of one duty cycle. This cycle is divided into an *active duration* (AD), where the radio transceiver is always on, followed by an *inactive duration* (ID), where the radio, MCU and sensors enter low power modes (if available) to save energy.

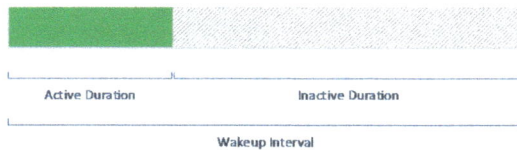

Figure 3. Time structure of IEEE 802.15.5 energy-saving mechanisms.

The length of a *wakeup interval* is defined by the *wakeup order* (WO) parameter according to the following expression:

$$WI = meshBaseActiveDuration \times 2^{WO}$$
$$0 \leqslant WO \leqslant 14 \tag{1}$$

Similarly, the *active order* (AO) parameter controls the *active duration*:

$$AD = meshBaseActiveDuration \times 2^{AO}, \ 0 \leqslant AO \leqslant WO \leqslant 14 \tag{2}$$

where *meshBaseActiveDuration* is a constant defined by the IEEE 802.15.5 standard with a 5 ms default value.

3.2.2. Asynchronous Energy Saving (ASES)

In the ASES mode, nodes are configured with the same *wakeup interval* but may have different *active durations*. Also, there is no synchronization among the *active durations* of network nodes. In order to enable communications, one or some of the nodes, depending on the case, have to delay their *active durations*. For instance, unicast communication requires that every potential receiver periodically sends a broadcast frame advertising its *active duration*. When any other node wants to transmit, upon receiving these broadcast frames, it sends its pending data during the *active duration* of the receiver using CSMA/CA (Carrier Sense Multiple Access with Collision Avoidance) for medium access. For further details about the operation of the ASES mode, the following references give more insight [28,30].

The ASES mode is intended for networks with low data sampling and data transmission rate requirements and it supports node mobility. However, it is not appropriate for video transmission because it is not able to provide an adequate QoS for more intensive data flows.

3.2.3. Synchronous Energy Saving (SES)

Unlike ASES, the Synchronous Energy Saving (SES) mechanism is intended for static networks. It is based on a precise synchronization mechanism of the *active* and *inactive durations* of the nodes, which potentially decreases delays in packet transmissions. It is well-suited for delay-sensitive applications, ensuring quick reporting of events of monitoring interest.

SES defines two transmission methods, namely the *contention-based method* and the *reservation-based method*. In the *contention-based method*, devices can only carry out data transmission within the *active duration*, competing with other nodes for medium access. In the *reservation-based method*, nodes can transmit data within the *inactive duration* when other nodes are in sleep mode. In this way, it is possible to take advantage of the available transmission bandwidth to relay data, making it potentially suitable for reliable transmission of multimedia data.

In the *reservation-based method*, the *inactive duration* is divided into slots of a fixed length. Every time a node wants to transmit, it sends a *reservation request* message during the *active duration*, competing with other nodes for accessing the medium through the CSMA-CA mechanism. If the reservation succeeds, a slot in the *inactive period* is assigned to transmit to the next hop in the source-destination route. The reservation process is repeated for each hop in the route during the *active duration*. Once the *inactive duration* starts, all the successfully reserved slots are used to transmit packets from the source to the following nodes in the route. If the *active duration* has ended and there are no slots reserved for all the hops in the route, the reservation for the remaining hops of the route will continue in the next *active duration*.

The format of the *reservation request* message can be observed in Figure 4. As the names of the fields suggest, the *end address* is the address of the destination or sink node, the *next address* is the address of the next hop of the route, selected by the TDLS routing protocol, and the *reservation slot number* identifies the requested slot. The purpose of the *previous address* field is to save bandwidth, since the same message used to request the reservation of a slot can also be used to inform the previous node that its reservation was accepted.

Command Frame Identifier	Previous Address	Next Address	End Address	Reservation Slot Number

Figure 4. Format of the reservation request message.

As an example of the *reservation-based* method, Figure 5 illustrates a scenario where node/device A transmits to device D through a route consisting of devices B, C and D. First, A sends a *reservation request* message (identified in the figure by label 1) setting the *end address* to D, the *next address* to B and the *reservation slot* to zero (first slot). The message is sent to the broadcast address and received by all neighbor nodes but only B processes it. In a second step, B sends a *reservation request* (label 2) for the second slot to C (*next address* of the message) and acknowledges A that it has successfully reserved the first slot by means of the *previous address* field. Next, device C sends a reservation request (label 3) for the third slot to D. Finally, D sends a *reservation reply* message (label 4) to C confirming the allocation of the third slot. After the reservation in the *active duration*, the data transmission starts in the *inactive duration*, each device using the slot that was assigned (labels 5–7) and receiving acknowledgement frames from the receivers during the same slot. This example shows, as mentioned previously, how the *reservation-based method* can forward a data frame from the source to the destination, using a single *inactive duration*.

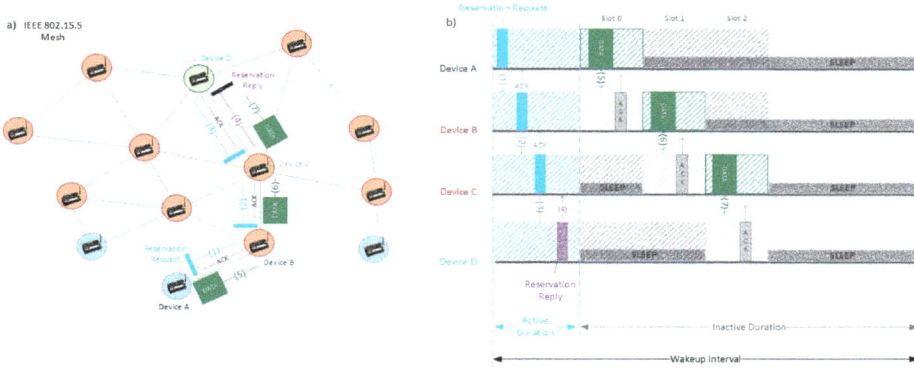

Figure 5. Data transmission in SES. (**a**) Topological view; (**b**) Time structure view.

4. Simulation

4.1. Simulation Environment

In order to carry out the simulation of multimedia transmission in an IEEE 802.15.5 mesh network, ns-2 simulator and EvalVid software are used. ns-2 [31] is a popular discrete-event network simulator, which provides support for assessing many different protocols and network technologies, including IEEE 802.15.4.

The other component used for the simulations, EvalVid, is a tool-set for evaluating the quality of video transmitted over a real or simulated network [32]. It can be used to measure several QoS metrics of the underlying communication network such as packet loss, delay, jitter and standard video quality metrics like PSNR. Unlike other available tools, EvalVid works properly even if there are packet drops, making it suitable for unreliable networks such as WSN.

In order to provide researchers with an integrated multimedia tool for simulating WSNs with mesh topology, a new module was added to ns-2. This new module, accessible via [33], includes the IEEE 802.15.5 SES mode and its integration with EvalVid. In order to perform the simulation experiments, the video is encoded and some information on how to fragment each video frame for its transmission in an IEEE 802.15.5 network is generated. This information is passed to ns-2, which simulates the transmission of packets. It is also used by EvalVid, in conjunction with data about the arrival of packets, to calculate the aforementioned QoS metrics of interest. Figure 6 shows the steps involved in the simulation and detailed in the next paragraph.

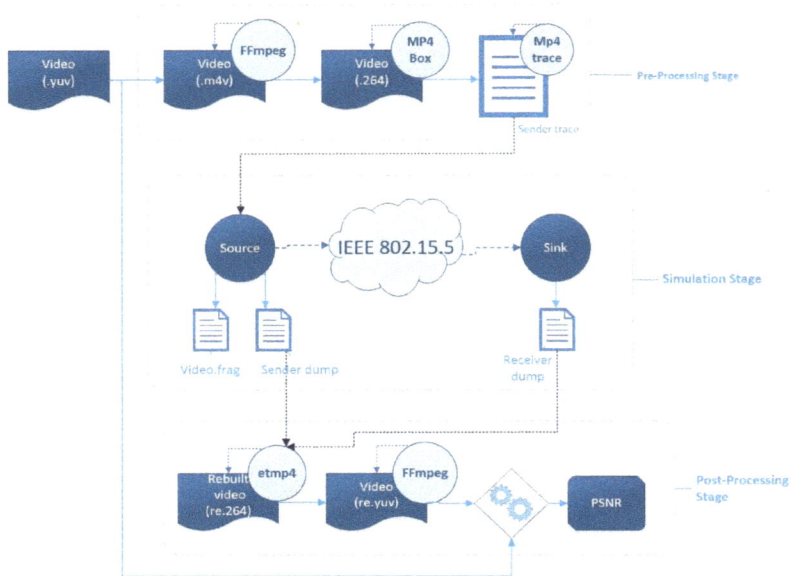

Figure 6. Simulation workflow.

The original file stores information using the YUV (Y luminance component, U and V chrominance components) color space as it allows for greater compression than Red, Green, Blue (RGB). Since the human eye has little spatial sensitivity to color, the bandwidth of the UV chrominance channels can be considerably reduced with little impact on visual perception. The information in this file is sequentially processed by the *FFmpeg* tool to encode the video and the *MP4Box* tool to create an H.264 flow with video frames and a *hint track*. This track describes how to fragment each video frame that is going to be transmitted. From the resulting file, EvalVid's *MP4trace* tool creates a text file, the *sender trace*, which contains the information about every frame shown in Table 3.

Table 3. Sender trace file structure.

Frame	Type	Size (bytes)	Packets	Sender Time
1	I	2281	23	0.081
2	P	288	3	0.158
3	P	76	1	0.159
4	P	238	3	0.199
...	...	...	...	...

The information of the *sender trace* file is used for generating the *video1.frag* file (Table 4) in order to adjust the file to the requirements of the ns-2 framework and to perform the simulation. This file has all the necessary information to allow a modified User Datagram Protocol (UDP) agent to transmit video fragments at the appropriate time. This agent will be in charge of passing the fragments to the

IEEE 802.15.5 mesh layer for their transmission in the simulated network, where packet delays or losses are detected and computed. The agent, in addition, stores the generation time of each packet in a file known as the *sender dump*. Upon reception, another modified UDP agent generates another file, the *receiver dump*, with the arrival times of each packet.

Table 4. Video1.frag file structure.

Time Interval	Size (bytes)	Type	Priority	Maximum Size [1]
81,000	2281	1	0	100
+77,000	288	2	0	100
+1000	76	2	0	100
+40,000	238	2	0	100
. . .	. . .	. . .	. . .	. . .

[1] The Maximum Size, in bytes, corresponds to the maximum payload defined by the 802.15.5 standard, obtained as aMaxMACPayloadSize − meshcMaxMeshHeaderLength = 118 − 18 = 100, where aMaxMACPayloadSize = aMaxPHYPacketSize − aMinMPDUOverhead = 127 − 9 = 118.

Finally, EvalVid, by means of the ET (Evaluate Trace, *etmp4*) tool, makes use of the encoded video and the *sender trace*, *sender dump* and *receiver dump* files to generate reports with the most relevant metrics: latency, jitter and number of packets and video frames lost. In addition, after decoding the video with the *ffmpeg* tool, the PSNR quality metric can also be obtained.

4.2. Simulation Scenarios

Simulations consisted of the transmission of commonly used video sequences [34]: Akiyo QCIF, Foreman QCIF and Mobile QCIF. These video sequences were selected because of their different motion, sorted in ascending order according to their motion. Five simulation scenarios were chosen with different numbers of nodes. In the scenarios, nodes were arranged in a regular grid layout, ranging from a 2 × 2 network to a 6 × 6 network. For each scenario, simulations were performed for various configurations of the WO and AO parameters, resulting in different lengths of the *active duration*, the *inactive duration* and the number of transmission slots in the *inactive duration*. Table 5 shows the WO–AO combinations used and the length of each period of the SES time structure according to Equations (1) and (2).

Table 5. Active and inactive durations according to WO–AO.

Wakeup Order	Active Order	Active Duration (ms)	Wakeup Interval (ms)	Inactive Duration (ms)	Slots	AD Assignment	ID Assignment
5	1	10	160	150	15	6.25%	93.75%
5	2	20	160	140	14	12.50%	87.50%
5	3	40	160	120	12	25.00%	75.00%
5	4	80	160	80	8	50.00%	50.00%
6	1	10	320	310	31	3.13%	96.88%
6	2	20	320	300	30	6.25%	93.75%
6	3	40	320	280	28	12.50%	87.50%
6	4	80	320	240	24	25.00%	75.00%
6	5	160	320	160	16	50.00%	50.00%
7	1	10	640	630	63	1.56%	98.44%
7	2	20	640	620	62	3.13%	96.88%
7	3	40	640	600	60	6.25%	93.75%
7	4	80	640	560	56	12.50%	87.50%
7	5	160	640	480	48	25.00%	75.00%
7	6	320	640	320	32	50.00%	50.00%

The length of each slot within the *inactive duration* was chosen to be 10 ms [35]. This was regarded as sufficient for carrying out the transmission of a single data message between two consecutive nodes.

Table 6 shows the set of network simulation parameters common to all the simulations performed. With regard to the video encoding parameters, the GOP size was set to 25, CRF to 24 and frames per second (FPS) to 25.

Table 6. Simulation parameters.

Parameter	Value
Channel Type	Channel/WirelessChannel
Radio-propagation model	Propagation/TwoRayGround
Physical layer	Phy/WirelessPhy/802_15_4
Medium Access Layer (MAC)	Mac/802_15_4
Antenna model	Antenna/OmniAntenna
Frequency	2.4 GHz
CS Threshold	2.13643×10^{-7}
RX Threshold	2.13643×10^{-7}
Coverage range	30 m
Distance between two consecutive nodes	25 m
Routing protocol	TDLS
Energy-saving mechanism	SES: Reservation-based method
Simulation time limit	1500 s (25 min)
Traffic	Akiyo, Foreman and Mobile QCIF (176 × 144 resolution)
Video generation rate/service rate	53 kbps bitrate/1 packet per *wakeup interval*

5. Performance Evaluation

The simulation results for every scenario (Figure 7) are presented below. Different metrics of interest have been plotted, including latency, jitter, throughput, message delivery ratio, power consumption/network lifetime and PSNR. These metrics are commonly assessed in other studies related to multimedia traffic or WSNs. It should be noted that, for all scenarios, power consumption and lifetime were computed for node 1, as will be explained later in the text. To conclude this section, a scalability study is also included. It is intended to show the impact of the aforementioned metrics in networks composed of a large number of nodes and different node densities.

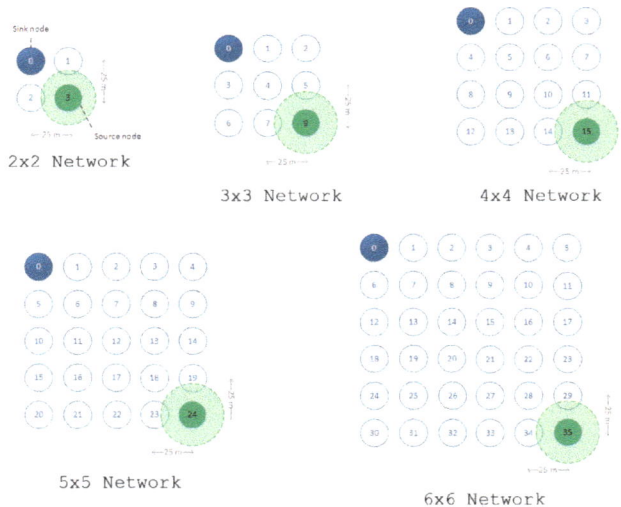

Figure 7. Simulation scenarios.

5.1. Latency

Latency has been computed as the average time between the generation of data messages and their delivery to the destination node (*i.e.*, data sink). This includes the time dedicated by all the intermediate nodes belonging to the source–destination path for accomplishing tasks such as the data processing, temporary storage in the node's memory or transmission/reception procedures, among others.

The latency for all the packets that have successfully reached the destination is shown in Figure 8. It can be observed that the latency shows a considerable variation for different values of WO. It should be noted that while the generation rate remains constant at 53 kbps, only one packet is transmitted per *wakeup interval*, whose length is defined by this parameter. Consequently, as WO (and subsequently the *wakeup interval*) increases, the capacity of the network to accommodate data traffic decreases and packets are stored in the queue of the first node waiting for transmission, therefore increasing the latency measured. Some simulations were also performed transmitting two packets per *wakeup interval*, but the obtained performance was really poor. For example, for a 6 × 6 configuration, an increase in the latency of around 50% was measured. In this case, increasing the amount of data sent led to a higher contention for medium access, which worsened the performance of the network.

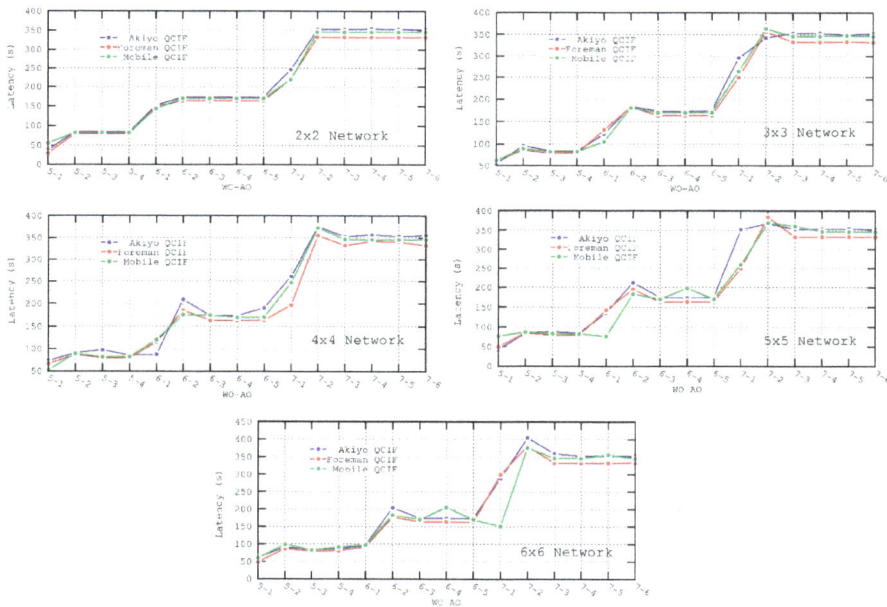

Figure 8. Latency metric for the different scenarios.

Under these circumstances, for any given WO–AO configuration, the latency increases linearly with respect to the number of transmitted packets (see Figure 9) because of the packet service rate of the network (one packet per WI). The figure shows the evolution of the latency calculated after the arrival of each packet for a 6 × 6 network with a 5–3 (WO–AO) configuration. It has to be noted that latency values of zero in the figure represent packet loss.

In Figure 8 there is a case that requires further explanation, specifically for AO = 1, where the latency measured is smaller. However, these low values are obtained because of high contention for medium access and the resulting packet loss. These figures must be analyzed in conjunction with the Message Delivery Ratio figures. For AO = 1, it can be observed that packet loss increases considerably since the *active duration* is too short to successfully reserve all transmission slots. In particular, as the congestion of the network grows with time, most of the packets that are successfully received correspond to the first packets sent, which have a lower latency. On the contrary, packets sent subsequently, which have a higher latency as shown in Figure 9, have a higher loss probability. Since the displayed latency is averaged over all the successfully received packets, this results in a deceptively short latency.

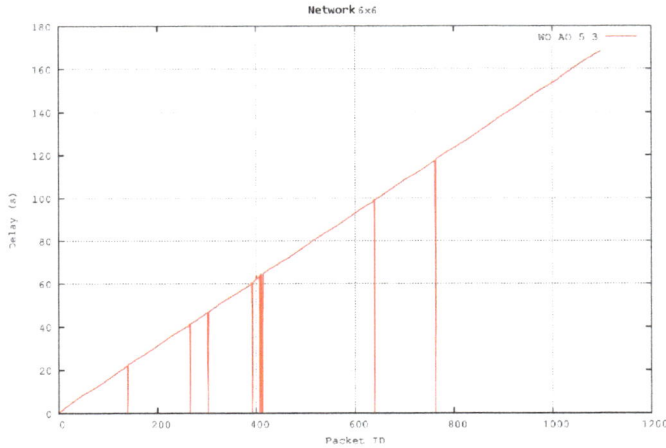

Figure 9. Packet latency.

Besides, it can be observed that, for a fixed value of WO, the latency remains almost constant in the interval $3 \leqslant AO < WO$. Thus, a further increase of the *active duration* will not have noticeable effects on the latency. This holds true even for $AO = WO - 1$, where the *inactive duration* will still be long enough. According to Table 5, for $AO = WO - 1$, the *inactive duration* is as big as the active duration and thus adequate for transmission.

In conclusion, for a fixed value of WO, the latency remains almost constant except for the two following cases:

- For $AO = 1$ the *active duration* is too short and, given that the contention for the medium is too high, few nodes are able to reserve a slot for transmission and soon the network experiences congestion collapse. Packets are stored in queues (of infinite size) and many of them, due to the failure of the CSMA-CA mechanism (no successful transmission after the maximum number of retries), are discarded and will not reach the destination. The latency measured corresponds to the few packets that are successfully received.
- For $AO = 2$, the contention is still high (the *inactive duration* is not long enough yet) but packet loss is not as important as in the previous case (though still present). Therefore, the average delay represented is higher since the final packets (with higher latency) experience less congestion than for $AO = 1$. However, the latency measured is also slightly higher than for AO values greater than 2 because of the network congestion, as the transmission of packets may require more than one *wakeup interval*.

5.2. Inter-Packet Arrival Time

Inter-packet arrival time refers to the time that elapses from the receipt of a packet at the data sink to the instant at which the next packet is received.

According to the operation of the SES *reservation-based method*, it can be expected that the time between the receipt of two consecutive packets at the destination node is determined by the time between slots that one particular node has available for transmission. This time is set by the length of the *wakeup interval*, which, in turn, depends on the WO parameter. The theoretical inter-packet arrival time is:

$$WI = meshBaseActiveDuration \times 2^{WO} = 5 \ ms \times 2^5 = 160 \ ms, \ if \ WO = 5 \qquad (3)$$

$$WI = meshBaseActiveDuration \times 2^{WO} = 5 \ ms \times 2^6 = 320 \ ms, \ if \ WO = 6 \qquad (4)$$

$$WI = meshBaseActiveDuration \times 2^{WO} = 5 \ ms \times 2^7 = 640 \ ms, \ if \ WO = 7 \qquad (5)$$

$$\overline{it}_{P_n} = \begin{cases} 160 \ ms, \ WO = 5 \\ 320 \ ms, \ WO = 6 \\ 640 \ ms, \ WO = 7 \end{cases} \qquad (6)$$

Increasing the WO by one doubles the inter-packet arrival time, causing a decrease in throughput. This has been tested through simulations of a 6 × 6 network with WO–AO configurations in Figure 10.

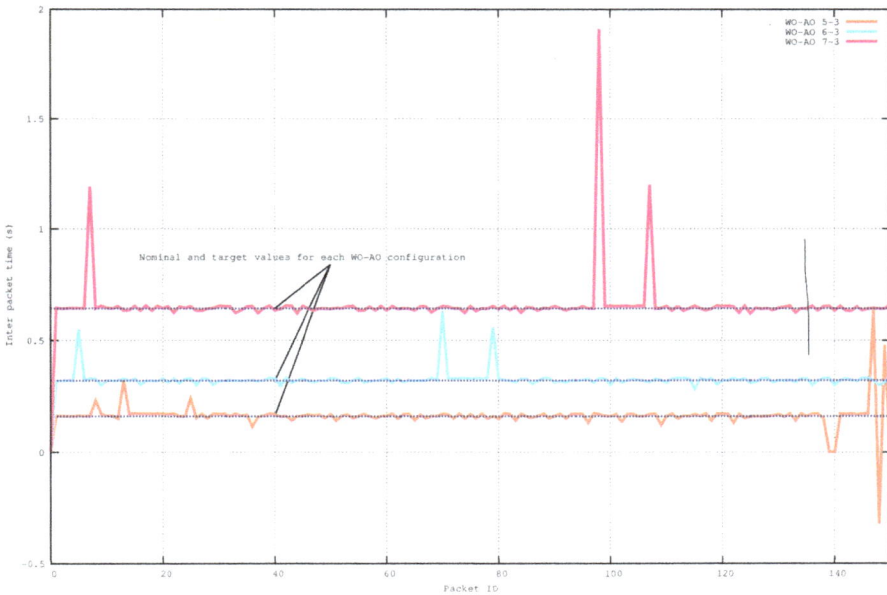

Figure 10. Inter-packet arrival times for a 6 × 6 network.

It can be observed that the obtained measurements are consistent with the theoretical values, with values around 0.16 s for WO = 5, 0.32 s for WO = 6 and 0.64 s for WO = 7. There are, however, small variations in the inter-packet arrival times, also known as jitter, which will be discussed in the next section. These variations are mainly due to the imperfect synchronization process performed in the SES mechanism and the CSMA-CA mechanism. In addition, larger variations, also caused by the CSMA-CA mechanism, may indicate packet delays through several *wakeup intervals* and, when packets arrive out of order, negative values may be found.

5.3. Jitter

Jitter is calculated as the variation in the inter-packet time caused by network congestion, packet loss, loss of synchronization, or by the use of multiple paths from source to destination. It must be noted that due to the use of the TDLS routing protocol, which by default selects the shortest paths between source and destination and performs load balancing across them, all the paths used for data forwarding will have the same number of hops. Therefore, in spite of the use of multiple paths, all the packets will traverse the same number of hops. Consequently, the jitter will not be affected.

However, there are situations where jitter is more likely to be affected. Analyzing Figure 11, when AO = 1 or AO = 2 the jitter increases since the *active duration* is insufficient to perform all slot reservations, resulting in more congestion and consequently in more unpredictable inter-packet arrival times.

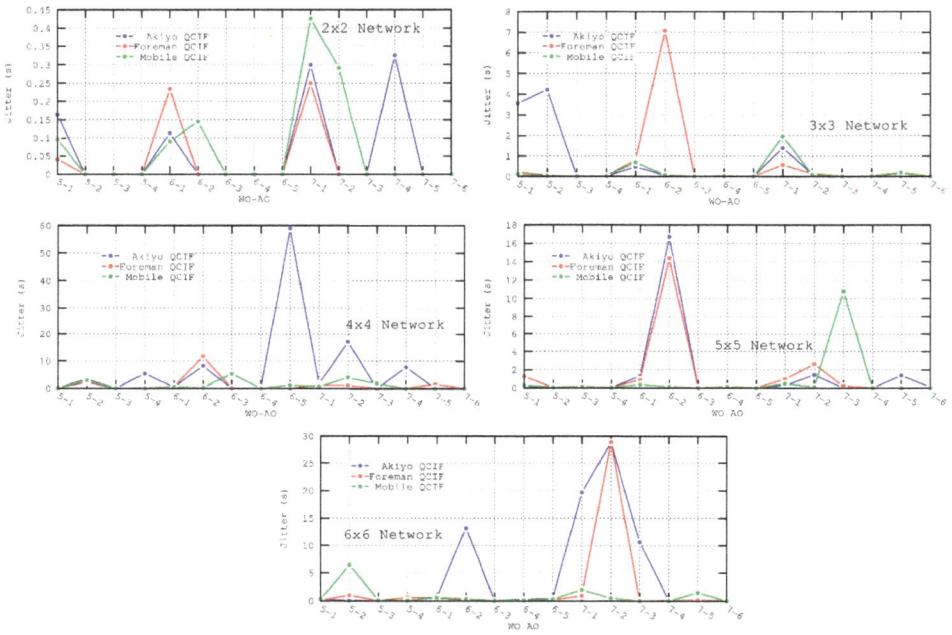

Figure 11. Jitter metric for the different scenarios.

5.4. Throughput and Message Delivery Ratio

Throughput is defined as the amount of raw information (header and payload) received in a given period. From this definition, it is clear that throughput is directly related to the *message delivery ratio metric*, that is, to the rate of successfully received messages. Consequently, these two metrics are discussed below jointly.

In order to evaluate the *message delivery ratio*, an entry is created in a text file for each message transmitted or received, as explained before, letting EvalVid check the number of packets sent and received.

It should be noted that UDP is used at the transport layer over the IEEE 802.15.5 mesh network layer and IEEE 802.15.4 MAC and PHY layers. Therefore, if a packet does not arrive at the destination node for whatever reason, the packet will not be retransmitted.

Since the size of queues has been assumed to be infinite, it can be deduced that packet losses are not related to the size of buffers. Consequently, packet losses are mainly caused by the occupation of the channel and the failure when accessing the medium through the CSMA-CA mechanism (no successful transmission after the maximum number of retries). In addition to existing medium contention, the exposed node problem and the hidden nodes problems also affect transmissions in the *active duration*, since CSMA-CA does not avoid collisions by hidden nodes or unnecessary backoff delays by exposed nodes [30]. The exposed node problem prevents a sender from transmitting to a receiver if the sender detects a signal from another node, even if that signal is not interfering at the receiver's location. This increases the probability of CSMA failures. Besides, even if a node manages to transmit a message, it may be lost due to the hidden node problem. In this regard, the reservation process that takes place during the *active duration* will be affected by both primary (two hidden nodes transmitting to the same receiver) and secondary (receiver in range of another node transmitting to a different node) collisions. On the other hand, the *inactive duration* will be affected to a lesser extent by the hidden terminal problem, as discussed in Section 5.6.

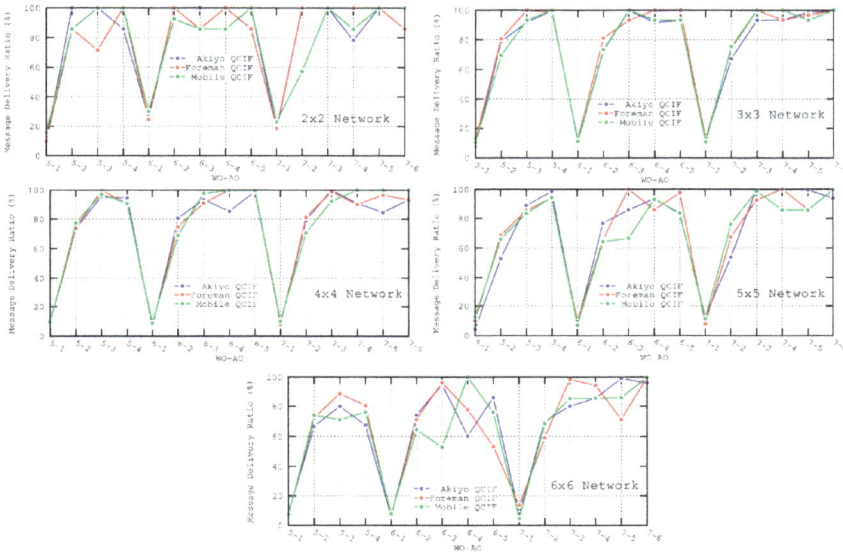

Figure 12. Throughput metric for the different scenarios.

As observed in Figures 12 and 13 AO values of 1 and, to a lesser degree, 2 present the worst performance in both the throughput and message delivery ratio. In addition, throughput is also affected by the selection of WO values, since it determines the separation between consecutive packets sent. That is, a longer *inactive period* (or WO values) implies lower throughput. In particular, the best results for the throughput metric are obtained for 5–3 (WO–AO) configuration. In this case, the lengths of the *Wakeup Interval* and Active Duration are found to be suitable for forwarding data messages without incurring long node delays.

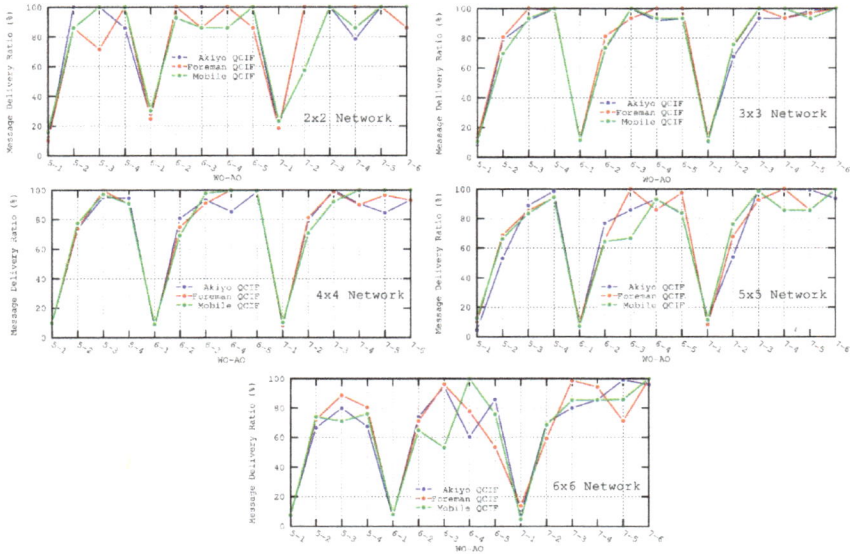

Figure 13. Message delivery ratio metric for the different scenarios.

5.5. Power Consumption and Lifetime

The SES mechanism provides an efficient energy-saving mechanism because the devices are sleeping most of the time, turning on their transceivers during short periods of time for transmitting or receiving video packets. In addition, according to the type of device in an IEEE 802.15.5 network (*i.e.*, whether the node acts as *mesh coordinator*, *mesh or end device*), it can be intuitively deduced which of them will have a higher power consumption during the *inactive period* of the *SES reservation-based* method:

- *Mesh coordinator*: Most of its power consumption is due to the continuous performance of synchronization tasks [28], which are out of the scope of this paper. Regarding data transmission power consumption, which is the focus of this section, the mesh coordinator will not have to perform data transmission and the only significant consumption is from packet reception during the allocated slots (it changes the radio transceiver state from *sleep* to *listening*).
- *End device*: This type of device corresponds to the source node that sends packets continuously through the network. Therefore, its power is consumed during the transmission in one slot per SES time structure, that is, one slot for each *wakeup interval*.
- *Mesh device (router)*: Unlike the other two types of nodes, mesh devices require two consecutive slots of the *inactive period*, one for receiving from the previous node and another for transmission to the next node. Consequently, they are the devices that have the highest power consumption.

On the other hand, all nodes have additional power consumption during the whole *active duration* since their transceivers are either in the *listening* state or in the *transmit* state. Consequently, the *active duration* affects to a greater extent the power consumption metric than the *inactive duration*, where nodes remain asleep for most of the time (their state is switched only to perform the scheduled transmissions/receptions in the reserved slots).

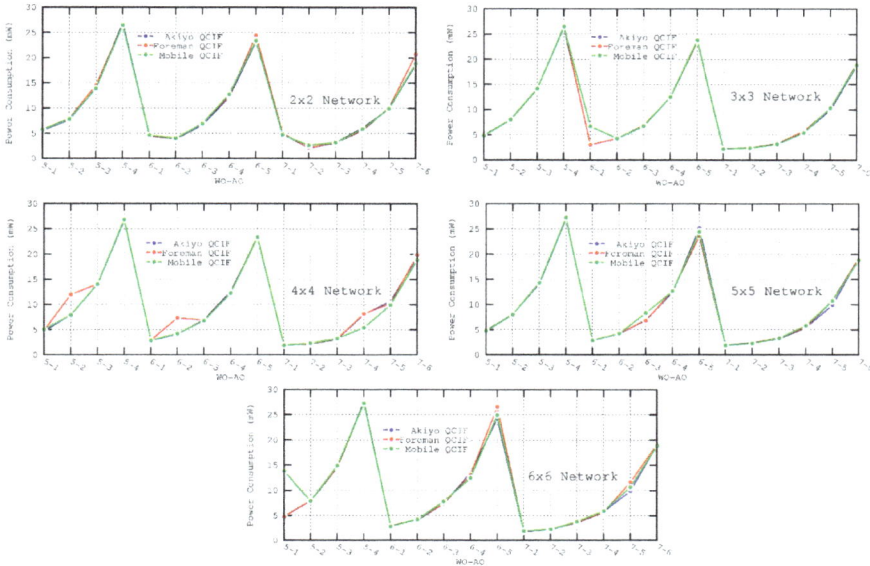

Figure 14. Power Consumption metric for the different scenarios.

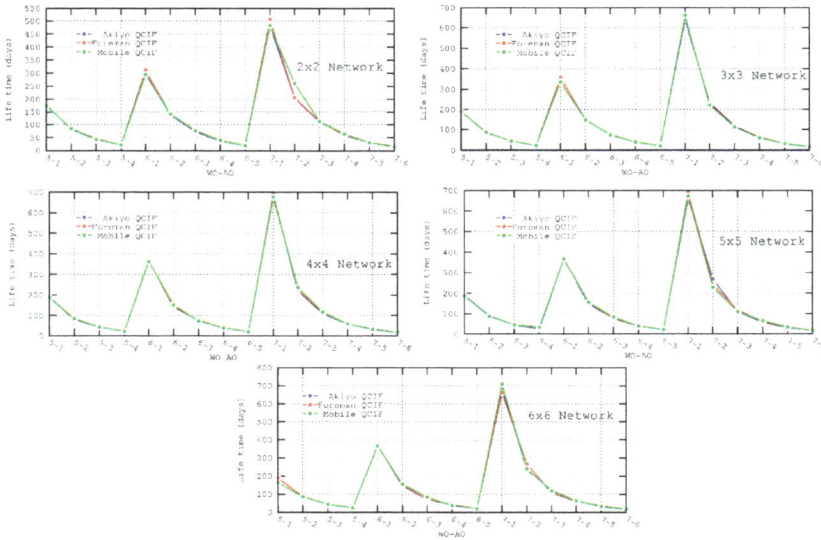

Figure 15. Lifetime metric for the different scenarios.

It can be observed in all the evaluated scenarios—Figures 14 and 15—how the AO has a considerable impact on power consumption (regardless of the WO, nodes configured with a low AO value consume less energy). It should be noted that both power consumption and lifetime metrics were calculated for node 1, which has a high power consumption as it is a mesh device and it is located close to the destination node where all the paths generated by the TDLS algorithm converge.

5.6. Peak Signal-to-Noise Ratio

The most widespread method to determine the quality of a video sequence is the *Peak Signal-to-Noise Ratio* (PSNR), which represents the ratio between the maximum possible power of a signal and the power of the distorting noise that affects the fidelity of its representation. Because many signals have a wide dynamic range, PSNR values are usually expressed in logarithmic scale.

In order to compute the PSNR, the *Mean Squared Error* (MSE) of each frame must first be obtained. The MSE represents the cumulative squared error between the compressed frame and the original frame. For a monochrome image, it is calculated as:

$$MSE = \frac{1}{M * N} \sum_{i=0}^{M-1} \sum_{j=0}^{N-1} \| I(i,j) - K(i,j) \|^2 \tag{7}$$

where $I(i,j)$ corresponds to the value of the (i,j) pixel of the transmitted frame and $K(i,j)$ is the value of the same pixel in the received frame.

The PSNR is defined as:

$$PSNR = 10 * \log_{10} \left(\frac{MAX^2}{MSE} \right) = 20 * \log_{10} \left(\frac{MAX}{\sqrt{MSE}} \right) \tag{8}$$

where MAX denotes the maximum possible pixel value of the image. If B bits are used per pixel (luminance component), $MAX = 2^B - 1$.

Considering a video sequence composed of a series of frames, the MSE and PSNR are calculated for each of the frames and the PSNR of the complete video sequence is the mean value of the PNSR of the frames.

EvalVid uses the PSNR as an evaluation metric of the quality of compressed videos. When the difference between the frames of an original video sequence and the frames of the received and decoded video sequence is high, the PSNR is low. Therefore, higher values of PSNR denote better quality of the video. Acceptable values for wireless communication systems are considered to be about 20 dB to 25 dB.

On the other hand, for the subjective evaluation of video quality, the MOS (*Mean Opinion Score*) [36] is usually employed. This technique is based on the perceived quality from the users' perspective of the received video (or any other kind of signal), giving a score to video sequences from 1 (worst) to 5 (best). In [37] a possible mapping between PNSR and MOS values is provided. The conversion between the two metrics can be seen in Table 7.

Above, in Figure 16, the results of the PSNR for each of the simulated scenarios are shown. In view of the presented results, a degradation in the quality of the video can be observed as the network grows. Besides, the values of WO seem to have some impact for larger networks, 5 × 5 and 6 × 6, but not for small networks. In particular, the PSNR value deteriorates for lower WO values due to the hidden terminal problem. With fixed AO values and therefore fixed *active duration* and contention for medium access, the only reason for this drop in performance is packet loss in the *inactive duration*. Since there is no contention for medium access during this period, packet loss is due to the hidden terminal problem. More particularly, only secondary collisions from hidden nodes' transmissions can take place, whereas primary collisions cannot happen because of the slot reservation process performed in the *active duration*.

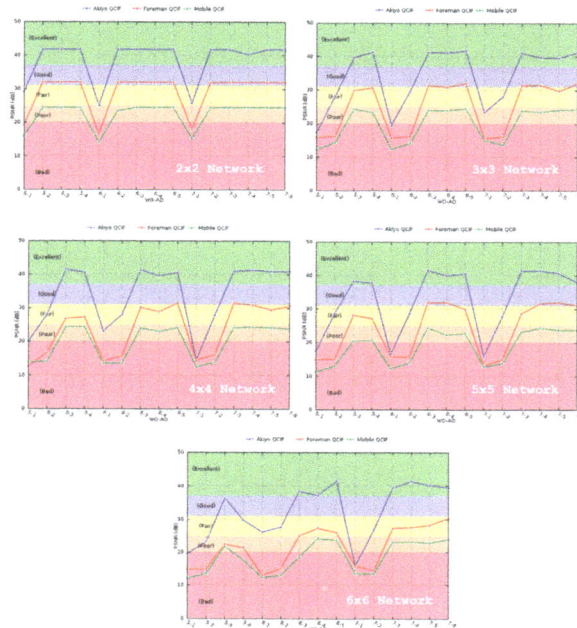

Figure 16. PSNR metric for the different scenarios.

As a consequence of the above-mentioned, the size of the network has to be taken into account for the appropriate selection of the WO, finding a balance between the quality of the video and throughput (high WO values reduce the throughput considerably).

With respect to the AO parameter, the worst PSNR values are obtained for AO = 1 and AO = 2, since the large amount of packet losses distorts the quality of the decoded video. For $3 \leqslant AO < WO$ there is little variation in the PSNR, implying a good or excellent video reproduction.

Comparing the three different video sequences, it can be easily deduced that, regardless of the network size, the PSNR gets worse for sequences with higher motion. However, this can be misleading, since the drop in quality is not as high as it may seem. This is due to the use of the CRF encoding method. CRF compresses different frames by different amounts, taking the motion into account. Therefore, frames with higher motion will have a greater compression, discarding more information from them. The reason for doing so is that the human eye perceives more detail in still objects than in moving objects. Consequently, in spite of having a worse PSNR because of greater compression, subjectively, the quality of the video sequences with high motion will still be acceptable. To this extent, the video sequences with higher motion, Foreman and Mobile QCIF, obtained after the simulation using a 5–3 (WO–AO) configuration for the 6×6 scenario can be seen [38], where, although its quality is rated as "very annoying" according to the PSNR to MOS conversion, the video can be considered as acceptable from a subjective point of view.

Table 7. PSNR to MOS mapping [37].

PSNR	MOS	Impairment
>37	5 (Excellent)	Imperceptible
31–37	4 (Good)	Perceptible but not annoying
25–31	3 (Fair)	Slightly annoying
20–25	2 (Poor)	Annoying
<20	1 (Bad)	Very annoying

5.7. Scalability

The evolution of the obtained performance metrics according to the size of the networks is shown in Figure 17, ranging from 2 hops for a 2 × 2 network to 10 hops for a 6 × 6 network. All metrics remain within acceptable values for any network size, proving the scalability of the IEEE 802.15.5 SES mechanism for video transmission. For the largest network sizes, though, there is a small drop in performance. In these cases, increasing the value of the WI parameter will improve the quality of the received video at the expense of throughput.

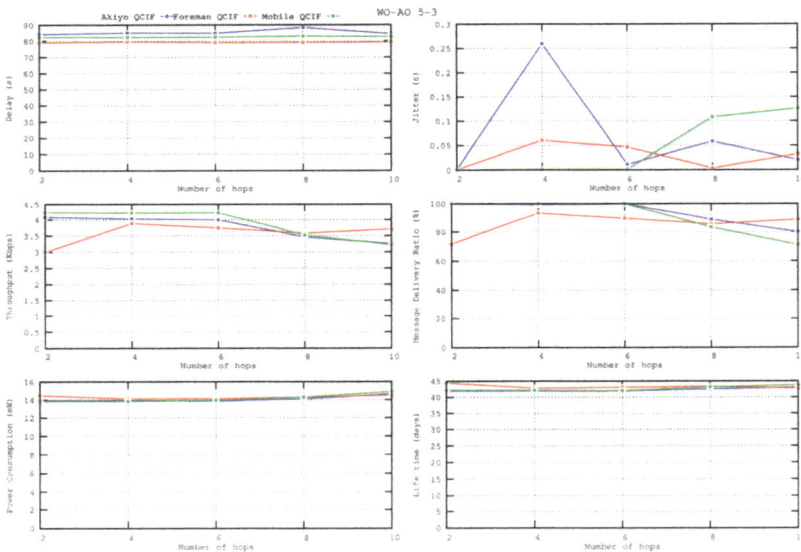

Figure 17. Performance metrics according to the size of the network, WO–AO = 5–3.

6. Discussion

The previous section has shown the performance of IEEE 802.15.5 mesh networks transmitting with a packet generation rate of 53 kbps and dispatching one packet per *wakeup interval*, since transmitting more than one packet per *wakeup interval* was observed to lead to congestion and therefore worsen the performance of the network. In particular, the effect of using different WO–AO configurations for several network sizes has been studied.

In spite of considering that just a single node is transmitting video, the previous metrics show that there is some drop of performance because different packets of the same stream may be routed by different paths and compete for transmission. In this regard, the performance drop observed is due to repetitive failures to gain access to the medium while it is busy (CSMA-CA failure), aggravated by the exposed and hidden nodes problems. To cope with these problems, the reservation-based method of the SES mode proposes the use of a relatively short *active duration* period, where all contention for reserving transmission slots takes place, and the *inactive duration*, where nodes have slots where transmission is guaranteed. In data-intensive applications, though, this inactive period can also be affected by the hidden node problem to some degree, as mentioned.

During the active period, nodes have the opportunity to reserve slots for subsequent transmissions in the inactive period. Since the active duration of all neighbor nodes is synchronized, it is likely that several nodes will try to perform simultaneous reservations. In order to avoid collisions, the CSMA-CA scheme allows that, if some reservation message cannot be sent because the channel is busy, several subsequent retransmission attempts (up to a configurable maximum number in the IEEE 802.15.4 standard, *macMaxCSMABackoff*, 4 by default) can be performed. However, according to the simulations, AO values of 1 and 2 result in active durations not long enough to perform all CSMA-CA retries. For example, the maximum delay (in a worst case scenario) due to the CSMA-CA backoff is 36.8 ms (calculated as $\sum_{backoff=0}^{backoff=4}(2^{BE} - 1) * \text{aUnitBackoffPeriod} * \text{symbol\_time}$ where $BE = \min(\text{macMinBE} + \text{backoff}, \text{macMaxBE})$) using the IEEE 802.15.4 default values at 2.4 GHz, whereas for AO = 1 the active duration is only 10 ms long. This implies that several active durations may be necessary in order to reserve one transmission slot. However, during that time, new packets would also require to be transmitted, increasing the contention for medium access (if the packets are received by neighbor nodes) or the queue size of the node, and therefore the performance of the network would get worse as time goes on.

Given all the factors that affect negatively the transmissions during the *active duration*, it is clear that it is very important to select an AO value that allows transmission with an appropriate QoS. It has been observed for all the previous QoS metrics that for AO values equal to or greater than 3, the performance has been quite satisfactory. Taking into account that the AO parameter is detrimental to power consumption and network lifetime, a value of 3 can be used for the scenarios studied without affecting latency, jitter, throughput and message delivery ratio.

Regarding the selection of the *wakeup interval* length, and consequently the WO parameter, it is important to find a balance between power consumption (which improves with high WO values) and throughput (which improves with low WO values). In addition, it has also been observed that the size of the network may affect the quality of the received video due to the hidden terminal problem, and more particularly secondary collisions. Other negative factors such as contention for medium access, the exposed node problem or primary collisions due to the hidden node problem are avoided by the use of transmission in reserved slots of the inactive period. Thanks to this fact, the drop of performance in the *inactive duration* is not important and it is only noticeable for networks of size 5×5 and greater. As the size of the networks grows, the number of packet retransmissions in the mesh network increases, and therefore the probability of secondary collisions. For the scenarios studied, a WO value of 5 has shown to provide good video quality without worsening the throughput. This can be better observed by watching the received videos since, as mentioned in the previous section, values of PSNR values may be misleading for videos with high motion. It has to be noted that, due to the tendency of worsening PSNR and packet loss with network size, for networks larger than the studied ones, higher values of WO should be considered. In this case, increasing the WO by one implies a more than twofold increase of the *inactive duration* and the number of slots (see Table 5). Therefore transmissions in the *inactive duration* are sparser and the probability of collision decreases.

In view of the simulation results, it can be affirmed that it is feasible to transmit video using the IEEE 802.15.5 standard. Using appropriate WO–AO values, it can achieve the necessary reliability for a satisfactory video quality. On the other hand, the main drawback of the standard is that it does not achieve high throughput values, which hinders its applicability to real-time video applications, although this is anyway very challenging for multi-hop networks. In order to improve throughput, some enhancements are necessary that allow more than one packet per *wakeup interval* to be reliably sent. This can be achieved by tackling the diverse problems that have been discussed in this section in order to arrange transmissions so that multimedia flows can use more than one slot per *wakeup interval*, but also by applying enhancements in the directions given in the Related Work section. In this line, the standard has an important feature, the possibility of allowing transmissions to any other node of the network, which facilitates the development of collaborative or multi-sink applications.

7. Conclusions

In this paper, the transmission of video sequences using the low rate part of the IEEE 802.15.5 standard has been evaluated by means of a new simulation tool which is available in [33]. This tool implements the SES reservation-based method as energy-saving mechanism, following faithfully the standard specification. Furthermore, the simulation framework integrates the SES method and the EvalVid software, which generates compressed QCIF video sequences with different motion, encoded using H.264/AVC. These sequences are transmitted through mesh networks of different sizes. The main results of the simulations have been presented and discussed, highlighting the most important metrics such as latency, jitter, time between packets, message delivery ratio, power consumption, lifetime and PSNR. The importance of the selection of appropriate values for the *active order* and *wakeup order* parameters has been emphasized, observing that for $3 \leqslant AO < WO$ video transmission is feasible. A 5–3 (WO–AO) configuration has been found to be the best for the scenarios studied, although, for larger networks, higher WO values can be considered. The video sequence obtained after the simulation of 5–3 and 7–2 (WO–AO) configurations for Akiyo, Foreman and Mobile QCIF can be found at [38–43].

In addition, the developed tool can be used to evaluate different enhancements over the IEEE 802.15.5 standard. In spite of the advantages of this standard reported in this work, such as fault tolerance or the possibility that any node transmits to any other node, there is still room for improvement for data-intensive applications and, more particularly, WMSN applications. In this regard, it is possible to introduce enhancements to the original standard, with no detriment to the interoperability with other IEEE 802.15.5 devices, which take into account the particularities of WMSN and increase its performance for this type of application.

Acknowledgments: This research has been supported by the MINECO/FEDER project grant TEC2013-47016-C2-2-R (COINS).

Author Contributions: Antonio-Javier Garcia-Sanchez and Felipe García-Sánchez conceived the idea and participated in the elaboration of the manuscript and in the analysis of results. Fernando Losilla was involved in the writing of the manuscript and in the analysis of the results. David Rodenas-Herraiz developed the SES module of the simulation tool. Felipe Cruz-Martínez took part in the elaboration of the simulation tool and the execution of simulations. All authors revised the paper.

Conflicts of Interest: The authors declare no conflict of interest.

References

1. Akyildiz, I.F.; Su, W.; Sankarasubramaniam, Y.; Cayirci, E. Wireless sensor networks: A survey. *Comput. Netw.* **2002**, *38*, 393–422. [CrossRef]
2. Akyildiz, I.F.; Melodia, T.; Chowdhury, K.R. A survey on wireless multimedia sensor networks. *Comput. Netw.* **2007**, *51*, 921–960. [CrossRef]
3. Rodenas-Herraiz, D.; Garcia-Sanchez, A.-J.; Garcia-Sanchez, F.; Garcia-Haro, J. Current trends in wireless mesh sensor networks: A review of competing approaches. *Sensors* **2013**, *13*, 5958–5995. [CrossRef] [PubMed]
4. Lee, M.; Zhang, R.; Zhu, C.; Park, T.R.; Shin, C.-S.; Jeon, Y.; Lee, S.-H.; Choi, S.-S.; Liu, E.Y.; Park, S.-W. Meshing wireless personal area networks: Introducing IEEE 802.15. 5. *IEEE Commun. Mag.* **2010**, *48*, 54–61. [CrossRef]
5. Ali, A.; Ahmed, M.E.; Piran, M.J.; Suh, D.Y. Resource Optimization Scheme for Multimedia-Enabled Wireless Mesh Networks. *Sensors* **2014**, *14*, 14500–14525. [CrossRef] [PubMed]
6. Pham, C. Communication performances of IEEE 802.15. 4 wireless sensor motes for data-intensive applications: A comparison of WaspMote, Arduino MEGA, TelosB, MicaZ and iMote2 for image surveillance. *J. Netw. Comput. Appl.* **2014**, *46*, 48–59. [CrossRef]

7. MICAz Datasheet. Available online: http://www.memsic.com/userfiles/files/Datasheets/WSN/micaz_datasheet-t.pdf (accessed on 3 February 2016).
8. Farooq, M.O.; Kunz, T. Wireless sensor networks testbeds and state-of-the-art multimedia sensor nodes. *Appl. Math. Inf. Sci.* **2014**, *8*, 935–940. [CrossRef]
9. Ben Ammar, A.; Bouattay, O.; Dziri, A.; Terre, M.; Youssef, H. Performance analysis of AODV and AOMDV over SMAC and IEEE 802.15. 4 in Wireless Multimedia Sensor Network. In Proceedings of the 2015 International Wireless Communications and Mobile Computing Conference (IWCMC 2015), Dubrovnik, Croatia, 24–28 August 2015.
10. Andrea, P.; Scavongelli, C.; Orcioni, S.; Conti, M. Performance analysis of JPEG 2000 over 802.15. 4 wireless image sensor network. In Proceedings of the 2010 8th Workshop on Intelligent Solutions in Embedded Systems, Heraklion, Greece, 8–9 July 2010.
11. Pekhteryev, G.; Sahinoglu, Z.; Orlik, P.; Bhatti, G. Image transmission over IEEE 802.15. 4 and ZigBee networks. In Proceedings of the 2005 IEEE International Symposium on Circuits and Systems (ISCAS 2005), Kobe, Japan, 23–26 May 2005.
12. Garcia-Sanchez, A.-J.; Garcia-Sanchez, F.; Garcia-Haro, J.; Losilla, F. A cross-layer solution for enabling real-time video transmission over IEEE 802.15. 4 networks. *Multimed. Tools Appl.* **2011**, *51*, 1069–1104. [CrossRef]
13. Farooq, M.O.; St-Hilaire, M.; Kunz, T. Cross-layer architecture for QoS provisioning in wireless multimedia sensor networks. *KSII Trans. Internet Inf. Syst.* **2012**, *6*, 176–200.
14. Alanazi, A.; Elleithy, K. Real-Time QoS Routing Protocols in Wireless Multimedia Sensor Networks: Study and Analysis. *Sensors* **2015**, *15*, 22209–22233. [CrossRef] [PubMed]
15. Razaque, A.; Elleithy, K. Modular energy-efficient and robust paradigms for a disaster-recovery process over wireless sensor networks. *Sensors* **2015**, *15*, 16162–16195. [CrossRef] [PubMed]
16. Kandris, D.; Tsagkaropoulos, M.; Politis, I.; Tzes, A.; Kotsopoulos, S. Energy efficient and perceived QoS aware video routing over wireless multimedia sensor networks. *Ad Hoc Netw.* **2011**, *9*, 591–607. [CrossRef]
17. Zaidi, S.M.A.; Song, B. Prioritized Multipath Video Forwarding in WSN. *J. Inf. Process. Syst.* **2014**, *10*, 176–192. [CrossRef]
18. Felemban, E.; Lee, C.-G.; Ekici, E. MMSPEED: Multipath Multi-SPEED Protocol for QoS Guarantee of Reliability and Timeliness in Wireless Sensor Networks. *IEEE Trans. Mob. Comput.* **2006**, *5*, 738–754. [CrossRef]
19. Teo, J.-Y.; Ha, Y.; Tham, C.-K. Interference-minimized multipath routing with congestion control in wireless sensor network for high-rate streaming. *IEEE Trans. Mob. Comput.* **2008**, *7*, 1124–1137.
20. Maimour, M. Maximally radio-disjoint multipath routing for wireless multimedia sensor networks. In Proceedings of the 4th ACM Workshop on Wireless Multimedia Networking and Performance Modeling, Vancouver, BC, Canada, 27–31 October 2008; pp. 26–31.
21. Li, B.-Y.; Chuang, P.-J. Geographic energy-aware non-interfering multipath routing for multimedia transmission in wireless sensor networks. *Inf. Sci.* **2013**, *249*, 24–37. [CrossRef]
22. Bidai, Z.; Maimour, M. Interference-aware multipath routing protocol for video transmission over ZigBee wireless sensor networks. In Proceedings of the 2014 International Conference on Multimedia Computing and Systems (ICMCS 2014), Marrakech, Morocco, 14–16 April 2014; pp. 837–842.
23. Haenggi, M.; Puccinelli, D. Routing in ad hoc networks: A case for long hops. *IEEE Commun. Mag.* **2005**, *43*, 93–101. [CrossRef]
24. Memsic. *TelosB Datasheet*; Crossbow Technology, Inc.: San Jose, CA, USA, 2011.
25. Wiegand, T.; Sullivan, G.J.; Bjontegaard, G.; Luthra, A. Overview of the H. 264/AVC video coding standard. *IEEE Trans. Circuits Syst. Video Technol.* **2003**, *13*, 560–576. [CrossRef]
26. Ullah, S.; Ahmad, J.J.; Khalid, J.; Khayam, S.A. Energy and distortion analysis of video compression schemes for Wireless Video Sensor Networks. In Proceedings of the 2011 Military Communications Conference (MILCOM 2011), Baltimore, MD, USA, 7–10 November 2011; pp. 822–827.
27. Jung, D.; Teixeira, T.; Savvides, A. Sensor node lifetime analysis: Models and tools. *ACM Trans. Sens. Netw.* **2009**, *5*, 3. [CrossRef]
28. *IEEE Std 802.15.5™-2009: Mesh Topology Capability in Wireless Personal Area Networks (WPANs)*; IEEE Computer Society: Washington, DC, USA, 2009.

29.   Zheng, J.; Lee, M.J. A resource-efficient and scalable wireless mesh routing protocol. *Ad Hoc Netw.* **2007**, *5*, 704–718. [CrossRef]

30.   Rodenas-Herraiz, D.; Garcia-Sanchez, F.; Garcia-Sanchez, A.-J.; Garcia-Haro, J. On the influence of the hidden and exposed terminal problems on asynchronous IEEE 802.15. 5 networks. *Comput. Stand. Interfaces* **2015**, *42*, 53–70. [CrossRef]

31.   The Network Simulator (ns-2). Available online: http://www.isi.edu/nsnam/ns (accessed on 3 May 2016).

32.   Klaue, J.; Rathke, B.; Wolisz, A. Evalvid—A framework for video transmission and quality evaluation. In *Computer Performance Evaluation: Modelling Techniques and Tools*; Springer: Berlin, Germany, 2003; pp. 255–272.

33.   IEEE 802.15.5 Video Simulation Tool. Available online: http://labit501.upct.es/~{}agarcia/Project/ (accessed on 18 February 2016).

34.   YUV Video Sequences. Available online: http://trace.eas.asu.edu/yuv/ (accessed on 18 February 2016).

35.   Garcia-Sanchez, A.-J.; Garcia-Sanchez, F.; Rodenas-Herraiz, D.; Garcia-Haro, J. On the synchronization of IEEE 802.15. 5 wireless mesh sensor networks: Shortcomings and improvements. *EURASIP J. Wireless Commun. Netw.* **2012**, *2012*, 1–23. [CrossRef]

36.   Mean Opinion Score. Available online: https://www.itu.int/rec/T-REC-P.800.2-201305-I/en (accessed on 18 February 2016).

37.   Ohm, J.-R. *Bildsignalverarbeitung Fuer Multimedia-Systeme*; Skript Institut für Nachrichtentechnik und theoretische Elektrotechnik der TU: Berlin, Germany, 1999.

38.   Mobile QCIF WO-AO 5-3. Available online: https://www.youtube.com/watch?v=KVWEtJbNfoM (accessed on 18 February 2016).

39.   Akiyo QCIF WO-AO 5-3. Available online: https://www.youtube.com/watch?v=UMYEFdpfRXY (accessed on 18 February 2016).

40.   Akiyo QCIF WO-AO 7-2. Available online: https://www.youtube.com/watch?v=31avC1TEFo0 (accessed on 18 February 2016).

41.   Foreman QCIF WO-AO 5-3. Available online: https://www.youtube.com/watch?v=rU-83jD8rDI (accessed on 18 February 2016).

42.   Foreman QCIF WO-AO 7-2. Available online: https://www.youtube.com/watch?v=neM4w04LyNY (accessed on 18 February 2016).

43.   Mobile QCIF WO-AO 7-2. Available online: https://www.youtube.com/watch?v=CdgpNBUzFhQ (accessed on 18 February 2016).

sensors

MDPI

Review

State of the Art in LP-WAN Solutions for Industrial IoT Services

Ramon Sanchez-Iborra * and Maria-Dolores Cano

Departamento de Tecnologías de la Información y las Comunicaciones, Universidad Politécnica de Cartagena, Cartagena 30202, Spain; mdolores.cano@upct.es
* Correspondence: ramon.sanchez@upct.es; Tel.: +34-968-325-953

Academic Editor: Gonzalo Pajares Martinsanz
Received: 25 February 2016; Accepted: 9 May 2016; Published: 17 May 2016

Abstract: The emergence of low-cost connected devices is enabling a new wave of sensorization services. These services can be highly leveraged in industrial applications. However, the technologies employed so far for managing this kind of system do not fully cover the strict requirements of industrial networks, especially those regarding energy efficiency. In this article a novel paradigm, called Low-Power Wide Area Networking (LP-WAN), is explored. By means of a cellular-type architecture, LP-WAN–based solutions aim at fulfilling the reliability and efficiency challenges posed by long-term industrial networks. Thus, the most prominent LP-WAN solutions are reviewed, identifying and discussing the pros and cons of each of them. The focus is also on examining the current deployment state of these platforms in Spain. Although LP-WAN systems are at early stages of development, they represent a promising alternative for boosting future industrial IIoT (Industrial Internet of Things) networks and services.

Keywords: Low-Power Wide Area Networks (LP-WAN); Machine-to-Machine (M2M) communications; Industrial Internet of Things (IIoT); Internet of Things (IoT); wireless sensor networks

---

1. Introduction

Machine-to-Machine (M2M) networks and Industrial Internet of Things (IIoT) services are two key enabling approaches for future industrial networking [1]. As reflected from the forecast investments predicted in the IIoT field [2], the advent of low-cost, always-connected devices opens new and exciting opportunities involving many stakeholders from a wide range of sectors. Deploying well-structured and easily-accessible M2M networks will facilitate having a precise control over the production or company's installations, which could be translated into a smart strategy for saving logistic costs [3]. As an example, new services such as real-time event processing or 24/7 access to tracking information will be introduced into the supply chain. Having a thorough monitoring system deployed all along the manufacturing and supply chain allows enriching the complete value chain with precious information, minimizing losses against unexpected events, and hence improving both business processes and the information exchange among stakeholders (Business-to-Business (B2B) networks) [4]. In this case, smart metering (water, oil, *etc.*), goods and facilities monitoring, or smart farming are good examples of areas of activity for M2M/B2B networks.

M2M networks can be seen as a revamp of the widely-deployed Wireless Sensor Networks (WSN); we could also think that most of the aforementioned applications are already covered by this well-studied approach. It is true that we have survived so far with the existing WSN *classic* solutions such as ZigBee, Bluetooth, or even WiFi (short-range technologies), but the main point of industrial M2M networks is the huge increase in the number of devices composing them and the notable widening of the covered areas. Global device connections are estimated to be about 28 billion by 2020 (Figure 1) [5]. This enormous growth requires (i) minimized cost per unit; (ii) optimized edge-nodes' energy consumption; (iii) high network scalability; and (iv) wide network coverage. As discussed in the next sections, one or many of these points are the main weaknesses of traditional WSN technologies. In addition, as mentioned previously, lots of industrial applications need to operate over vast regions that are unaffordable for those *classic* WSN solutions. The need of rich coverage has been solved by means of existing cellular technologies (usually with low bandwidth), e.g., GSM (Global System for Mobile communications), GPRS (General Packet Radio Service), *etc.*, or satellite connectivity (long-range technologies), but the increased costs and the high level of power demanded by these systems make them unsuitable for long-term M2M networks composed by a massive number of devices.

Figure 1. Worldwide IoT connected devices and revenues forecast. Data extracted from [5].

A new paradigm called Low-Power Wide Area Networking (LP-WAN) has arisen recently, aimed at filling the existing gap for deploying overcrowded M2M networks [6]. The main foundation of these systems is the deployment of highly scalable systems, usually in an operated fashion, employing low-cost edge-devices with low battery consumption. Figure 2 presents the typical architecture of a LP-WAN system. Observe that, essentially, the network architecture is similar to that of cellular networks, where one or a series of base stations provides direct connectivity from edge-devices to the backhaul network and, then, to the cloud, where the data is stored and prepared to be accessed. Regarding the edge-network architecture, it is notably different from that employed by traditional WSN. Basically, instead of composing a local network and using a gateway for sending outside the collected data, end-nodes directly connect to the base station. This configuration allows simplifying the network management complexity and also reduces energy consumption given that routing tasks are avoided.

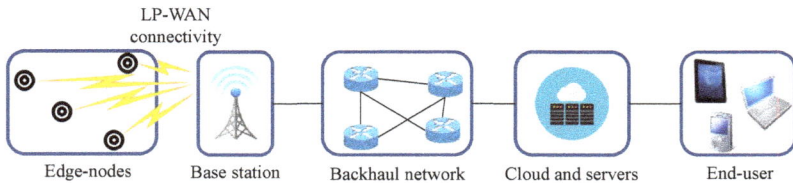

Figure 2. LP-WAN network architecture.

Different LP-WAN platforms have been proposed, each of them with their own particularities and individual features that make them more suitable for different types of IIoT services. This issue will be addressed in the next sections as follows. Section 2 identifies the limitations that the *classic* IIoT solutions present. A detailed overview of the LP-WAN paradigm, covering the key characteristics of the most prominent LP-WAN platforms, is developed in Section 3. Section 4 focuses on the deployment state of LP-WAN technology in Spain. Section 5 presents a thorough discussion about the reviewed LP-WAN proposals, exploring the answers given to the challenges previously identified. Finally, the paper ends outlining the main conclusions.

2. Limitations on Existing IIoT Solutions

Current enabling technologies for IIoT services can be divided into short-range and long-range approaches. The main impediments found to implement sustainable cost-effective IIoT solutions are related to: (i) network management costs; (ii) scalability and network organization; (iii) edge-nodes' dimensioning and power efficiency; and (iv) coverage. In the following, these points are identified and reviewed for different short-range and long-range technologies that have been employed so far for supporting IIoT applications. Please note that although the list of solutions provided in this section does not intend to be exhaustive, it permits us to identify the principal challenges in deploying these types of M2M networks.

2.1. Short-Range Connectivity

Systems with short-range connectivity were the first ones employed to manage WSN. Depending on the adopted wireless technology, which strongly determines the Physical (PHY) and Medium Access Control (MAC) layers, the network presents more suitable characteristics for supporting one application or another.

Regarding the network management costs, one typical characteristic for this kind of solution is the private ownership of a great part of the network. This fact should not be ignored because it causes an increase in both the expense and complexity of the operations. On the one hand, the owner is in charge of the complete deployment process, from the edge-device placement to the backhaul network management, in order to make data accessible from outside (including security issues). Besides, failures happening in the private part of the system should be handled by the owner company, which might not be always be able to cope with these tasks and would have to assume extra expenses by outsourcing this service. On the contrary, by employing public networks, there is a clear change in the business model and, hence, the deployment costs are shared: the subscriber assumes the edge-device costs, whereas the network operator bears the backhaul network deployment and maintenance expenses. During the operation stage, the subscriber pays a fee to the network operator for the system maintenance service, some kind of technical support, and, usually, for having a friendly back-end for data accessing. In the case that a traditional WSN adopted the public-architecture strategy, the direct communication between both extremes would not be feasible due to the limited transmission range of the edge-nodes [7]. Thus, additional equipment, *i.e.*, gateways, or sophisticated data-collection strategies, would be needed to connect the edge-nodes to the central base station.

In addition to network management, as a large-scale issue, there are other problems regarding the edge-nodes' functionality when they are managed by the existing short-range solutions. For example, the most employed technologies for operating WSNs, *i.e.*, the IEEE 802.15.4-based protocols ZigBee and 6loWPAN, present highly interesting features in terms of energy efficiency and the low cost of the edge-devices. However, the growth of this type of network is limited because the management complexity and interference issues could suffer a noticeable increase with the increment of the network size [7,8]. Although several routing algorithms based on different paradigms such as multi-hop routing, opportunistic networks, or delay-tolerant networks have been proposed, an important number of concentrators (or information collectors) might be still needed in relatively large networks, which could also increase the overall network power consumption [9].

As well as the possible effect in terms of higher network consumption in more dense scenarios with ZigBee or 6loWPAN protocols mentioned before, the use of other technologies such as WiFi and Bluetooth (not oriented to WSN in its inception, but widely used for this purpose) could have a negative impact on energy efficiency. The main issue presented by these solutions is that they were designed to support highly-bandwidth-demanding applications and, hence, transmission/reception tasks waste a lot of energy. Additionally, the management of a network composed by a significant number of nodes is also tricky as these networks are often based on the Internet Protocol (IP), so different topology-organization methods, e.g., clustering, are needed [10,11].

Another important issue, common to all short-range technologies mentioned so far (IEEE 802.15.4-based protocols ZigBee and 6loWPAN, WiFi, and Bluetooth), is the need for a connection to the Internet in order to upload all collected data to the cloud. While in urban or suburban areas this should not be a problem, in remote locations it could be difficult or, at least, expensive because these areas usually lack a preexisting infrastructure that could provide Internet access [12]. Additionally, special equipment such as bridges is needed for different reasons. Firstly, these nodes are employed as intermediate points between the backhaul network and the edge-nodes due to the limited coverage range of the latter. Besides, all the collected data need to be gathered and formatted before sending it to the storage servers. When talking in terms of Big Data, accomplishing an accurate dimensioning of the bandwidth and the temporal storage needs of these devices is not a trivial task. For all these reasons, other approaches based on long-range technologies have also been employed for deploying IIoT services.

2.2. Long-Range Connectivity

The first idea that comes to mind in order to solve the issues described above is cellular networks: they are based on public infrastructure, they are widely deployed and cover large areas, and they are operated employing well-known standards such as GSM, GPRS, or 3G/4G. Following this strategy, the edge-sensors collect the data of interest and, afterwards, send it to the cloud via a cellular data link, e.g., GPRS, 3G, *etc.* However, the main problem with these systems is that they were designed to fulfill different requirements than those of IIoT services. While in cellular networks the trend has been increasing the available bandwidth, aiming to accomplish the increasing demand of multimedia traffic by human users [13], in IIoT services the strategy should be optimizing bandwidth usage and decreasing energy consumption and costs [14]. Current cellular base stations are capable of hosting a small number of connected users (in comparison with the needs of sensorization services), with a relatively high bandwidth assured for each of them. In turn, what a machine-only network demands is a solution for supporting a huge number of low-throughput connected devices that send short messages only once in a while. Therefore, the current cellular solutions are clearly inefficient in terms of scalability and energy consumption. Regarding the former, one possible strategy for organizing and providing connectivity to independent systems is using femtocells [15] or picocells [16]. However, this solution notably increases the system cost as new equipment and connection infrastructure are required. Focusing on energy efficiency, cellular networks need a quasi-constant communication between edge-nodes and the base station for management tasks (protocol overhead), which is completely devastating for battery lifetime. Moreover, existing cellular networks work on scarce and expensive (licensed) frequency bands.

Another solution with even more drawbacks is satellite communications. Although they provide a good coverage worldwide, the energy consumed in each transmission is too much for IIoT applications. In addition, the high latency of these transmissions could be inadmissible for certain applications with strict temporal constraints. Finally, with respect to network costs, subscribing a satellite connection plan is still excessively expensive. Although cheaper, current cellular network operators have not substantially reduced their subscription fees. For all these reasons, Low-Power Wide Area Networks appear as an alternative long-range solution to give response to the IIoT services' demands.

3. LP-WAN Solutions for IIoT Services

Recently, a number of different platforms following the LP-WAN paradigm have arisen. These proposals aim at gathering both the long transmission range provided by cellular technologies and the low energy consumption of WSNs (Figure 3). Many LP-WAN proposals are at an early development stage and others have already begun their architecture deployment. LoRaWAN, Sigfox, and Ingenu are currently the LP-WAN platforms with the greatest momentum and they have been reviewed in recent works [17,18]. However, there are many other proprietary and standard platforms with interesting proposals that we also consider in the following sections. Although each of these LP-WAN solutions has its own particularities and protocols (many of them proprietary), there are some common foundations which all of them rely on.

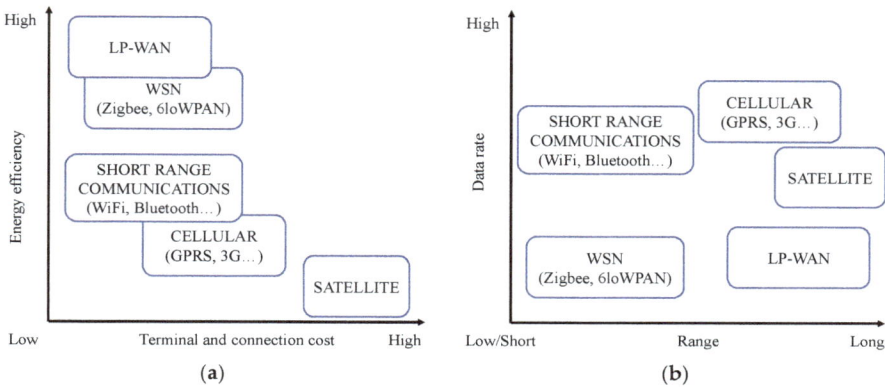

Figure 3. Principal characteristics of IIoT-enabling technologies. (**a**) Data rate and coverage range; (**b**) Energy efficiency and terminal and connection cost.

As shown in Figure 2, LP-WANs make use of a star topology, where all edge-nodes are directly connected to the base station; hence, the LP-WAN modem is directly installed in edge-devices. In some cases, concentrators/gateways can be used to connect a cluster of nodes to the base station (star-of-stars topology). The base station and the backhaul network are usually public and operated by the service provider. As discussed above, this fact liberates subscribers from deployment, maintenance tasks, and operational costs related to this part of the system. Regarding the edge-network connectivity with the base station, most of the proposed platforms employ ISM (Industrial, Scientific, and Medical) frequency bands; concretely, the most employed frequencies are those within the sub-GHz bands, namely 868 MHz in Europe, 915 MHz in the US, and 920 MHz in Japan. In comparison with the 2.4 GHz band, transmitting in a lower-frequency band leads to a deeper wave penetration and range, which are highly valued characteristics in order to provide indoor connectivity. Furthermore, electronic circuits are more efficient at lower frequencies.

Another common characteristic in these systems is the asymmetric connectivity provided to edge-nodes. Aimed at reducing energy consumption, most of the solutions focus on the uplink connection; thus, the downlink is severely limited, hence reducing the necessary "listening" time needed for receiving data. It is clear that most data flow from the edge-network to the core, but in the case of having not only sensors but also actuators, an effective downlink would be also highly appreciated. It would be useful for updating the edge-nodes' software, too. To deal with these issues, different strategies have been adopted to provide a base station-to-edge-nodes downlink, as discussed later.

In summary, the main advantages that all LP-WAN platforms claim to own are: (i) high scalability and range, necessary for super-crowded networks deployed in vast areas; (ii) roaming, useful for goods-delivery tracking; (iii) real-time event alerts, which are set up by the customer and automatically triggered from the LP-WAN operator's management system; and; (iv) low edge-node energy consumption and cost. In the following, a brief review about the most prominent LP-WAN platforms arisen so far is provided.

3.1. LoRaWAN

This platform is promoted by the LoRa Alliance [19], composed by IBM, Semtech, and Actility, among others. It proposes a star-of-stars topology with dedicated gateways serving as transparent bridges between edge-nodes and the central network, where the data is stored and made available to the subscriber. The edge-nodes connect to the access points via one-hop links by using the LoRa (Long Range) modulation. This is Semtech's proprietary Chirp Spread Spectrum (CSS) radio scheme that employs a wide channel of up to 250/500 kHz (Europe/North America) and provides adaptive data rate capabilities by means of a variable processing gain. Please note that this concept represents the ratio between the chip rate and the baseband information rate, and is usually known as the Spreading Factor (SF). LoRaWAN presents a SF from 7 to 12. Using this last characteristic, edge-nodes can tune the transmission power and bitrate to the real network conditions, allowing a reduction in energy consumption. Moreover, LoRaWAN defines three types of edge-devices depending on their downloading needs: Class A devices have a scheduled downloading window just after each uplink connection (Receiver-Initiated Transmission strategy, low power consumption), Class B devices have additional scheduled downlink windows (Coordinated Sampled Listening strategy, medium power consumption), and Class C devices can receive messages almost at any time (Continuous Listening strategy, large power consumption). In its specification sheets, LoRaWAN claims a Class A edge-node's battery lifetime is over five years.

Originally, LoRaWAN was designed to work in ISM bands but it can be also adapted for supporting the licensed spectrum. Under these conditions, LoRaWAN claims to demodulate signals 19.5 dB below the noise floor, hence achieving greater ranges than those provided by cellular base stations. In both communication directions, the adaptive data-rate ranges from 0.25 kbps (0.98 kbps in North America due to FCC (Federal Communications Commission) limitations) up to 50 kbps, with a maximum payload length of 256 bytes. Finally, security issues have been thoroughly considered, so that end-to-end AES (Advanced Encryption Standard) encryption security, including the use of a unique network, application, and device keys for encrypting data at different OSI (Open Systems Interconnection) levels, is provided.

3.2. Sigfox

This is the platform in the most advanced deployment state in Europe. By means of agreements with local cellular network operators, Sigfox [20] claims to have covered most of the territory of France, Russia, and Spain, among others. Technically speaking, this solution is quite different from the LoRaWAN approach. Instead of using bidirectional spread spectrum channels, Sigfox employs proprietary ultra-narrow band modulation (Differential Binary Phase Shift Keying, DBPSK) with a heavily limited uplink connection. Using this modulation, a maximum data rate of 100 bps can be achieved by transmitting messages with a maximum payload length of 12 bytes. Meanwhile, using this low bitrate permits large ranges of 10 km and beyond with very low transmission power, which allows saving energy at edge-nodes. Sigfox's technical sheets claim a typical stand-by time of 20 years with a 2.5 Ah battery.

Sigfox's star topology is similar to a cellular architecture, with a wide deployment of base stations aimed at covering entire countries by employing ISM bands. This base station structure permits edge-nodes to upload the gathered data directly to Sigfox servers, which makes it accessible to subscribers through a web-based API (Application Programming Interface). The use of ISM bands together with Sigfox's medium access strategy, namely without collision-avoidance techniques, leads to a stringent bandwidth-occupancy limitation suffered by edge-nodes. For example, a duty cycle of 1% is established in the Europe regulations; hence, a maximum of 140 messages per edge-node per day are allowed. In the case of the USA regulations, Sigfox's limited data rate of 100 bps shows that transmitting single messages usually takes 2–3 s, which is outside the FCC's maximum message transmission time in ISM bands of 0.4 s. Although originally designed as a unidirectional system,

Sigfox has lately included a limited downlink window (four messages of eight bytes per edge-node per day) similar to the strategy adopted by LoRaWAN's Class A devices (please see previous sub-section).

Regarding security issues, Sigfox implements frequency-hopping and anti-replay mechanisms in their servers, but no encryption techniques are used between end-nodes and base stations. Additionally, the payload format is undefined. Therefore, Sigfox's security strategy relies on the fact that an intercepted message cannot be interpreted unless the attacker is able to understand the particular subscriber's system.

3.3. Weightless

Weightless is the alliance name for a set of three LP-WAN open standards: Weightless-W, Weightless-N, and Weightless-P [21]. The three Weightless flavors work in sub-GHz bands, but each of them has its own particularities.

The original Weightless-W standard makes use of the TV whitespace spectrum and provides a wide range of modulation schemes, spreading factors, and packet sizes. Considering all these features, and depending on the link budget, Weightless-W claims to achieve two-way data rates from 1 kbps to 10 Mbps with very low overhead. Due to the extensive feature set provided by Weightless-W, the edge-node's battery lifetime is limited to three years and the terminal cost is higher than that of its competitors. The communication between the edge-nodes and the base station can be established along 5 km, depending on the environmental conditions.

In turn, Weightless-N uses a class of low-cost technology, very similar to that employed by Sigfox. Thereby, ultra-narrow band (DBPSK) modulation is adopted in order to provide unidirectional-only connectivity of up to 100 bps, exploiting ISM bands. This scheme is based on nWave's technology [22], which was donated as a template for the Weightless-N standard. Because of the simplicity of this solution, Weightless-N allows a battery duration of up to 10 years, very low cost terminals, and a long connection range similar to that reached by Weightless-W.

Finally, the newest Weightless-P open standard is derived from the M^2Communication's Platanus protocol [23]. This version gathers together the most proper characteristics of the previous standards, and it claims to be specifically focused on the industrial sector. Using a narrow-band modulation scheme (Gaussian Minimum Shift Keying, GMSK, and Offset Quadrature Phase Shift Keying, OQPSK) operating in 12.5 kHz channels, Weightless-P implements bi-directional communication with an adaptive data rate from 200 bps to 100 kbps. It supports both ISM and licensed spectrum operation. Aimed at providing the reliability demanded by some industrial applications, Weightless-P includes, by default, valued characteristics such as acknowledged transmissions, auto-retransmission, frequency and time synchronization, and channel coding, among others. Compared with the other Weightless standards, Weightless-P provides a more limited range of 2 km and its advanced features in comparison with Weightless-N permit a shorter battery lifetime of three years.

Regarding security, the three Weightless versions provide end-to-end network authentication and 128 bit AES encryption.

3.4. Other Alternatives

Besides the three solutions mentioned so far, there are other alternatives that, up to the date of preparing this article, either are in a less advanced deployment state or their technical insights are not yet available. For example, Ingenu (formerly known as On-Ramp) is a LP-WAN platform currently beginning its deployment in the USA. It is based on its proprietary RPMA (Random Phase Multiple Access) technology, which has the particularity of working in the 2.4 GHz band. In addition, it permits both star and tree topologies by using different network hardware. Although Ingenu has raised high expectations regarding the range, edge-device's battery lifetime, and available bandwidth [24], these promising figures should be confirmed in real deployments as they have been extracted so far only from simulation studies.

Mostly focused in the Smart Cities market, Telensa [25] has also developed its own bi-directional ultra-narrow-band technology. Telensa's PLANet (Public Lighting Active Network) and PARKet are focused on street lighting control and smart parking enhancement, respectively. Both of them are defined as end-to-end systems, from edge-nodes (telecells) to the end-user interface, including base stations. By using their proprietary technology, Telensa claims to reach 2–3 km (urban) and 5–8 km (rural) real ranges. They have already deployed their solutions in different big cities worldwide.

In turn, Dash7 is an open standard promoted by the Dash7 Alliance [26], which has its origin in the ISO/IEC 18000-7. Unlike the afore-reviewed solutions, Dash7 proposes a two-hops tree topology composed by hierarchized devices, namely endpoints, sub-controllers, and gateways. Notice that this topology is similar to the traditional WSN architecture instead of the long-range systems described in this article. The main advantages provided by the Dash7 protocol are the extended range in comparison with other pure-WSN solutions due to the use of sub-GHz bands (433 MHz and 868/915 MHz), the possibility of direct device-to-device communication, which is not currently available in any of the LP-WAN platforms described above, and its compatibility with Near Field Communication (NFC) radio devices. However, this proposal has not been widely adopted yet, and only some pilot projects have been carried out so far [27].

Finally, it is worth mentioning other solutions such as those proposed by Helium [28], M2M Spectrum Networks [29] (recently joined the LoRa Alliance), or Amber Wireless [30] which, although less expanded, could bring more competence to this growing market in the future.

3.5. Standardization Bodies' Efforts

Besides the platforms reviewed above, there are different solutions proposed by well-recognized standardization bodies that are currently under study. For example, the IEEE has proposed the P802.11ah [31] and 802.15.4k [32] standards. The former presents a series of modifications at the 802.11 PHY and MAC layers aimed at adapting them to sub-GHz bands (excluding TV white space). Using the well-studied Orthogonal Frequency Division Multiplexing (OFDM), it is intended to reach a minimum data rate of 100 kbps and a transmission range up to 1 km [33]. In this standard, the co-existence with other technologies, such as all those based on the IEEE 802.15.4 PHY-layer specifications, is being considered. In turn, the IEEE 802.15.4k standard presents MAC and PHY layer specifications to facilitate Low Energy Critical Infrastructure Monitoring (LECIM) applications. This standard defines two PHY modes: Direct-Sequence Spread Spectrum (DSSS) and Frequency Shift Keying (FSK). The former permits links of up to 20 km in line of sight (5 km in non–line of sight) with data rates of up to 125 kbps. The proposed architecture is a point-to-multipoint network by means of a star topology composed by two types of nodes, namely a PAN (Personal Area Network) coordinator and the edge-devices. The communication between the collector and the sensors is asymmetric, aimed at limiting the "listening" time of the battery-powered sensors. This standard permits employing both sub-GHz and 2.4 GHz bands using Binary Phase Shift Keying (BPSK) and OQPSK modulations.

In turn, the 3GPP group (3rd Generation Partnership Project) is working on the development of the LTE-MTC (Long Term Evolution-Machine-Type Communications) standard [34]. In the LTE Release 12, the Cat 0 speed of 1 Mbps was defined, but in order to reduce the chipset's complexity and power consumption, there is a plan to define an even lower speed of about 200 kbps (referred to as Cat M) in the next release, Release 13. Although the standard is still being developed, it has been decided to make use of 1.4 MHz channels within the cellular band (450 MHz) in order to provide bi-directional connectivity between edge-nodes and the base station. Finally, aimed at presenting a comprehensive comparison among all the reviewed LP-WAN platforms, tab:sensors-16-00708-t001 shows their most relevant characteristics. Please note that the presented values have been extracted from the platform's specification sheets and some of them could be provisional figures due to the ongoing evolution of the different solutions.

Table 1. LP-WAN platforms summary.

	LoRaWAN	Sigfox	Weightless -W	Weightless -N	Weightless -P	Ingenu	Telensa	Dash7	IEEE 802.15.4k (DSSS)	IEEE P802.11ah	LTE-MTC
Band	433/868/780/915 MHz	868/915 MHz	TV whitespace	Sub-GHz	Sub-GHz	2.4 GHz	Sub-GHz	Sub-GHz	Sub-GHz/2.4 GHz	Sub-GHz	Cellular
Max. data-rate	50 kbps	100 bps	10 Mbps	100 bps	100 kbps	19 kbps/MHz	346 Mbps	-	125 kbps	346 Mbps	200 kbps
Range (urban)	5 km	10 km	5 km	5 km	2 km	15 km	1 km	3 km	5 km	1 km	5 km
Packet-size	Max. 256 B	12 B	Min. 10 B	Max. 20 B	Min. 10 B	Max. 10 kB	Max. 65 kB.	-	Max 32 B	Max. 65 kB.	-
Downlink	Yes. Different plans	Yes (not sym.)	Yes (sym.)	No	Yes (sym.)	Yes (not sym.)	Yes (sym.)	Yes (sym.)	Yes (not sym.)	Yes (sym.)	Yes (sym.)
Topology	Star-of-stars	Star	Star	Star	Star	Star/Tree	Star/Tree	Star	Star	Star/Tree	Star
Roaming	Yes	Yes	Yes	Yes	Yes	Yes	Yes	Yes	-	Yes	Yes
Security	Fully addressed	Partially addressed	Fully addressed	Fully addressed	Fully addressed	Fully addressed	In development	-	Partially addressed	In development	In development
Protocol ownership	Partially proprietary	Proprietary	Standard	Standard	Standard	Proprietary	Standard	Proprietary	Standard	Standard	Standard

4. Current Deployment State of LP-WAN Solutions in Spain

As in the rest of the world, the rollout of LP-WAN platforms in Spain is in its beginning stages. Currently, there is one solution with a clear advantage over the rest: Sigfox. After reaching an agreement with the network operator Cellnex Telecom [35,36], Sigfox has reached a count of more than 1300 base stations covering the Spanish territory. Thus, Sigfox employs the already-deployed Cellnex (previously known as Abertis Telecom) infrastructure. This strategy of partnering with a big network operator has been also adopted by Sigfox in other countries such as France (TDF [37]) and the Netherlands (Aerea [38]). Regarding the Spanish case, Sigfox has focused on security services (e.g., to connect alarm systems to the cloud) and is beginning its expansion to other niche markets (e.g., in smart farming and precision agriculture).

Although far from the Sigfox network's deployment state, other platforms have begun their landing in Spain, too. For example, a LoRaWAN pilot network is planned to be deployed in the city of Malaga by the Swiss company iSPHER [39,40]. Therefore, by rolling out their SPHER NET, an operational end-to-end LoRa IoT network solution, the full city territory will be covered. Up to the date of writing this article, this project is still at an early stage of development.

Regarding the standard solutions, the deployment of the LTE-MTC technology will permit current cellular carriers to take advantage of their deployed infrastructure. LTE-MTC will be compatible with the normal construct of LTE networks, so the network operators only will have to update their systems' software. In Spain, several cellular carriers have already deployed their own infrastructure; thus, more competitors will arise with the advent of this promising standard.

Aimed at providing a specific scenario of applicability for LP-WAN solutions, in the following the case of irrigation water smart metering is discussed; this is a greatly valued good in the southern regions of Spain [41,42]. Due to the shortage of water and its expensive price, both water companies and farmer associations are highly interested on having a thorough control of water consumption [43]. The main obstacle found until now is the remote location of the fields, which in many cases lack of any kind of connectivity or even electricity. Therefore, having a centralized control of water consumption is greatly challenging in this scenario. Due to the great distances among fields, it is not feasible to deploy an interconnected WSN with the aim of routing the collected to data to a gateway connected to the Internet. Even more, as explained in previous sections, the gateway's Internet connection would be difficult and expensive to establish. In such remote locations, it is usual to not have GSM/GPRS coverage, so employing cellular networks is not a valid strategy either. Therefore, this is a good example of the applicability of LP-WAN solutions. Given the great coverage range of base stations, especially in free space, one of these stations can provide connectivity to several water meters, which can directly submit their readings to the base station, making them accessible almost in real time. Thus, abusive consumption, water theft, or pipe losses can be easily detected, increasing the whole system's efficiency with an inexpensive investment [44].

5. Discussion (All that Glitters Is Not Gold)

We are witnessing the dawn of LP-WAN solutions for wide and overcrowded M2M networks and IIoT services. There are differentiating characteristics such as the data rate, power consumption, or cost that work against each other. Consequently, none of the existing platforms provides the best performance for all of these requirements. Thus, once the needs of the service to be deployed are specified, the LP-WAN solution that matches best will be chosen. For that reason, there is not a clear dominant platform yet among all the arisen platforms that could completely fulfill the key challenges identified in Section 2:

- Focusing on management costs, most platforms offer the same model to their customers: the subscriber assumes the expenses of deploying the edge-network and pays a fee to the LP-WAN operator for managing and making all the collected data accessible. This is an adequate solution, as the issues and expenses related to the information management process are avoided by the subscriber.

- In terms of network organization and the edge-nodes' dimensioning, it seems that the star topology allows an easy and straight connection from each end-node to the base station. However, although all the cited solutions claim high system scalability with base station capacities of thousands of simultaneously connected nodes, other topologies such as star-of-stars or tree architectures could improve this scalability at the expense of employing special nodes (concentrators) and increasing the edge-network complexity.

- Regarding power efficiency, every reviewed platform ensures edge-node lifetimes of some years. Of course, these figures depend on the number of messages transmitted per day, the transmission bitrate, and other factors such as the edge-node's downlink strategy.

- Concerning the area covered by the system, the explored solutions claim connectivity ranges of at least 1 km from the base station. Those platforms operating at the sub-GHz band take advantage of greater transmission distances and wave penetration in comparison with those systems employing the 2.4 GHz band. In addition, solutions adopting a hierarchized architecture, e.g., Dash7, could also extend the network coverage at the expense of needing more hops between the edge-nodes and the backhaul network.

Furthermore, there are other points regarding the service reliability and security that seem important for the proper operation of IIoT applications and represent challenges not fully covered yet. Focusing on reliability, it is clear that outdoor or industrial environment conditions are not the most favorable for sensor (edge-device) deployment. They are sometimes installed in extreme temperature and moisture conditions, near potential noise (acoustic and electromagnetic) sources, or under other hostile scenarios. Considering that M2M networks are self-regulated and that one unheard or non-transmitted message could provoke loss of revenues, the reliability of these systems should be heavily ensured. In addition, most of the cited platforms avoid using the 2.4 GHz band because of its "current saturation" [45]. However, in the near future the forecasted billions of connected things will be transmitting in the sub-GHz band; hence, the impact of the interferences among all the co-existing technologies will not be negligible either. As another relevant point, the sending and processing time for each transmission should not be ignored in applications with severe timing constraints or in the case of messaging between sensors and actuators. Besides, an effective downlink should be ready to transmit the proper message back to the edge-network if necessary. In architectures where direct device-to-device communication is allowed, e.g., Dash7, this issue could be easily solved, but in the more common star topology, messages should be firstly processed by the LP-WAN operator's systems.

Regarding security, for mission-critical or high-security applications, the use of private data storage or servers would be more convenient than using third-party (e.g., LoRaWAN, Sigfox, *etc.*) servers. In the last case, the data owner could lose control of the information management process; this could be risky or even unacceptable in certain applications. Additionally, as the ISM bands are freely accessible, they are vulnerable to a broad range of security threats; therefore, including extra functionality to support the functions of confidentiality, authentication, authorization, or even accounting would be very welcome. Of course, all these new features would be against the edge-device's power consumption, so a balance between the edge-nodes; functionality and energy use would be necessary.

Besides these important issues more focused on the network's technical insights, the business model emerges as another key challenge for taking advantage against the competitors. Having the best technological solution does not always lead to success. For example, we have seen that ultra-narrow-band technology presents a series of drawbacks in comparison with other modulation schemes that offer better connectivity. However, Sigfox seems to be very attractive to potential customers due to its simplicity and its higher degree of deployment. It is on this last point where LP-WAN companies have to make the biggest economical effort and some of them have focused on different specific regions. While Sigfox seems to be more focused, for the moment, in Europe, with several countries fully covered, LoRa-WAN and Ingenu are focused on the North American market. Regarding territory coverage plans, they are commonly designed regarding the territory's population; thus, the major urban areas are usually mostly covered but there is often a lack of connectivity in rural sites. Precisely, many big factories and farms are isolated in these emplacements, so quasi-dedicated base stations will be needed to provide services to these customers.

To sum up, we are currently in a highly dynamic scenario, with all the different platforms positioning themselves in the market. The diverse technological and business solutions offered by each of them will determine their success or failure, but there is no doubt that the LP-WAN is a rising technology that will play an important role in the forthcoming expansion of IIoT services.

6. Conclusions

This article discussed different enabling solutions for the imminent IIoT era. Taking advantage of these technologies will make companies ready to tackle future large-scale challenges, improving business productivity at several levels. In addition, the new networking solutions presented here are also focused on reducing power consumption in order to construct more efficient and sustainable architectures. The LP-WAN paradigm seems to be a promising response to the limitations showed by current technologies, but we are just at the very beginning of the IIoT explosion, so it will be necessary to remain vigilant to the new challenges that the upcoming M2M-based services will pose.

Acknowledgments: This work was supported by the MINECO/FEDER project grant TEC2013-47016-C2-2-R (COINS).

Conflicts of Interest: The authors declare no conflict of interest.

Abbreviations

The following abbreviations are used in this manuscript:

LP-WAN	Low-Power Wide Area Networking
M2M	Machine-to-Machine
IIoT	Industrial Internet of Things
B2B	Business-to-Business
WSN	Wireless Sensor Networks
GSM	Global System for Mobile communications)
GPRS	General Packet Radio Service
MAC	Medium Access Control
ISM	Industrial, Scientific, and Medical
CSS	Chirp Spread Spectrum
SP	Spreading Factor
FCC	Federal Communications Commission
DBPSK	Differential Binary Phase Shift Keying
GMSK	Gaussian Minimum Shift Keying
OQPSK	Offset Quadrature Phase shift Keying
RPMA	Random Phase Multiple Access
LECIM	Low Energy Critical Infrastructure Monitoring
3GPP	3rd Generation Partnership Project
LTE-MTC	Long Term Evolution-Machine-Type Communications

References

1. Anton-Haro, C.; Dohler, M. *Machine-to-Machine (M2M) Communications. Architecture, Performance and Applications*; Elsevier: Waltham, MA, USA, 2015.
2. Manyika, J. *The Internet of Things: Mapping the Value Beyond the Value*; McKinsey Global Institute: San Francisco, CA, USA, 2015.

3. McFarlane, D.; Giannikas, V.; Lu, W. Intelligent logistics: Involving the customer. *Comput. Ind.* **2016**. in press. [CrossRef]

4. Stock, T.; Seliger, G. Opportunities of sustainable manufacturing in Industry 4.0. *Procedia CIRP* **2016**, *40*, 536–541. [CrossRef]

5. Worldwide and Regional Internet of Things (IoT) 2014–2020 Forecast: A Virtuous Circle of Proven Value and Demand. Available online: https://www.business.att.com/content/article/IoT-worldwide_regional_2014-2020-forecast.pdf (accessed on 10 May 2016).

6. Xiong, X.; Zheng, K.; Xu, R.; Xiang, W.; Chatzimisios, P. Low power wide area machine-to-machine networks: Key techniques and prototype. *IEEE Commun. Mag.* **2015**, *53*, 64–71. [CrossRef]

7. Wang, F.; Liu, J. Networked wireless sensor data collection: Issues, challenges, and approaches. *IEEE Commun. Surv. Tutor.* **2011**, *13*, 673–687. [CrossRef]

8. Chen, D.; Brown, J.; Khan, J.Y. An interference mitigation approach for a dense heterogeneous wireless sensor network. In Proceedings of the 9th International Conference on Signal Processing and Communication Systems (ICSPCS), Cairns, Australia, 14–16 December 2015; pp. 1–7.

9. Mekikis, P.-V.; Antonopoulos, A.; Kartsakli, E.; Lalos, A.S.; Alonso, L.; Verikoukis, C. Information Exchange in Randomly Deployed Dense WSNs With Wireless Energy Harvesting Capabilities. *IEEE Trans. Wirel. Commun.* **2016**, *15*, 3008–3018. [CrossRef]

10. Vambase, S.V.; Mangalwede, S.R. Cooperative clustering protocol for mobile devices with bluetooth and Wi-Fi interface in mLearning. In Proceedings of the 3rd IEEE International Advance Computing Conference (IACC), Ghaziabad, India, 22–23 February 2013; pp. 505–510.

11. Younis, O.; Krunz, M.; Ramasubramanian, S. Node clustering in wireless sensor networks: Recent developments and deployment challenges. *IEEE Netw.* **2006**, *20*, 20–25. [CrossRef]

12. Wolff, R.; Andrews, E. Broadband access, citizen enfranchisement, and telecommunications services in rural and remote areas: A report from the american frontier (Topics in Wireless Communications). *IEEE Commun. Mag.* **2010**, *48*, 128–135. [CrossRef]

13. Nasralla, M.M.; Ognenoski, O.; Martini, M.G. Bandwidth scalability and efficient 2D and 3D video transmission over LTE networks. In Proceedings of the IEEE International Conference on Communications Workshops (ICC), Budapest, Hungary, 9–13 June 2013; pp. 617–621.

14. Lin, F.; Liu, Q.; Zhou, X.; Chen, Y.; Huang, D. Cooperative differential game for model energy-bandwidth efficiency tradeoff in the Internet of Things. *China Commun.* **2014**, *11*, 92–102.

15. Zhang, H.; Jiang, C.; Mao, X.; Chen, H.-H. Interference-limited resource optimization in cognitive femtocells with fairness and imperfect spectrum sensing. *IEEE Trans. Veh. Technol.* **2016**, *65*, 1761–1771. [CrossRef]

16. Zhang, H.; Liu, H.; Jiang, C.; Chu, X.; Nallanathan, A.; Wen, X. A practical semidynamic clustering scheme using affinity propagation in cooperative picocells. *IEEE Trans. Veh. Technol.* **2015**, *64*, 4372–4377. [CrossRef]

17. Centenaro, M.; Vangelista, L.; Zanella, A.; Zorzi, M. Long-range communications in unlicensed bands: The rising stars in the IoT and smart city scenarios. Available online: http://arxiv.org/pdf/1510.00620v1.pdf (accessed on 16 May 2016).

18. Margelis, G.; Piechocki, R.; Kaleshi, D.; Thomas, P. Low throughput networks for the IoT: Lessons learned from industrial implementations. In Proceedings of the 2015 IEEE 2nd World Forum on Internet of Things (WF-IoT), Milan, Italy, 14–16 December 2015; pp. 181–186.

19. LoRa-Alliance. *A Technical Overview of LoRa and LoRaWAN*; LoRa-Alliance: San Ramon, CA, USA, 2015.

20. *Sigfox M2M and IoT Redefined through Cost Effective and Energy Optimized Connectivity*; CBS Interactive Inc.: San Francisco, CA, USA, 2014; Available online: https://lafibre.info/images/3g/201302_sigfox_whitepaper.pdf (accessed on 16 May 2016).

21. Poole, I. Weightless Wireless M2M White Space Communications. Available online: http://www.radio-electronics.com (accessed on 10 May 2016).

22. nWave. Available online: http://www.nwave.io (accessed on 10 May 2016).

23. M2Communication. A Cellular-Type Protocol Innovation for the Internet of Things. Available online: http://www.theinternetofthings.eu/sites/default/files/%5Buser-name%5D/M2C%20Whitepaper%20for%20IoT%20Connectivity.pdf (accessed on 10 May 2016).

24. An educational guide. *How RPMA Works*; Ingenu: San Diego, CA, USA, 2015.

25. *Public Lighting Active Network*; Telensa Ltd.: Essex, UK, 2015.

26. Alliance, D. DASH7 Alliance Protocol Specification v1.0. Available online: http://95.85.41.106/wp-content/uploads/2014/08/005-Dash7-Alliance-Mode-technical-presentation.pdf (accessed on 16 May 2016).

27. Cetinkaya, O.; Akan, O.B. A DASH7-based power metering system. In Proceedings of the 12th Annual IEEE Consumer Communications and Networking Conference (CCNC), Las Vegas, NV, USA, 9–12 January 2015; pp. 406–411.

28. Helium. Available online: https://helium.com (accessed on 10 May 2016).

29. M2M Spectrum Networks. Available online: http://m2mspectrum.com (accessed on 10 May 2016).

30. Amber Wireless. Available online: https://www.amber-wireless.de (accessed on 10 May 2016).

31. *IEEE P802.11ah/D5.0—IEEE Draft Standard for Information Technology-Telecommunications and Information Exchange Between Systems-Local and Metropolitan Area Networks-Specific Requirements-Part 11: Wireless LAN Medium Access Control and Physical Layer Specification*; IEEE: New York, NY, USA, 2015.

32. *IEEE 802.15.4k—IEEE Standard for Local and metropolitan area networks—Part 15.4: Low-Rate Wireless Personal Area Networks (LR-WPANs). Amendment 5: Physical Layer Specifications for Low Energy, Critical Infrastructure Monitoring Networks*; IEEE: New York, NY, USA, 2013.

33. Khorov, E.; Lyakhov, A.; Krotov, A.; Guschin, A. A survey on IEEE 802.11ah: An enabling networking technology for smart cities. *Comput. Commun.* **2015**, *58*, 53–69. [CrossRef]

34. Flore, D. Evolution of LTE in Release 13. Available online: http://www.3gpp.org (accessed on 10 May 2016).

35. Cellnex Telecom. Available online: https://www.cellnextelecom.com/noticia-8 (accessed on 10 May 2016).

36. eSMARTCITY. Available online: https://www.esmartcity.es/noticias/abertis-y-sigfox-lanzaran-en-espana-una-red-de-internet-de-las-cosas (accessed on 10 May 2016).

37. TDF. Available online: http://www.tdf.fr (accessed on 28 December 2015).

38. Aerea. Available online: http://www.aerea.nl (accessed on 31 January 2016).

39. iSPHER. Available online: http://www.ispher.com (accessed on 20 December 2015).

40. Europa Press. Available online: http://www.europapress.es/andalucia/malaga-00356/noticia-sostenible-ayuntamiento-malaga-estudia-implantacion-proyecto-internet-cosas-20151025122230.html (accessed on 10 May 2016).

41. Martínez-Alvarez, V.; García-Bastida, P.A.; Martin-Gorriz, B.; Soto-García, M. Adaptive strategies of on-farm water management under water supply constraints in south-eastern Spain. *Agric. Water Manag.* **2014**, *136*, 59–67. [CrossRef]

42. Navarro-Hellín, H.; Martínez-del-Rincon, J.; Domingo-Miguel, R.; Soto-Valles, F.; Torres-Sánchez, R. A decision support system for managing irrigation in agriculture. *Comput. Electron. Agric.* **2016**, *124*, 121–131. [CrossRef]

43. Lopez-Gunn, E.; Zorrilla, P.; Prieto, F.; Llamas, M.R. Lost in translation? Water efficiency in Spanish agriculture. *Agric. Water Manag.* **2012**, *108*, 83–95. [CrossRef]

44. Hsia, S.-C.; Chang, Y.-J.; Hsu, S.-W. Remote monitoring and smart sensing for water meter system and leakage detection. *IET Wirel. Sens. Syst.* **2012**, *2*, 402–408. [CrossRef]

45. Zhang, H.; Chu, X.; Guo, W.; Wang, S. Coexistence of Wi-Fi and heterogeneous small cell networks sharing unlicensed spectrum. *IEEE Commun. Mag.* **2015**, *53*, 158–164. [CrossRef]

sensors

MDPI

Article

Evaluation of MPEG-7-Based Audio Descriptors for Animal Voice Recognition over Wireless Acoustic Sensor Networks

Joaquín Luque *, Diego F. Larios, Enrique Personal, Julio Barbancho and Carlos León

Department of Electronic Technology, University of Seville, Seville 41011, Spain; dlarios@us.es (D.F.L.);
epersonal@us.es (E.P.); jbarbancho@us.es (J.B.); cleon@us.es (C.L.)
* Correspondence: jluque@us.es; Tel.: +349-545-528-38; Fax: +349-545-528-33

Academic Editor: Gonzalo Pajares Martinsanz
Received: 7 March 2016; Accepted: 12 May 2016; Published: 18 May 2016

Abstract: Environmental audio monitoring is a huge area of interest for biologists all over the world. This is why some audio monitoring system have been proposed in the literature, which can be classified into two different approaches: acquirement and compression of all audio patterns in order to send them as raw data to a main server; or specific recognition systems based on audio patterns. The first approach presents the drawback of a high amount of information to be stored in a main server. Moreover, this information requires a considerable amount of effort to be analyzed. The second approach has the drawback of its lack of scalability when new patterns need to be detected. To overcome these limitations, this paper proposes an environmental Wireless Acoustic Sensor Network architecture focused on use of generic descriptors based on an MPEG-7 standard. These descriptors demonstrate it to be suitable to be used in the recognition of different patterns, allowing a high scalability. The proposed parameters have been tested to recognize different behaviors of two anuran species that live in Spanish natural parks; the *Epidalea calamita* and the *Alytes obstetricans* toads, demonstrating to have a high classification performance.

Keywords: sensor network; habitat monitoring; audio monitoring

1. Introduction

Nowadays, there is a growing interest in studying the evolution of certain environmental parameters associated with climate change. The scientific studies of the evidence of climate change through animals is done through phenology [1]. A clear example of these effects can be observed in animal behavior [2].

One of the most common methods to evaluate animal behavior is using an acoustic study. It consists of recording and analyzing animal voices in an area of interest. These studies can help us to understand key issues in ecology, such as the health status of an animal colony. Consequently, this is an important research area in ethology.

With the purpose of acquiring animal sound patterns, ethologists have traditionally deployed audio recording systems over the natural environment where their research was being developed. However this procedure requires human presence in the area of interest at certain moments. To avoid these events, some data loggers have been proposed in the literature [3,4]. Although they reduce the monitoring effort, they still require collecting the acquired information *in-situ*. In this regard, Banerjee *et al.* [5], Dutta *et al.* [6] and Diaz *et al.* [7] proposed different remotely accessible systems in order to minimize the impact of the presence of human beings in the habitat that is being monitored.

Once the data is collected through any of the described systems, several audio processing techniques have to be developed in order to extract useful information about the animals' behavior. This task represents a heavy workload for biologists. Therefore, automated techniques have to be considered. In this sense, some processing systems have been proposed in the literature. There are many strategies that can be followed in order to extract features from the audio records. Some examples are [8,9] where automatic audio detection and localization systems are described. Other examples are [10,11], where the authors presented an automatic acoustic recognition system for frog classification; Chunh *et al.* [12] proposes an automatic detection and recognition system for pig diseases, based on the evaluation of Mel Frequency Cepstrum Coefficients; and Kim *et al.* [13] details an automatic acoustic localization system that detect events using direct waveform comparison with acoustic patterns.

Automatic acoustic recognition systems in natural environments are not used only for animal detection. There are other applications that are based on this scheme of information extraction. This is the case of [14], which describes a detection and location system for an illegal presence in a natural park. This system is based on the calculation of the differences between the time of arrival of the sound waves emitted by multiple microphones. However, all of these *ad-hoc* solutions are specific for an application. Consequently, they cannot be scaled well to study other phenomena, such as the recognition of a new animal or a new behavior.

To overcome these limitations, this paper proposes a different approach based on the use of the MPEG-7 descriptors. MPEG-7 [15] is a multimedia content description standardized by ISO and IEC as the ISO/IEC 15938 standard. MPEG-7 was designed initially to propose a family of multimedia descriptors that allow a fast and efficient search for material that is of interest to the user.

In this paper, the MPEG-7 low level audio descriptors [16] are called primary descriptors. There are some studies that have demonstrated that these audio descriptors are suitable to be used to detect audio patterns over not well defined or variable scenarios [17], such as music identification [18], voice diseases [19], speech emotion recognition [20] or bird detection [21].

One important audio characteristic of animal voices is reverberation which can be used in animal detection and recognition. However, the basic primary descriptors do not depict these characteristics well. Due to that and taking into account applications for animal monitoring, we propose the addition of other parameters to characterize reverberations. These parameters are called secondary descriptors and are based on long-term analysis of the variations of primary descriptors.

Therefore, we propose the use of a Wireless Acoustic Sensor Network (WASN) to monitor animals' voices. In the proposed architecture, primary descriptors are only sent through the networks, reducing the required bandwidth. Thanks to that, it is possible to use Low-Power Wireless radio communication devices, improving the network lifetime. Furthermore, the base station only requires storing primary and secondary descriptors, saving a high amount of dedicated storage space for long term environmental analysis. This information is usually enough to solve animals' song recognition problem, as is described in Section 3.

The proposed recognition algorithm allows us to search and detect animal voices, even for animals not initially considered. In this regard, an adequate classifier must be selected and used.

The proposed distributed monitoring architecture has been mainly designed to be spread out in Spanish natural parks. This is the case of Doñana Natural Park, a park that is internationally well known for its research in the application of novel sensor technologies for environmental monitoring [22]. In this way, the proposed descriptors have been tested using a Spanish anuran audio database, obtaining promising results in the detection of common anuran species that live in these environments. This fact makes the work of the biologists easier, letting them avoid tedious tasks such as audio collection and analysis of large sets of audio patterns.

This work has been done in collaboration with the Doñana Biological Station and the National Museum of Natural Sciences in Madrid, Spain. According to our results, the proposed classifier can not only be used to distinguish an individual of Spanish species among other ones, it even detects different behaviors of the same species with accuracy. These results can be even obtained only using a small group of descriptors.

The rest of the paper is organized as follows: Sections 2 and 3 describe the architecture of the proposed system for audio monitoring and its audio descriptors. These descriptors have been evaluated in a real case study with real anuran sounds in Section 4. Finally, Section 5 sums up the conclusions, final remarks and presents future work.

2. Acquisition Architecture of the Recognition System

Wireless Acoustic Sensor Networks (WASNs) represent a new paradigm for acoustic signal acquisition. They typically consist of nodes with microphones, signal processing and communication capabilities. WASNs find use in several applications, such as hearing aids, ambient intelligence, hands-free telephony, and acoustic monitoring [23].

WASN has several advantages. For example, this type of network improves the coverage area in comparison with single recording devices due to the limited operative area of a microphone. Thanks to this, this technology allows us to monitor large areas of interest, such as big natural parks like Doñana. It can also be solved using microphone arrays, but the deployment cost of a WASN is generally lower than the deployment cost of a microphone arrays.

Moreover, the redundancy of the acquired information that appears in WASNs adds extra advantages to the recognition systems. In a WASN it is possible to use pattern classifiers that consider information from multiple acoustic nodes to improve the pattern recognition and localization in environmental applications [24].

All these reasons inspired the use of a WASN for animals' voice recognition in environmental monitoring scenarios. In this case, a system with the architecture depicted in Figure 1 is proposed. As it can be seen, it is based on two different devices: the Acoustical Nodes (*i.e.*, the smart measurement devices) and the Base Station, which collaborates amongst them to solve the animals' voice recognition problem. Both types of devices are described briefly in the following subsections.

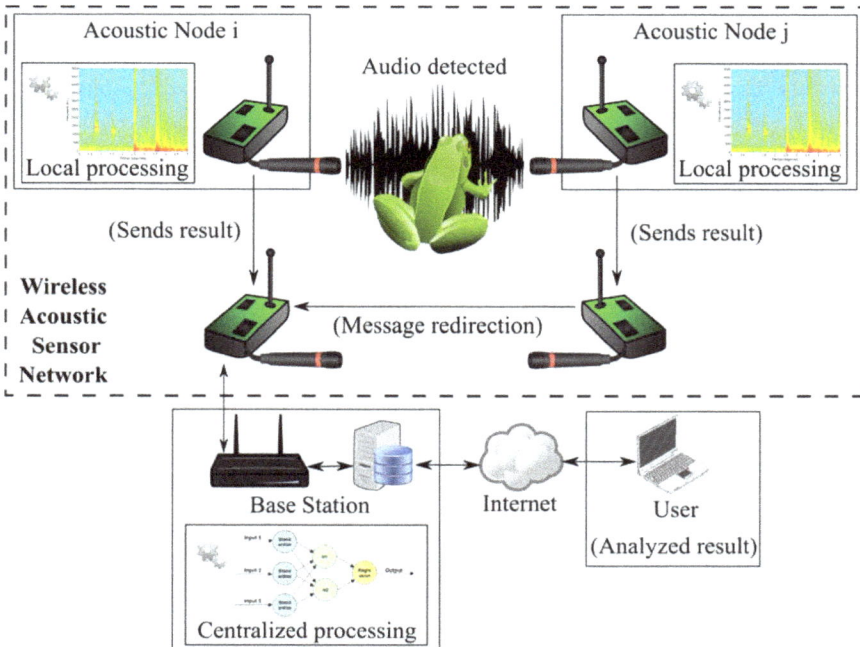

Figure 1. Architecture of the proposed Wireless Acoustic Sensor Network.

2.1. Wireless Acoustic Nodes

The proposed acoustic nodes, are based on a small low-power ARM architecture, with a radio communication module and an audio interface. This node and the enclosure for its use in environmental applications is depicted in Figure 2.

(a)   (b)

Figure 2. Current prototype of the acoustic sensor: (**a**) External enclosure; (**b**) Sensor board.

Essentially, this acoustic node uses a low-cost development board based on ARM Cortex A7 with SIMD (NEON) instruction support, which can efficiently execute audio processing. This development board is credit card sized, with low power consumption. It has 1 GB of RAM and can execute a fully functional Linux OS. The board also includes multiple I/O ports that can be used to connect external sensors, such as microphones and communication interfaces.

In this way, the proposed radio module is based on the IEEE802.15.4 standard [25]. This radio transceiver is very common in Wireless Sensor Network (WSN) nodes, following the reduction of its power consumption. Moreover, this standard is able to use the 6LoWPAN stack (an IPv6-based specification especially designed for low power consumption radios), which is also used to implement a communication layer [26]. This implementation allows multi-hop communication, which extends the basic communications amongst the nodes (in the 100–300 m range).

The use of a WASN has some drawbacks, mainly, the maximum bandwidth available and the power consumption of the radio device. This is why it is necessary to reduce, as much as possible, the amount of information sent by the radio transceiver.

Sending the entire raw audio pattern using wireless communications typically requires the use of high bandwidth radio transceivers, such as IEEE 802.11 ones. However, these radio transceivers have large power consumption. Instead of using these kinds of devices, this application proposes sending a reduced amount of data per each pattern, based on the primary descriptors described below. Thanks to this feature, we can use these low power radio devices in this application.

Dealing with the audio interface being used, we have selected a USB soundcard connected to an environmental microphone, capturing its surrounding sound. In acoustic monitoring, the type of microphone and acquisition system will be selected according to the animals under study. However, due to its limited acquisition frequencies [27], in most common cases of terrestrial observations, it is enough to use audio devices that can only sense within the human audible range. This is the case in anuran surveillance, with voices in the 20–20,000 Hz range [28].

If only one microphone is used in an environmental monitoring application, it needs to be highly sensitive. It needs to be very linear and it requires a complex calibration [29]. This is why these microphones are expensive (over $200). Furthermore, we have to consider the cost of its protection to work in harsh scenarios.

Fortunately, this restriction is relaxed if an array or a network of different microphones is used, due to the redundancy of the information that appears in different nodes. Thanks to that, it is possible to use low-cost electret microphones [30] in WASNs. Figure 3 depicts the frequency response of some of these low cost (less than $20) microphones. As it can be seen, even being cheaper, they have a good frequency response in the range required for our application.

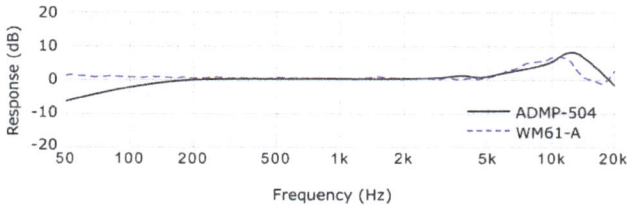

Figure 3. Frequency response of different microphones.

2.2. Base Station

The base station (BS), also called the information service, is a PC-based device with two communication modules: one to acquire information from the wireless acoustic sensor network, and another to allow researchers to obtain the gathered information. Therefore, this device acts typically as a database, where the nodes save the information, and the user can retrieve it as needed.

However, in the proposed animal voice recognition system, the required information is not the stored one. As described in the processing architecture, we use a classifier in the BS to retrieve the audio voice recognition service. It increases the computational requirements of the base station, but allows us to detect voices of animals not considered at the moment of the deployment.

3. Processing Architecture of the Recognition System

The processing architecture of the animal voice detection system is depicted in Figure 4. In this implementation, audio patterns are acquired from the monitoring area by the acoustic sensors and processed locally to obtain primary descriptors in wireless acoustic nodes.

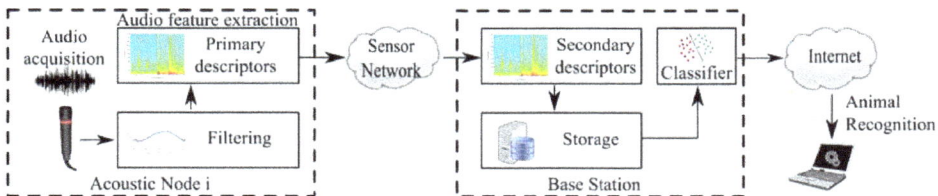

Figure 4. Processing architecture of the system.

After that, these primary descriptors are sent to the BS through the wireless radio transceiver of the nodes. Then, when the BS receives this data, secondary parameters are obtained. Only primary and secondary descriptors are stored in the database, where it can be used by a classifier for animal voice detection. The next subsections describe briefly the main blocks of the processing architecture:

3.1. Filtering

The main objective of the filtering stage consists of reducing the amount of noise in the patterns. This goal is obtained using a band-pass filter that eliminates frequencies of the acquired audio that do not contain relevant information. This filter can be adjusted for each application. For most animal voice analysis, in the human audible range, a 0.3–10 kHz band-pass is a good tradeoff between noise elimination and information loss.

3.2. Primary Descriptors

This subsystem is responsible for obtaining the primary descriptors. These descriptors receive the name of Low Level Descriptors (*LLDs*) and they are defined in MPEG-7 specification.

According to this specification, *LLDs* can be obtained from the audio spectrogram. It consists of a time–frequency analysis based on obtaining the amplitude (in dB) of the frequency response of each frame. In this case, the specifications recommend using a temporal window of 10 ms as a frame.

As shown in Figure 5, the spectrograms are commonly depicted as an X-Y level graph, where X is the time, Y the frequency and the darkness of the image represents the relative amplitude of the power of the signal.

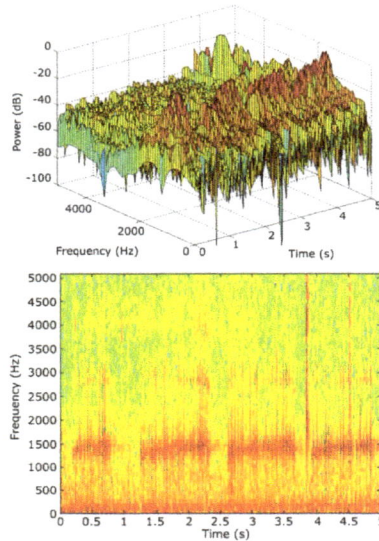

Figure 5. Audio spectrogram example.

This spectrogram can be used by a trained user to distinguish animal audio patterns (in a human manual processing step). Nevertheless, *LLDs* can be used by an automatic system (without human intervention) to obtain similar conclusions. In our research, 18 primary descriptors have been considered. The cost associated with sending the information expressed with these descriptors represents, in the worst case (using single precision floating point), a data payload of only 72 bytes each 10 ms, instead of the 882 bytes per 10 ms required to send a standard 44.1 kHz raw audio pattern.

Therefore, this way of representing the information provides a huge data bandwidth reduction, especially if we consider that these parameters can be expressed using only 2 bytes, with a fixed point representation. This feature makes this kind of representation suitable for being used in low power radio modules with restricted bandwidth, as the ones used in typical WSN nodes (IEEE 801.15.4) [31]. *LLDs* characterize spectrograms with the following parameters:

3.2.1. Frame Power

This parameter, expressed in dB, is obtained as the maximum energy of the frame, with respect to its minimum. Two different frame power indicators are defined in MPEG-7 specification:

- Relevant Power (PR): The relevant spectral power of the frame, defined in a frequency range. It is usually used normalized to the maximum power of the sound. By default, a 0.5–5 kHz band is used, due to the fact that most useful audio information for the human ear is in this range.
- Total Power (Pt): The normalized total spectral power, considering all the spectral dynamic ranges of the frame.

If the previous filter of the architecture is well fitted, both indicators may offer similar results.

3.2.2. Frame Power Centroid (C_P)

In a particular frame, the power centroid briefly represents the shape of the frequency spectrum. The centroid offers information about the frequencies where most of the energy of the frame is distributed. It can be obtained as shown in Equation (1):

$$C_P = \frac{\sum_i [f_i \cdot P_i]}{\sum_i P_i} \tag{1}$$

where f_i is the frequency of the spectrogram, and P_i the acoustic power of the associated frequency.

3.2.3. Spectral Dispersion (D_E)

It represents approximately the shape of the frequency spectrum of a frame. It can be obtained as shown in Equation (2):

$$D_E = \sqrt{\frac{\sum_i \left[(f_i - C_p)^2 \cdot P_i\right]}{\sum_i P_i}} \tag{2}$$

3.2.4. Spectrum Flatness (P_l)

It represents the deviation of the audio spectrum in relation to a flat spectrum (*i.e.*, a white noise). It can be obtained with Equation (3):

$$P_l = \frac{\sqrt{\prod_i P_i}}{\frac{1}{n} \sum_i P_i} \tag{3}$$

3.2.5. Harmonicity Ratio (R_a)

R_a is the ratio of harmonic power to total power. According to MPEG-7 specifications, it can be obtained from the autocorrelation of the audio signal $s\,(n)$ into the frame, as expressed in Equation (4):

$$R_a = \max_k \left(\frac{\sum_j s\,(j) \cdot (j - k)}{\sqrt{\sum_j s\,(j)^2 \cdot \sum_j s\,(j - k)^2}} \right) \tag{4}$$

where k is the length of the frame.

3.2.6. Fundamental Frequency (*Pitch*)

The fundamental frequency (or sound pitch) is the frequency that best describes the periodicity of a signal. It can be obtained with the Linear Prediction Coding (*LPC*) algorithm [32]. *LPC*, models the spectral power as sum of the polynomial A (z) defined by Equation (5):

$$A\,(z) = 1 - \sum_{k=1}^P \left[a_k \cdot z^{-k} \right] \tag{5}$$

where the polynomial degree p represents the model precision. The coefficients a_k represent the formants of the audio signal, *i.e.*, the spectral maximum produces by the audio resonances. According to [32], these coefficients can be obtained reducing the error between the polynomial prediction and the spectral power of the audio frame using a minimum least squares estimation.

For the sound s (n) the error function $\varepsilon\,(n)$ is obtained according to Equation (6):

$$\varepsilon\,(n) = s\,(n) - \sum_{k=1}^P a_k s\,(n - k) \tag{6}$$

For $\varepsilon\,(n)$, the autocorrelation function is obtained by Equation (7):

$$\varphi\,(k) = \sum_{m=-\infty}^{+\infty} \varepsilon\,(m)\ \varepsilon\,(m + k) \tag{7}$$

Using this information, the *Pitch* can be defined as the first maximum of the $\varphi(k)$ autocorrelation function.

Due to the execution complexity of *LPC* a tradeoff between time of computation and accuracy is necessary. According to our tests, a polynomial of size 30 is enough to obtain *Pitch* estimations with good accuracy.

3.2.7. Upper Limit of Harmonicity (F_{la})

As defined in Section 5.3.13.3.1 of the MPEG standard [33], this is the frequency beyond which the spectrum cannot be considered harmonic. It can be obtained using the comb filter of Equation (8):

$$c(j) = s(j) - \left(\sum_{j=m}^{m+n-1} [s(j) \cdot s(j-K)] / \sum_{j=m}^{m+n-1} \left[s(j-K)^2 \right] \right) \cdot s(j-K) \tag{8}$$

where $s(n)$ is the audio signal and K is the delay related to R_a (the Harmonicity Ratio).

Considering $P(f)$ and $P'(f)$ as the spectral power of $s(n)$ and the filtered signal $c(j)$, the ratio for each frequency is defined as Equation (9):

$$\alpha(f_{\lim}) = \sum_{f=f_{\lim}}^{f_{max}} P'(f) / \sum_{f=f_{\lim}}^{f_{max}} P(f) \tag{9}$$

Using it, F_{la} can be obtained as the maximum frequency where $\alpha(f_{\lim})$ is higher than a threshold, 0.5 in this case, as is commonly used.

3.2.8. Harmonic Peaks

The harmonic peaks are the n highest peaks of the spectrum located around the multiple of the fundamental frequency of the signal. The terms that surround the fundamental frequency are used in order to take into account the slight variations of harmonicity of some sounds.

MPEG-7 describes harmonic peaks that we propose to form the following descriptors:

- Frequency of the harmonic peaks (FrFi).
- Bandwidth of the harmonic peaks (AbFi).

Both parameters can be obtained using the aforementioned *LPC* algorithm. In most practical scenarios, it is usual to consider the three first harmonic peaks to characterize the frame.

Figure 6 depicts the results of an *LPC* analysis of two different frames. The red shape is the original audio spectrum and the blue shape represents the *LPC* algorithms. Blue and red bands are the bandwidth of the harmonic peaks, whose frequency is depicted as a vertical blue line.

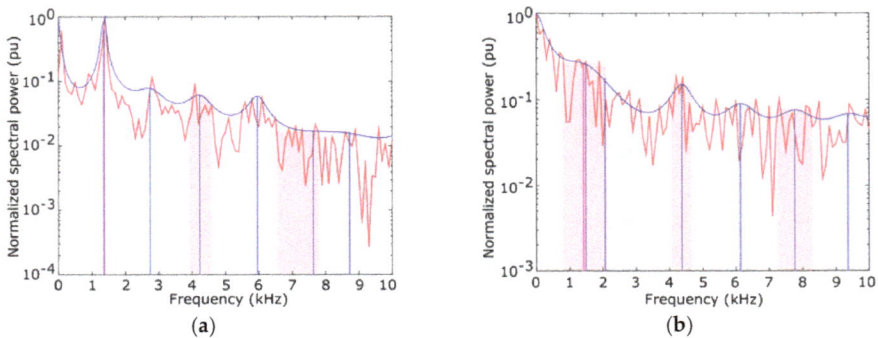

Figure 6. Example of results obtained with LPC analysis. (a) $t = 0.80$ $f_f = [36, 810, 682]$ Hz; (b) $t = 2.00$ $f_f = [1316, 2191, 647]$ Hz.

3.2.9. Harmonic Centroid (C_A)

It is a weighted average that selects a frequency that represents the harmonic peaks of the frame. It can be obtained with Equation (10):

$$C_H = \frac{\sum_{i=1}^{n_f} [f_i \cdot v_i]}{\sum_{i=1}^{n_f} v_i} \tag{10}$$

where f_i is the ith frequency of the harmonic peak and v_i its peak value.

3.2.10. Harmonic Spectral Deviation (D_{eA})

It is computed as the deviation between harmonic peaks and the signal spectrum of the frame. It can be obtained with Equation (11):

$$D_{eA} = \max_k \left(\frac{\sum_{i=1}^{n_f} \left| \log(v_i) - \log\left(\frac{1}{3} \cdot \sum_{k=-1}^{1} [v_i + k]\right) \right|}{\sum_{i=1}^{n_f} \log(v_i)} \right) \tag{11}$$

3.2.11. Harmonic Spectral Spread (D_{iA})

It is computed as the typical deviation between harmonics peaks and the harmonic centroid. It can be obtained with Equation (12):

$$D_{iA} = \frac{1}{C_a} \cdot \sqrt{\frac{\sum_{i=1}^{n_f} \left[v_i^2 \cdot (f_i - C_a)^2 \right]}{\sum_{i=1}^{n_f} v_i^2}} \tag{12}$$

3.2.12. Harmonic Spectral Variation (V_A)

It is obtained as a normalized correlation between the harmonic peaks of two adjacent frames. It can be obtained with Equation (13):

$$V_A = 1 - \frac{\sum_{i=1}^{n_f} \left[v_{i,j} \cdot v_{i,j-1} \right]}{\sum_{i=1}^{n_f} \sqrt{v_{i,j}^2} \cdot \sum_{i=1}^{n_f} \sqrt{v_{i,j-1}^2}} \tag{13}$$

where $v_{i,j}$ is the amplitude of the ith harmonic of the frame j.

3.3. Secondary Descriptors

One shortcoming in the use of MPEG-7 *LLDs* for animal voice detection is that they do not consider reverberation effects. This is due to the reduced size of the frame (only 10 ms), which is smaller than the duration of these effects. However, many animal characteristic acoustic patterns have an important reverberation component, which can be used to distinguish between species.

To describe these effects we propose the use of secondary descriptors. These descriptors are obtained with a methodology based on the same methodology used with the primary descriptors. The dispersion between longer time frames is analyzed. These time frames are called micro-segments and have a duration of 100 ms (*i.e.*, 10 frames). This time is chosen considering that, audio delays higher than 100 ms are identified by human being as echo, instead of a reverberation.

We define these secondary descriptors as the interquartile range of each one of the primary parameters, in a 100 ms micro-segment.

According this procedure, we obtained the next secondary parameters:

- Dispersion of relevant Power (PR~).
- Dispersion of total Power (Pt~).

- Dispersion of centroid of frame power (CP~).
- Dispersion of spectral dispersion (DE~).
- Dispersion of spectrum flatness (Pl~).
- Dispersion of harmonicity ratio (Ra~).
- Dispersion of fundamental frequency (Pitch~).
- Dispersion of upper limit of harmonicity (Fla~).
- Dispersion of Frequency of the harmonic peaks (FrFi~).
- Dispersion of Bandwidth of the harmonic peaks (AbFi~).
- Dispersion of harmonic centroid (CA~).
- Dispersion of harmonic spectral deviation (DeA~).
- Dispersion of harmonic spectral spread (DiA~).
- Dispersion of harmonic spectral variation (VA~).

The set of secondary descriptors does not need to be computed (and sent) by the acquisition system. Due to the fact that they are obtained based on variations of primary descriptors, they can be obtained by the BS before storing the information in the storage server.

Figure 7 depicts examples with the values of all proposed parameters in two different samples: an *Epidalea calamita* (natterjack toad) and an *Alytes obstetricans* (common midwife toad) one. In this figure, the red circles represent the current value of each parameter in a particular frame, while the vertical line indicates its range during a full segment of the audio file.

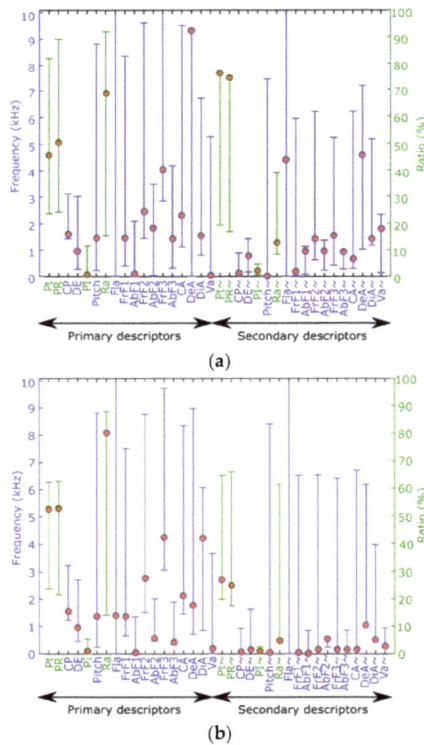

Figure 7. Examples of primary and secondary descriptors for two different frames. (**a**) *Epidalea calamita* (0501) *t* = 0.5; (**b**) *Alytes obstetricans* (1325) *t* = 0.8.

On the other hand, the color of the axis indicates the units of the parameters. As it can be seen, red ones are defined in Hertz, while green ones are in percentages.

This example graphically shows that the proposed parameters have very different range and values for different audio samples. Therefore, they can be effectively used by an automatic classification system in order to distinguish different patterns.

3.4. Data Storage

This subsystem is dedicated to storing the primary and secondary descriptors. This information must be tagged with a timestamp, and can be stored in a regular SQL database.

3.5. Classifier

This subsystem plays an important role in the whole system. It is responsible for the implementation of the classification method. Its main goal is to classify the audio patterns (*i.e.*, its descriptors) into one of the proposed classes. In this sense, hundreds of supervised classifiers have been proposed in the literature. The selection for the best method is a decision that has to be done considering the characteristics of the problem to solve [34]. However, in our real case analysis, even simple classifiers offer good class separations using the aforementioned descriptors.

4. A Case Study: Anuran classification system

The proposed animal audio detection descriptors have been evaluated considering the different behavior of two anuran species: *Alytes obstetricans* (Figure 8a) and *Epidalea calamita* (Figure 8b).

(a)   (b)

Figure 8. Two specific anuran species: (**a**) *Alytes obstetricans* toad, carrying egg strings [35]; (**b**) *Epidalea calamita* toad [36]. Photographs by Ursina Tobler.

Ethology studies focused on the anuran species, such as the described ones, are especially interesting for the biologist. The reason for this interest is based on the fact that in some amphibian species, climate change and UV radiation have important effects on their population (largely determined by their habitat) [1,37]. More concretely, an interesting phenomenon can be observed in anurans (frogs and toads). These kinds of amphibians present peaks in the timing of reproductive calling (songs), depending on each species and temperature evolution [38]. This is why an anuran classifier is used as a case study of the proposed descriptors.

4.1. Sound Database Information

In order to analyze the performance of the proposed descriptors, this research has based its evaluation on a relevant database of anuran songs. This database is composed by a set of audio samples of *Alytes obstetricans* and *Epidalea calamita* provided by *"Fonozoo: Fonoteca Zoológica del Museo Nacional de Ciencias Naturales"*, a Spanish phonotheque based on recordings of animal sounds [39].

This organization provided a scientific collection of animal sounds. Fonozoo was created with the aim of supporting the study of acoustic communication in animals.

The collection includes more than 45,000 sound recordings belonging to more than 12,000 different species, all of which makes Fonozoo a valuable tool in the systematic study of animal behavior. The main goal of using a public audio database is that these files can be used by other researchers all over the world for comparative performance analysis. In the case of the studied anurans, the collection is composed of 63 audio samples, according to the characteristics described in Table 1.

Table 1. Analysis of the sound collections.

Caption	Vocalization	Files
Epidalea calamita	standard	20
	chorus	3
	oviposit	10
Alytes obstetricans	standard	29
	distress call	1

This collection of anuran audio files offers a total of 6053 s of recorded sounds, with a mean duration of 96 s and a median of 53 s. Therefore, our test collection has 605,300 frames, where each frame has a total amount of 36 associated values that are its primary and secondary descriptors.

As can be seen, the number of chorus vocalization samples of *Epidalea calamita* toads is not enough to train a system. Moreover, it sounds similar to the standard vocalization. Due to this, both cases are going to be considered as the same classification category.

In the case of the *Alytes obstetricans* toads, there are not enough distress call vocalization samples to train a system. Due to this, in the test of the proposed descriptor, we have considered the next number of samples by categories:

- Type 1: 23 samples of *Epidalea calamita* toad with standard or chorus vocalization.
- Type 2: 10 samples of *Epidalea calamita* toad with oviposit vocalization.
- Type 3: 30 samples of *Alytes obstetricans* toad with standard or distress call vocalization.

Due to the fact that the samples have been obtained from environmental scenarios, not only animal vocalization is recorded. In fact, there is a high percentage of background sound, composed of wind, rain, traffic, human voices, and so forth. Therefore, there are intervals of time in each sample that do not correspond with animal vocalization, and which need to be detected and rejected by the classification system. With this purpose, a fourth type of audio has been considered, that corresponds with noise, *i.e.*, sounds in which we are not interested in.

4.2. Classification Results

Supervised classifier techniques require the use of sound models for each class. These models are required in the training and adjusting phase of the classifier. In this case, two files have been chosen as a pattern from both *Epidalea calamita* and *Alytes obstetricans* toads with standard vocalization, and one file for *Epidalea calamita* toad with oviposit vocalization. These models have been selected empirically, looking for files in which the sound is easily distinguishable. In this case, files "0501" and "0707" from Fonozoo collection are chosen as type 1 pattern, "0503" as type 2, and "1325" and "1330" as type 3.

Once the classification criterion has been selected, it is necessary to evaluate its value. In this sense, a classification can only offer good accuracy if there exists a surface in the input parameters (*i.e.*, the descriptors) space, where the different classes can be isolated completely or with a reduced error. This feature can be observed in Figure 9, where the value of each parameter (centroid and dispersion *versus* power in this example) for each frame is summarized. A color code has been chosen to indicate its class type. Blue for type 1 (*Epidalea calamita* toad with standard or chorus vocalization), green for type 2 (*Epidalea calamita* toad with oviposit vocalization), red for type 3 (*Alytes obstetricans* toad with standard or distress call vocalization) and purple for type 4 (noise).

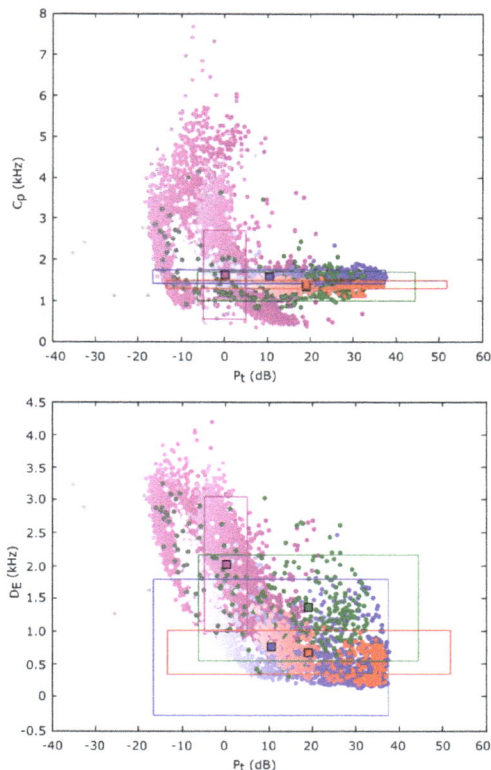

Figure 9. Point cloud of C_p and D_E descriptors *versus* its power in patterns.

It is important to take into account that although there are 36 parameters (18 primary and 18 secondary), it is not necessary to impose the use of all of them in the classifier task. Furthermore, in order to improve the response time, it is better to choose a representative subgroup of descriptors as inputs for the classifier in function of the voice animal to be recognized. In this case, we empirically chose the next descriptors for the recognition of the different voices of the *Epidalea calamita* and the *Alytes obstetricans* toads:

- Harmonicity ratio (Ra).
- Bandwidth of the first harmonic peaks (AbF1).
- Dispersion of total Power (Pt~).

As it can be seen, only three descriptors are used, two primary and one secondary. Under these conditions, the following classifiers have been tested:

4.2.1. Minimum Distance Classifier

The minimum distance classifier [40] is one of the simplest, classic classification algorithms. In spite of its simplicity, it offers good accuracy in many situations. Some animal voice classification systems, such as [11] for anuran species, are based on this classifier.

The classification is based on obtaining the distance d_{ij} between the input vector to classify i with each one of the representative class j. This distance can be obtained with the Euclidean distance, according to Equation (14):

$$d_{ij} = \sqrt{\sum_{k=1}^{P} \left[x_{ik} - x_{jk} \right]^2} \tag{14}$$

where x_{ik} is the value of the k-th parameter in the i-th frame.

According to it, the input can be classified as the class that minimizes the distance with it. However, the units of each parameter of the proposed system are clearly different. Therefore, a parameter normalization is needed. In this case, we propose the use a of spread range of R_k for each parameter, which can be obtained as Equation (15) describes:

$$R_k = \max_i (x_{ik}) - \min_i (x_{ik}) \tag{15}$$

Therefore, the normalized distance nd_{ij} can be obtained according to Equation (16):

$$nd_{ij} = \sqrt{\sum_{k=1}^{P} \left[\frac{x_{ik} - x_{jk}}{R_k} \right]^2} \tag{16}$$

If this classifier is applied to each frame of an audio pattern, the final classification of the audio sample can be obtained as the most likely class to appear in the pattern. Moreover, it can be used as a qualify indicator Q. If we consider n_a as the number of frames correctly classified, and n_j the total number of frames, Q can be obtained as expressed by Equation (17):

$$Q = \frac{n_A}{\sum_j n_j} \tag{17}$$

Figure 10 depicts the classification distance from each frame of the "0560" sample to the four classes. Blue indicates type 1 (*Epidalea calamita* toad with standard or chorus vocalization), green for type 2 (*Epidalea calamita* toad with oviposit vocalization), red for type 3 (*Alytes obstetricans* toad with standard or chorus vocalization) and purple for type 4 (noise). The top band of the figure indicates the minimum distance class. As can be seen, most patterns are classified correctly. In this example the quality indicator is 84.36%.

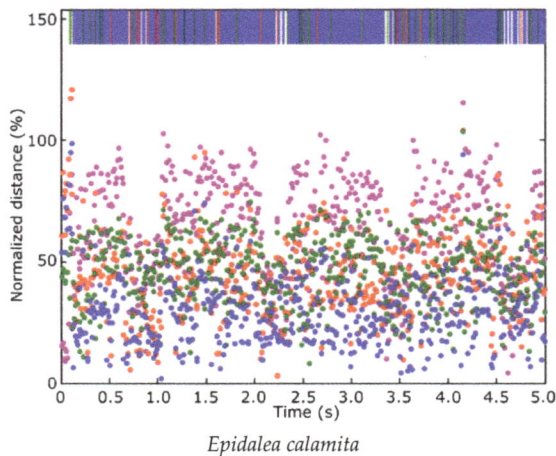

Epidalea calamita

Figure 10. Frame classification of "0560" sample (*Epidalea calamita* in standard vocalization).

The classification results, obtained with all the frames, are depicted in Figure 11, where the quality indicator Q, has been represented. In this figure the blue area indicates type 1 (*Epidalea calamita* toad with standard or chorus vocalization), green area type 2 (*Epidalea calamita* toad with oviposit vocalization) and red type 3 (*Alytes obstetricans* toad with standard or chorus vocalization). Vertical colored lines indicate the estimated classification. Therefore, a pattern corresponding to a file is correctly classified if its line color matches the band color.

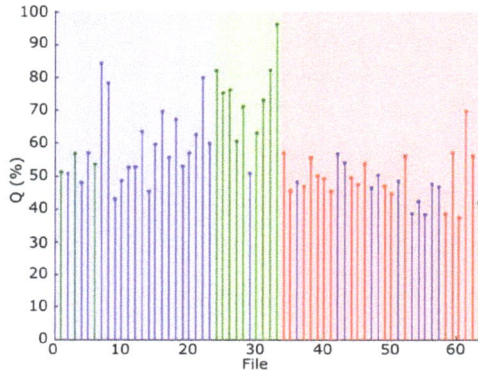

Figure 11. Classification result of all the samples of the audio collection.

As it can be seen, correctly classified patterns have high Q values. Then, Q can be used to determine a confidence degree in the classification result.

Figure 12 depicts the performance of the minimum distance classifier, according to the results obtained with the audio collection. As it can be seen, even a simple classifier as the minimum distance offers an acceptable classification result.

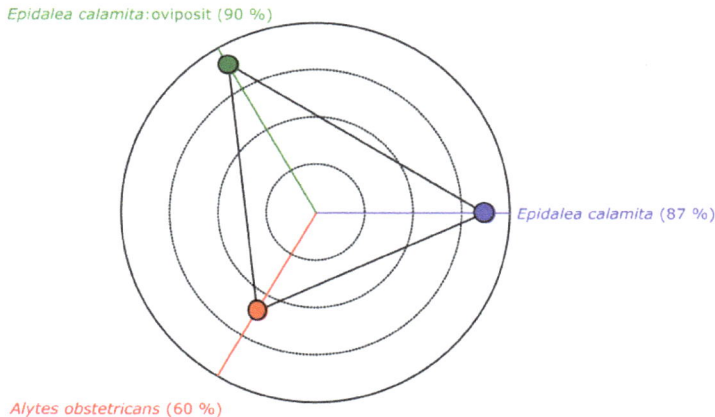

Figure 12. Performance of minimum distance classifier.

4.2.2. Modified Maximum Likelihood Classification

Maximum likelihood classification [41] is another classic supervised classification algorithm which has been widely used in the literature. This classification algorithm is divided into two stages: training and execution.

The training stage is executed only once, as a previous step to classification. In this stage the statistical information to execute the classification algorithm is obtained. In the case of the maximum likelihood classification consists of generating several probability density functions $f_\theta(x_k)$ for each one of the θ classes to be detected. $f_\theta(x_k)$ is a multivariable function that is going to be used to obtain a numeric probability for each class and frame to analyze in the execution stage.

In this case, we propose using a Gaussian mixture distribution (GMD) [42] of two Gaussians as $f_\theta(x_k)$, obtained from the audio models. It is depicted in Figure 13, where the blue line indicates class type 1 (*Epidalea calamita* toad with standard or chorus vocalization); green type 2 (*Epidalea calamita* toad with oviposit vocalization); red type 3 (*Alytes obstetricans* toad with standard or distress call vocalization) and purple type 4 (noise). Continuous lines depict raw data from the models, while dotted lines are the GMD ones.

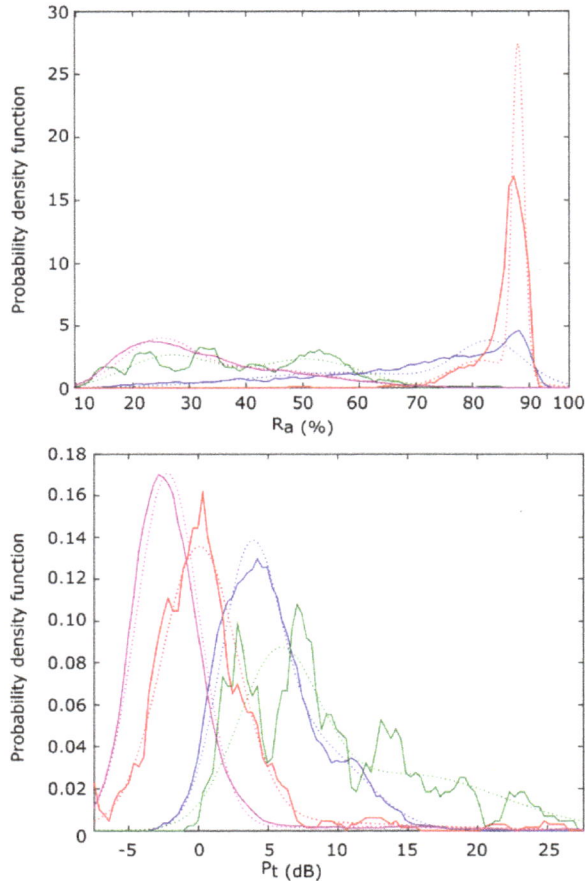

Figure 13. *Cont.*

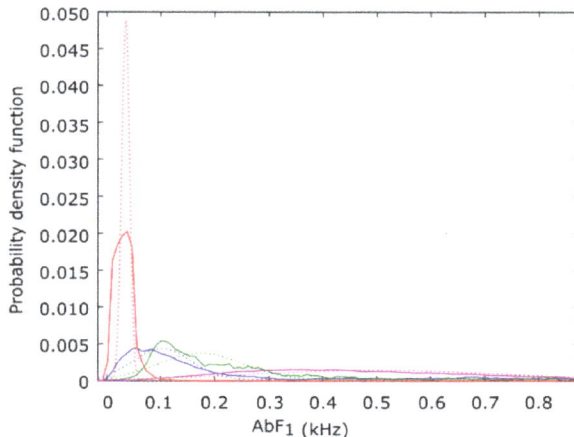

Figure 13. Probability density functions for each descriptor.

In execution time, maximum likelihood classification uses these $f_\theta(x_k)$ functions to obtain the probability of each frame to be one of the proposed θ classes. To do that, this algorithm uses the likelihood of the descriptors of a frame, in relation to the classes to recognize $\mathcal{L}(\theta|x)$. It can be obtained using the Expression (18):

$$\mathcal{L}(\theta|\mathbf{x}) = P(\mathbf{x}|\theta) = f_\theta(\mathbf{x}) \tag{18}$$

This expression give us a numeric likelihood of each class for all the frames in a sample to analyze. But to reduce the calculus complexity in maximum likelihood classification, the *log* of the likelihood $V(\theta|x)$ is commonly used. It can be done with the Equation (19):

$$V(\theta|\mathbf{x}) = log\left[\mathcal{L}(\theta|\mathbf{x})\right] = log\left[P(\mathbf{x}|\theta)\right] = log\left[f_\theta(\mathbf{x})\right] \tag{19}$$

In the modified maximum likelihood algorithm, we propose using to classify the differential likelihood D_{ij} instead of $V(\theta|\mathbf{x})$. D_{ij} of a frame i is the difference of the likelihood of belonging to that class j (type 1 to 3), rather than the likelihood it being noise r (type 4). It can be obtained with the Equation (20):

$$D_{ij} \equiv V(\theta_j|\mathbf{x}_i) - V(\theta_r|\mathbf{x}_i) = log\left[f_{\theta j}(\mathbf{x}_i)\right] - log\left[f_{\theta r}(\mathbf{x}_i)\right] \tag{20}$$

In the proposed algorithm, the temporal evolution of D_{ij} of a sample to analyze is smoothed out using a size 10 moving average filter to reduce spurious changes in D_{ij} evolution. For classification purposes, only the maximum of D_{ij} for each frame is considered. Figure 14 summarizes the results of this step for the "0560" file, an *Epidalea calamita* example. As it can be seen, in this step most of the frames are yet to be correctly classified.

To increase the detection accuracy, only filtered D_{ij} higher than a threshold β are considered for classification proposes. β is adjusted empirically at 30% of the relative maximum of the sample (approx. 0.5 for this example). Figure 15 summarizes the result of this step for the "0560" file. As it can be seen, in this case all the frames higher than the threshold are correctly classified.

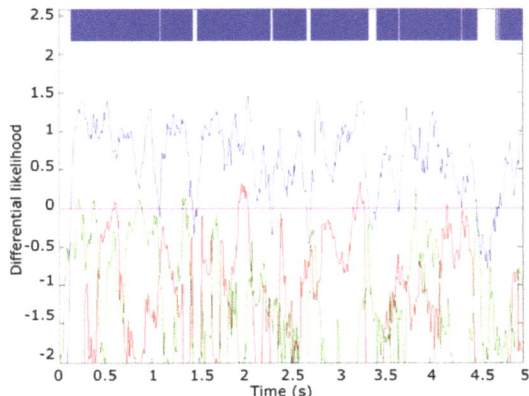

Figure 14. Result of the smoothed differential likelihood of "0560" sample.

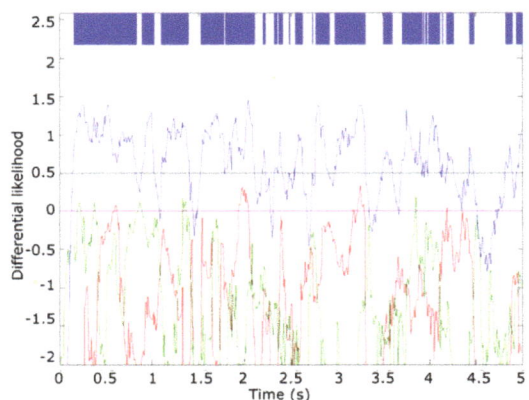

Figure 15. Result of the threshold application.

To estimate the class θ of the audio sample, a qualifier indicator Q_θ is defined, according to Equation (21):

$$Q_\theta = \frac{A_\theta}{\sum_j A_j} \tag{21}$$

Q_θ does not use only the maximum of the smoothed D_{ij} count, but the area A_θ of the curve higher than the threshold β, According to our analysis, it increases the classification accuracy. Considering this information, the final estimation of the class of the sample is the one with the maximum Q_m indicator, according to Equation (22):

$$Q = \max_j (Q_j) \tag{22}$$

As summarized, the proposed modified maximum likelihood classificatory is based on a $log[f_\theta(\mathbf{x})]$ functions obtained from the model of the defined four classes θ. Using this information, the algorithm consist of the execution of the next steps:

- Obtain $D_{ij} = log[f_{\theta j}(\mathbf{x})] - log[f_{\theta r}(\mathbf{x})]$ for the three possible classes of sound.
- Filter D_{ij} with a size 10 moving average filter.

272

- Select the class for each frame, as the maximun D_{ij} for each class j, if its value is higher than the threshold β. Otherwise the frames are discarded from the classification.
- Obtain the final class estimation as the class with maximum area over the threshold β.

The classification results obtained with all the frames, using this classification method is depicted in Figure 16. In this figure, the blue area indicates type 1 (*Epidalea calamita* toad with standard or chorus vocalization), green area type 2 (*Epidalea calamita* toad with oviposit vocalization) and red type 3 (*Alytes obstetricans* toad with standard or distress call vocalization). Vertical color lines indicate the estimated classification. Therefore, a pattern is correctly classified if its line color matches the band color.

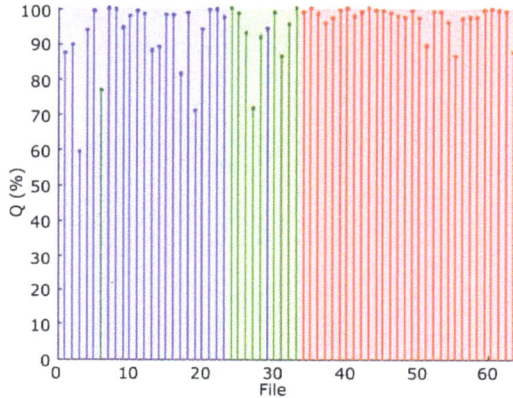

Figure 16. Classification result of all the samples of the audio collection.

As it can be seen, most of the correctly classified patterns have high Q values. Therefore, Q can be used to determine a confidence degree in the classification result.

Figure 17 depicts the performance of the modified maximum likelihood classifier, according to the results obtained with the audio collection. As it can be seen, this classifier offers a very good classification, with a success rate higher than 98%.

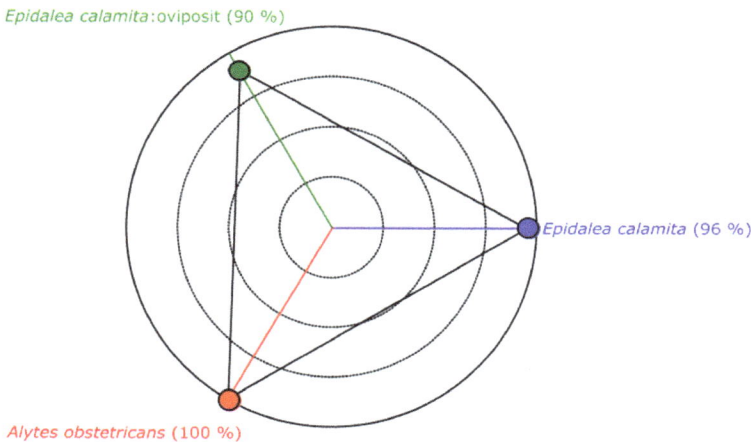

Figure 17. Performance of the modified maximum likelihood classifier.

5. Conclusions and Future Work

This paper describes a classification architecture for animal voice recognition. The main goal of the proposed architecture is first to reduce biologists' effort to manually classify audio patterns, and second, to provide a method that can be used to classify classes not previously considered. This classification architecture is based on the use of MPEG-7 low level descriptors, plus a family of derived descriptors used to detect animal voice reverberations. These descriptors have been tested in real scenarios detecting anuran voices stored in a public sound database.

The obtained results allow us to conclude that the proposed generic descriptors, using adequate classifiers, can be used to classify animal voices with a high degree of accuracy. In addition, the proposed architecture does not have the most common drawbacks of audio classification architectures.

The future work that is being considered by the authors focuses on two strategies: first, the deployment of this animal voice detection system in several natural environments; and secondly, the development of other classifiers using the proposed descriptors.

Acknowledgments: This work has been supported by the Consejería de Innovación, Ciencia y Empresa, Junta de Andalucía, Spain, through the excellence eSAPIENS (reference number TIC-5705). The authors would like to thank Rafael Ignacio Marquez Martinez de Orense (Museo Nacional de Ciencias Naturales) and Juan Francisco Beltrán Gala (Faculty of Biology, University of Seville) for their collaboration and support.

Author Contributions: Joaquín Luque conceived and designed the experiments; Joaquín Luque, Diego F. Larios, Enrique Personal, Julio Barbancho and Carlos León performed the experiments, analyzed the data and wrote the paper.

Conflicts of Interest: The authors declare no conflict of interest.

References

1. Walls, S.; Barichivich, W.; Brown, M. Drought, deluge and declines: The impact of precipitation extremes on amphibians in a changing climate. *Biology* **2013**, *2*, 399–418. [CrossRef] [PubMed]
2. Parmesan, C.; Yohe, G. A globally coherent fingerprint of climate change impacts across natural systems. *Nature* **2003**, *421*, 37–42. [CrossRef] [PubMed]
3. Cambron, M.; Bowker, R. An automated digital sound recording system: The amphibulator. In Proceedings of the Eighth IEEE International Symposium on Multimedia (ISM'06), San Diego, CA, USA, 11–13 December 2006.
4. Cambron, M.E.; Emberton, A.C. Amphibulator II. In Proceedings of the IEEE Southeastcon, Jacksonville, FL, USA, 4–7 April 2013.
5. Banerjee, R.; Mobashir, M.; Bit, S.D. Partial DCT-based energy efficient compression algorithm for wireless multimedia sensor network. In Proceedings of the 2014 IEEE International Conference on Electronics, Computing and Communication Technologies (CONECCT), Bangalore, India, 6–7 January 2014.
6. Dutta, I.; Banerjee, R.; Bit, S.D. Energy efficient audio compression scheme based on red black wavelet lifting for wireless multimedia sensor network. In Proceedings of the 2013 International Conference on Advances in Computing, Communications and Informatics (ICACCI), Mysore, India, 22–25 August 2013.
7. Diaz, J.J.M.; Nakamura, E.F.; Yehia, H.C.; Salles, J.; Loureiro, A.A.F. On the use of compressive sensing for the reconstruction of anuran sounds in a wireless sensor network. In Proceedings of the 2012 IEEE International Conference on Green Computing and Communications, Besancon, France, 20–23 November 2012.
8. Salomons, E.; Havinga, P. A survey on the feasibility of sound classification on wireless sensor nodes. *Sensors* **2015**, *15*, 7462–7498. [CrossRef] [PubMed]
9. Okuyucu, C.; Sert, M.; Yazici, A. Audio feature and classifier analysis for efficient recognition of environmental sounds. In Proceedings of the 2013 IEEE International Symposium on Multimedia, Anaheim, CA, USA, 9–11 December 2013.
10. Croker, B.; Kottege, N. Using feature vectors to detect frog calls in wireless sensor networks. *J. Acoust. Soc. Am.* **2012**, *131*, EL400. [CrossRef] [PubMed]
11. Huang, C.J.; Yang, Y.J.; Yang, D.X.; Chen, Y.J. Frog classification using machine learning techniques. *Expert Syst. Appl.* **2009**, *36*, 3737–3743. [CrossRef]

12. Chung, Y.; Oh, S.; Lee, J.; Park, D.; Chang, H.H.; Kim, S. Automatic detection and recognition of pig wasting diseases using sound data in audio surveillance systems. *Sensors* **2013**, *13*, 12929–12942. [CrossRef] [PubMed]

13. Kim, Y.; Ahn, J.; Cha, H. Locating acoustic events based on large-scale sensor networks. *Sensors* **2009**, *9*, 9925–9944. [CrossRef] [PubMed]

14. González-Castaño, F.J.; Alonso, J.V.; Costa-Montenegro, E.; López-Matencio, P.; Vicente-Carrasco, F.; Parrado-García, F.J.; Gil-Castiñeira, F.; Costas-Rodríguez, S. Acoustic sensor planning for gunshot location in national parks: A pareto front approach. *Sensors* **2009**, *9*, 9493–9512. [CrossRef] [PubMed]

15. Quackenbush, S.; Lindsay, A. Overview of MPEG-7 audio. *IEEE Trans. Circuits Syst. Video Technol.* **2001**, *11*, 725–729. [CrossRef]

16. You, S.D.; Chen, W.H. Comparative study of methods for reducing dimensionality of MPEG-7 audio signature descriptors. *Multimed. Tools Appl.* **2013**, *74*, 3579–3598. [CrossRef]

17. Lin, C.H.; Tu, M.C.; Chin, Y.H.; Liao, W.J.; Hsu, C.S.; Lin, S.H.; Wang, J.C.; Wang, J.F. SVM-based sound classification based on MPEG-7 audio LLDs and related enhanced features. In *Convergence and Hybrid Information Technology*; Springer Science + Business Media: Berlin, Germany, 2012; pp. 536–543.

18. You, S.D.; Chen, W.H.; Chen, W.K. Music Identification system using MPEG-7 audio signature descriptors. *Sci. World J.* **2013**, *2013*, 1–11. [CrossRef] [PubMed]

19. Muhammad, G.; Melhem, M. Pathological voice detection and binary classification using MPEG-7 audio features. *Biomed. Signal Process. Control* **2014**, *11*, 1–9. [CrossRef]

20. Lampropoulos, A.S.; Tsihrintzis, G.A. Evaluation of MPEG-7 descriptors for speech emotional recognition. In Proceedings of the 2012 Eighth International Conference on Intelligent Information Hiding and Multimedia Signal Processing, Piraeus, Greece, 18–20 July 2012.

21. Tyburek, K.; Prokopowicz, P.; Kotlarz, P. Fuzzy system for the classification of sounds of birds based on the audio descriptors. In *Artificial Intelligence and Soft Computing*; Springer Science + Business Media: Berlin, Germany, 2014; pp. 700–709.

22. Larios, D.; Barbancho, J.; Sevillano, J.; Rodríguez, G.; Molina, F.; Gasull, V.; Mora-Merchan, J.; León, C. Five years of designing wireless sensor networks in the doñana biological reserve (Spain): An applications approach. *Sensors* **2013**, *13*, 12044–12069. [CrossRef] [PubMed]

23. Bertrand, A. Applications and trends in wireless acoustic sensor networks: A signal processing perspective. In Proceedings of the 2011 18th IEEE Symposium on Communications and Vehicular Technology in the Benelux (SCVT), Ghent, Belgium, 22–23 November 2011.

24. Wang, H.; Chen, C.E.; Ali, A.; Asgari, S.; Hudson, R.E.; Yao, K.; Estrin, D.; Taylor, C. Acoustic sensor networks for woodpecker localization. In Proceedings of the Advanced Signal Processing Algorithms, Architectures, and Implementations XV, San Diego, CA, USA, 2–4 August 2005.

25. The Institute of Electrical and Electronics Engineers (IEEE). *IEEE Standard for Local and metropolitan area networks. Part 15.4: Low-Rate Wireless Personal Area Networks (LR-WPANs)*; IEEE Standards Association: New York, NY, USA, 2011.

26. Ko, J.; Terzis, A.; Dawson-Haggerty, S.; Culler, D.E.; Hui, J.W.; Levis, P. Connecting low-power and lossy networks to the internet. *IEEE Commun. Mag.* **2011**, *49*, 96–101.

27. Blumstein, D.T.; Mennill, D.J.; Clemins, P.; Girod, L.; Yao, K.; Patricelli, G.; Deppe, J.L.; Krakauer, A.H.; Clark, C.; Cortopassi, K.A.; *et al.* Acoustic monitoring in terrestrial environments using microphone arrays: Applications, technological considerations and prospectus. *J. Appl. Ecol.* **2011**, *48*, 758–767. [CrossRef]

28. Acevedo, M.A.; Corrada-Bravo, C.J.; Corrada-Bravo, H.; Villanueva-Rivera, L.J.; Aide, T.M. Automated classification of bird and amphibian calls using machine learning: A comparison of methods. *Ecol. Inform.* **2009**, *4*, 206–214. [CrossRef]

29. Müller, G.; Möser, M. *Handbook of Engineering Acoustics*; Springer: Berlin/Heidelberg, Germany, 2012.

30. Fristrup, K.M.; Mennitt, D. Bioacoustical monitoring in terrestrial environments. *Acoust. Today* **2012**, *8*, 16. [CrossRef]

31. Akyildiz, I.F.; Melodia, T.; Chowdhury, K.R. A survey on wireless multimedia sensor networks. *Comput. Netw.* **2007**, *51*, 921–960. [CrossRef]

32. Rabiner, L.R.; Schafer, R.W. *Introduction to Digital Speech Processing*; Foundations and Trends in Signal Processing: Hanover, MA, USA, 2007; Volume 1, pp. 1–194.

33. International Organization for Standardization (ISO)/International Electrotechnical Commission (IEC). *Information Technology—Multimedia Content Description Interface—Part 4: Audio*; ISO/IEC FDIS 15938-4:2002; ISO/IEC Association: Geneva, Switzerland, 2002.

34. Fernández-Delgado, M.; Cernadas, E.; Barro, S.; Amorim, D. Do we need hundreds of classifiers to solve real world classification problems? *J. Mach. Learn. Res.* **2014**, *15*, 3133–3181.

35. Tobler, U.; Garner, T.W.J.; Schmidt, B.R. Genetic attributes of midwife toad (Alytes obstetricans) populations do not correlate with degree of species decline. *Ecol. Evol.* **2013**, *3*, 2806–2819. [CrossRef] [PubMed]

36. Oromi, N.; Richter-Boix, A.; Sanuy, D.; Fibla, J. Genetic variability in geographic populations of the natterjack toad (Bufo calamita). *Ecol. Evol.* **2012**, *2*, 2018–2026. [CrossRef] [PubMed]

37. Blaustein, A.R.; Walls, S.C.; Bancroft, B.A.; Lawler, J.J.; Searle, C.L.; Gervasi, S.S. Direct and indirect effects of climate change on amphibian populations. *Diversity* **2010**, *2*, 281–313. [CrossRef]

38. Walpole, A.A.; Bowman, J.; Tozer, D.C.; Badzinski, D.S. Community-level response to climate change: Shifts in anuran calling phenology. *Herpetol. Conserv. Biol.* **2012**, *7*, 249–257.

39. Solís, G.; Eekhout, X.; Márquez, R. *FonoZoo.com a New Resource in the Web for the Study of Animal Sounds*; Razprave IV. Razreda Sazu; Slovenska Akademija Znanosti in Umetnosti: Ljubljana, Slovenia, 2006; Volume XLVII, pp. 145–154.

40. Rudrapatna, M.; Sowmya, A. Feature weighted minimum distance classifier with multi-class confidence estimation. In *Lecture Notes in Computer Science*; Springer Science + Business Media: Berlin, Germany, 2006; pp. 253–263.

41. Beel, J.; Gipp, B. Academic search engine spam and google scholar's resilience against it. *J. Electron. Publ.* **2010**, *13*. [CrossRef]

42. Roweis, S.; Ghahramani, Z. A unifying review of linear gaussian models. *Neural Comput.* **1999**, *11*, 305–345. [CrossRef] [PubMed]

sensors

MDPI

Article

Accuracy and Resolution Analysis of a Direct Resistive Sensor Array to FPGA Interface

Óscar Oballe-Peinado [1,2], Fernando Vidal-Verdú [1,2,*], José A. Sánchez-Durán [1,2], Julián Castellanos-Ramos [1,2] and José A. Hidalgo-López [1,2]

[1] Departamento de Electrónica, Universidad de Málaga, Andalucía Tech, Campus de Teatinos, Málaga 29071, Spain; oballe@uma.es (O.O.-P.); jsd@uma.es (J.A.S.-D.); julian@elca.uma.es (J.C.-R.); jahidalgo@uma.es (J.A.H.-L.)

[2] Instituto de Investigación Biomédica de Málaga (IBIMA), Málaga 29010, Spain

* Correspondence: fvidal@uma.es; Tel.: +34-952-133-325; Fax: +34-952-133-324

Academic Editor: Gonzalo Pajares Martinsanz

Received: 27 December 2015; Accepted: 29 January 2016; Published: 1 February 2016

Abstract: Resistive sensor arrays are formed by a large number of individual sensors which are distributed in different ways. This paper proposes a direct connection between an FPGA and a resistive array distributed in M rows and N columns, without the need of analog-to-digital converters to obtain resistance values in the sensor and where the conditioning circuit is reduced to the use of a capacitor in each of the columns of the matrix. The circuit allows parallel measurements of the N resistors which form each of the rows of the array, eliminating the resistive crosstalk which is typical of these circuits. This is achieved by an addressing technique which does not require external elements to the FPGA. Although the typical resistive crosstalk between resistors which are measured simultaneously is eliminated, other elements that have an impact on the measurement of discharge times appear in the proposed architecture and, therefore, affect the uncertainty in resistance value measurements; these elements need to be studied. Finally, the performance of different calibration techniques is assessed experimentally on a discrete resistor array, obtaining for a new model of calibration, a maximum relative error of 0.066% in a range of resistor values which correspond to a tactile sensor.

Keywords: resistive sensor arrays; direct sensor-to-digital device interface; FPGAs; parallel analogue data acquisition

1. Introduction

An important number of sensors are based on the variation of resistance values shown due to the passage of an electric current, depending on a physical magnitude which is measured. These are resistive sensors, which can be classified according to their applications. They can be tactile sensors [1–10], temperature sensors [11,12], gas detectors [13–15] or for other substances [16,17]. In order to obtain the desired information, sometimes a single sensor can be enough. Several sensors are necessary in other occasions. For example, tactile sensors used in applications such as skin emulation or robotic hands (where large surfaces need to be scanned) are usually formed by an array with a large number of individual sensors. Substance detectors are also usually formed by a large number of sensors.

One of the main problems of resistive sensor arrays is that they need a large number of components and complicated wiring for the circuits which measure the different resistances. There are also numerous circuits carrying out signal conditioning and analog-to-digital conversion of the information for its subsequent processing. In principle, for a single access (SA) configuration $2 \cdot M \cdot N$ cables from the sensor to the circuit are needed, although one of them is normally shared by all sensors [10] so they are reduced to $M \cdot N + 1$. The time needed to extract the information of the whole array is practically the time needed to extract the information of a single sensor. However, each sensor needs its own circuit for conditioning and AD conversion. For this reason, for large size arrays this configuration can be very costly in terms of area and energy consumption.

With the aim of alleviating this problem, a resistive sensor array can be distributed in the shape of a two-dimensional matrix of M rows and N columns. There are several sensors sharing the rows and columns in this array access (AA) configuration, but in such a way that the information of a single sensor can be accessed by selecting a row and a column. Therefore, the number of cables from the array is reduced to $M + N$ and the number of necessary circuits to scan the information would be M or N depending on the type of array. In addition to this, additional circuits are needed to carry out sensor addressing. On the other hand, the time needed to extract the information of the whole array is the result of multiplying the time used in scanning a sensor by the number of N columns or M rows of the array plus the time required in multiplexing. This same solution has to be used if the construction of the sensor already has a structure of rows and columns and its output comes through $M + N$ wires, as in [2,4]. A second possibility to further reduce the circuitry which carries out AD conversion is the increase of the multiplexing of the information of the array in such a way that only one element is scanned at a time. This solution is obviously slower and the multiplexing circuits are more complicated.

The last two strategies to scan a resistive sensor array show, in addition to the increase of the time needed to access the information, the so-called crosstalk effect [11,18–23]. This effect happens when the resistance values of other elements of the array have an impact on the measurement of the resistance of the element which is being scanned. Because of this effect, resistance measurements show a certain degree of error when the structures mentioned above are used. For instance, errors of 30% are obtained for a 4×4 array with resistors in a range between 100 Ω and 1000 Ω (using buffers with internal resistance of 10 Ω to address the rows) [24], despite the use of four operational amplifiers to reduce crosstalk. This error can be reduced with more complex circuits including calibration techniques to improve the readout accuracy [25] or including analog switches and an additional OA [24]. Therefore, for accurate and fast readings without these complex circuits, the only solution is accessing each of the elements of the array individually and without multiplexing.

The solution used in this paper avoids resistive crosstalk, providing accuracy in the measurement as in a SA configuration, while having the intermediate speed and intermediate hardware cost of an AA configuration. The structure of the array is divided into rows and columns, but now, in each column, all sensors share a terminal while the other is individual (row access configuration, RA). Therefore, the sensor would have $(M + 1) \cdot N$ cables and the information of N sensors will be accessed simultaneously. The signal conditioning circuit is only one capacitor in each column, which is also part of the circuit for AD conversion (for our case a time-to-digital conversion). This is carried out by using an FPGA as in [26] instead of microcontrollers [27]. This configuration, with some modifications, can be applied to other sensors (capacitive or inductive) that provide a time-modulated signal that can be directly measured in the digital domain [28,29].

Another reason to use this array configuration lies in the need of calibrating the measurement of resistive sensors. There is extensive literature on calibration of single sensors [26,30–33], but not on calibration of resistive sensor arrays. In general, calibration is necessary to avoid errors in the measurement of parameters which are used in the determination of resistance values. In an AA type sensor, for best accuracy, $M \cdot N$ calibration resistors would also be needed, as the values of the capacitors which are used by each resistor of the array have to be calibrated as well. Calibrating the threshold voltage which makes digital devices go from detecting a 1-logic to a 0-logic would also be necessary if microcontrollers or FPGAs are used for AD conversion. This greatly complicates the total hardware of the system. However, only N calibration elements are necessary if an array with an RA configuration is used.

FPGAs have a large number of input/output ports which can be connected directly to the array, avoiding the need of additional circuits. They also allow for parallel processing of the digital information obtained simultaneously from the N sensors. For this reason, a circuit with a direct sensor-FPGA connection seems to be the best option. This type of architecture was shown in [34], however, the effects on precision and accuracy of having M sensors sharing the conditioning circuit and AD conversion were not studied, or the limitations this imposes on the number of rows that the array can have, or the hardware requirements of the FPGA to carry out sensing functions.

A characterization of this configuration of the sensor which takes into account all the facts mentioned above is carried out in this paper. The results are obtained experimentally using a circuit which will simulate the operation of a tactile sensor array, from discrete known resistors. Finally, the accuracy of the circuit will be analyzed with several techniques.

The structure of this paper is as follows: Section 2 describes the architecture and the operation mode of the proposed circuit of direct connection with an RA configuration; Section 3 studies what we will refer to as *RC crosstalk*, which derives from the RA configuration of the circuit; Section 4 explains how to assess all components which have an impact on uncertainty in time measurement of the resistors of the matrix; Section 5 describes the materials and procedures used for the implementation of the circuit; Section 6 shows the experimental results obtained and analyses the consequences of using different calibration techniques to estimate resistance values in the array. The conclusions section closes the paper.

2. Description of Architecture and Operation Mode

Figure 1 shows the proposed architecture as a direct interface to a tactile sensor with an RA configuration. As can be observed, there is no physical multiplexing circuit outside of the FPGA. The number of pins needed in the FPGA to address the sensor array is $(M + 1) \cdot N$. In the array, each R_{ij} resistor is connected to a P_{ij} pin in the FPGA which can be set as output (supplying voltages close to 0 V or to V_{DD}) or in high impedance (HZ). The common cable to all sensors of a column is connected to a C_j capacitor and to a PV_j pin of the FPGA which can be set as input/output (I/O). Additionally, each column has an extra R_{cj} resistor (not a sensor) connected to the P_{cj} pin, which is used for calibration purposes, as described further in this paper. Obviously, the maximum size of the sensor will be determined by the number of pins available in the FPGA, although this is not currently a severe limitation, as FPGAs with ranges of hundreds of pins can be found for use [35].

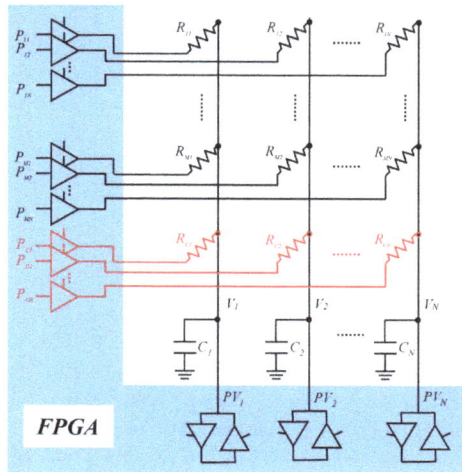

Figure 1. Direct interface resistive sensor array-FPGA.

The reading process of each row of the array has two phases. Firstly, in the charge phase, C_j capacitors are charged by setting PV_j pins as output to a '1' level while all P_{ij} pins are kept in HZ. Then, in the discharge phase, a whole row is selected by setting its corresponding I/O pins to '0'. For instance, pins $P_{k1}, \ldots P_{kN}$ are set to '0' while the remaining P_{ij} pins, are kept in HZ. Simultaneously, all PV_j pins are selected as input to be able to measure the V_j voltage of the C_j capacitor which is being discharged. A set of timers are started in the FPGA at the beginning of this discharging phase and their counts are stopped when the low threshold (VT_j) is reached at the related column pins. VT_j is the voltage in which the FPGA detects a '0' logic in PV_j. Therefore, a whole row is read in parallel. The process is repeated for the M rows of the array and the calibration row.

The purpose of the calibration resistor is to avoid having to know the C_j, VT_j and V_{DD} values, because these are normally difficult to measure. In addition, C_j and VT_j depend on power voltage supply or temperature and they can drift with time. As there is only one calibration resistor, it would seem that the method used is what is known in the literature as a single-point calibration [27]. According to this method, if when row i is selected, the discharge time measured for the C_j capacitor is Δt_{ij}, and when the calibration row is selected, the time is Δt_{cj}, then the R_{ij} resistance value is determined by:

$$R_{ij} = \frac{\Delta t_{ij}}{\Delta t_{cj}} R_{cj} \tag{1}$$

where R_{cj} is a value decided by the designer. However, this calibration method does not take into account the output resistance RB_j of the FPGA buffers (around 10 Ω for the FPGA of this paper) which are connected in series to R_{ij} and R_{cj}. For this reason, if Equation (1) is used, a mistake is made in the estimation of resistance value. More accuracy could be obtained if the two-point calibration method is used [27]. However, adding an additional row of calibration is then necessary, resulting in an increase of hardware in the circuit.

A calibration solution which takes into account RB_i resistances, and uses a single calibration resistor is proposed in [36]. The method consists on using off-time calibration (outside of the normal operation of the sensor) where, by using different resistors and measuring their discharge time, a relationship can be established between R_{ij} and R_{cj}. It is shown that the adjustment of the experimental data using a linear function is sufficiently accurate, resulting in the following relationship:

$$R_{ij} = \alpha \cdot \Delta t'_{ij} + B \tag{2}$$

with α and B as constants and where $\Delta t'_{ij}$ would be the time used for discharging C_j through R_{ij} during off-time calibration. $\Delta t'_{cj}$ must also be measured during off-time calibration. This is the time used in the discharge through R_{cj} (calibration resistor which will be used during the normal operation of the sensor). By using this time value, Equation (2) R_{ij} can be calculated as follows:

$$R_{ij} = A \cdot \frac{\Delta t'_{ij}}{\Delta t'_{cj}} + B \tag{3}$$

where $A = R_{cj} - B$. Using Equation (3), during the normal operation of the circuit, the value of R_{ij} can be calculated by measuring the Δt_{ij} and Δt_{cj} times:

$$R_{ij} = A \cdot \frac{\Delta t_{ij}}{\Delta t_{cj}} + B \tag{4}$$

This way of obtaining R_{ij} is similar to that of a single-point calibration but taking the RB_i buffer resistances into account and, therefore, with an accuracy which is similar to that of a two-point calibration. We will refer to this procedure as the *off-on time* method. This technique has the drawback of requiring prior pre-calibration, although the hardware is simpler, so the cost is reduced. Moreover, the time required to scan the array decreases when compared with that of the two-point calibration method, and the power consumption is also lower.

Although in principle there is no crosstalk in Equation (4), it must be noted that this expression does not take stray capacitors which appear in the FPGA pins into account, which are in HZ state when an array row is read. This originates what will be referred to as *RC crosstalk* in this paper, which will be studied in detail in the following section.

3. RC Crosstalk

Figure 2 shows the detail of the j column of the circuit when the discharge of C_j is being measured through the R_{ij} resistor. Cp_{kj} with $k \neq i$ are the capacitors which group all stray capacitors connected to P_{ij} pins in HZ state in the FPGA. By analyzing Figure 2, it can be observed how all stray capacitors have an impact on the discharge process of C_j. Moreover, the impact is not always the same, as the resistors which link them to C_j also vary. In Figure 2, it has been assumed that the discharge of the Cp_{ij} capacitor is very fast compared to that of C_j, (and it is true, as $RB_i \ll R_{ij}$ and $Cp_{ij} \ll C_j$).

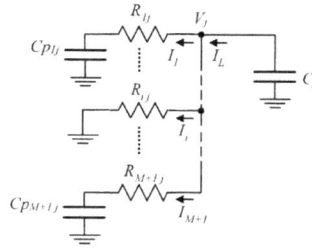

Figure 2. Detail of the j column of the circuit, showing the stray capacitors.

$V_j(s)$ is the voltage of C_j in the Laplace variable, during the discharge process. Its value is determined by:

$$V_j(s) = \frac{-I_L}{C_j s} + \frac{V_{DD}}{s} \tag{5}$$

where I_L is the current which discharges the capacitor and V_{DD} is its initial charging voltage. On the other hand, for each I_K with $k \neq i$:

$$V_j(s) = I_k \left(R_{kj} + \frac{1}{Cp_{kj}} \right) + \frac{V_{DD}}{s} \tag{6}$$

while:

$$I_i = \frac{V_j(s)}{R_{ij}} \tag{7}$$

Solving for I_L in Equation (5), I_k in Equation (6) and taking into account that:

$$I_L = \sum_{k=1}^{M+1} I_k \tag{8}$$

where $k = M + 1$ indicates the calibration row, the equation is:

$$- V_j(s) C_j s + C_j V_{DD} = \frac{V_j(s)}{R_{ij}} + \sum_{\substack{k=1 \\ k \neq i}}^{M+1} \frac{1}{R_{kj}} \frac{s V_j(s) - V_{DD}}{s + \dfrac{1}{R_{kj} Cp_{kj}}} \tag{9}$$

solving for $V_j(s)$:

$$V_j(s) = V_{DD} \cdot \frac{C_j + \sum\limits_{\substack{k=1 \\ k \neq i}}^{M+1} \frac{1}{R_{kj}} \frac{1}{s + \frac{1}{R_{kj} C p_{kj}}}}{C_j s + \frac{1}{R_{ij}} + \sum\limits_{\substack{k=1 \\ k \neq i}}^{M+1} \frac{1}{R_{kj}} \frac{1}{s + \frac{1}{R_{kj} C p_{kj}}}} \tag{10}$$

and finally:

$$V_j(s) = V_{DD} \cdot \frac{1 + \sum\limits_{\substack{k=1 \\ k \neq i}}^{M+1} \frac{C p_{kj}}{C_j} \frac{1}{R_{kj} C p_{kj} s + 1}}{s \left(1 + \sum\limits_{\substack{k=1 \\ k \neq i}}^{M+1} \frac{C p_{kj}}{C_j} \frac{1}{R_{kj} C p_{kj} s + 1} \right) + \frac{1}{R_{ij} C_j}} \tag{11}$$

Some facts are clear from Equation (11). Firstly, $V_j(s)$ is not only the discharge voltage of a capacitor with a single pole; it has $M + 1$ poles and M zeros. Equation (11) also shows crosstalk appearance, as $V_j(s)$ is not only a function of R_{ij} and C_j; it also depends on the rest of the resistors of the column (R_{kj}) and on the $C p_{kj}$ stray capacitors associated to each pin in the column (this is where the name RC crosstalk comes from).

Secondly, if $C p_{kj} << C_j$, Equation (11) is reduced to:

$$V_j(s) \approx \frac{V_{DD}}{s + \frac{1}{R_{ij} C_j}} \tag{12}$$

which is the discharge equation through an only capacitor. Therefore, the higher C_j is, compared to $C p_{kj}$, the lower the crosstalk will be. Reducing $C p_{kj}$ implies making a good design of the sensor and of the control circuit. However, there are terms composing $C p_{kj}$ which cannot be reduced, such as FPGA pin capacitance or that of their contacts with the wiring of the sensors (once the PCB technology of the sensors has been selected). The dependency with R_{kj} can also be observed in Equation (11). But this term cannot be varied as it is the resistance of the sensors themselves.

Although RC crosstalk has been analyzed in a sensor array, it will also appear if any of the calibration techniques mentioned in Section 2 are used. This situation occurs even when a single sensor is calibrated. Figure 3 shows a circuit for the measurement of an R_X resistor, with two calibration resistors, R_{C1} and R_{C2}. As shown in the figure, R_{C1} and R_{C2} resistors are associated with two parasitic capacitors, $C p_{C1}$ and $C p_{C2}$.

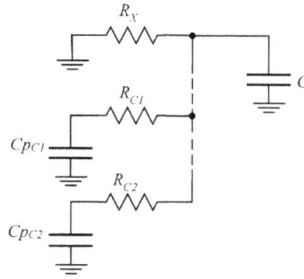

Figure 3. Two-point calibration circuit showing RC crosstalk due to stray capacitors.

4. Measurement Uncertainty Analysis

Equation (11) also helps us find the maximum $\Delta tmax_{ij}$ and the minimum $\Delta tmin_{ij}$ times used by the FPGA to detect a 0-logic in the discharge process through R_{ij}. Then, if R_{kj} tends to 0, Equation (11) would be:

$$V_j(s) = \frac{V_{DD}}{s + \dfrac{1}{R_{ij}C_{eq}}} \tag{13}$$

with:

$$Ceq_{ij} = C_j + \sum_{\substack{k=1 \\ k \neq i}}^{M+1} Cp_{kj} \tag{14}$$

On the other hand, if $R_{kj} \to \infty$ then, we find again that Equation (11) transforms into Equation (12). Then, the time needed to discharge C_j at a VT_j voltage using Equation (11) is limited by discharge times at that same voltage of two RC circuits with capacitors of different values and the same resistor. This is:

$$\Delta tmax_{ij} = R_{ij}Ceq_{ij} \cdot \ln \left(\frac{VT_j}{V_{DD}} \right) \tag{15}$$

$$\Delta tmin_{ij} = R_{ij}C_j \cdot \ln \left(\frac{VT_j}{V_{DD}} \right) \tag{16}$$

Therefore, $\Delta tmax_{ij} < \Delta t_{ij} < \Delta tmix_{ij}$ and, its exact value will depend on the rest of the resistors.

Another way of interpreting these results involves considering Δt_{ij} as a statistical variable with its corresponding probability density function and standard deviation, $\sigma(\Delta t_{ij})$. Depending on the process followed to find the value of Equations (15) and (16), the probability density function of Δt_{ij} will depend on the range of values of the R_{kj} resistors, in such a way that the smaller the range they can vary, the smaller $\sigma(\Delta t_{ij})$ will be. On the other hand, the standard deviation value will also be smaller, the smaller the difference between Ceq_{ij} and C_j is, or, in other words, the smaller the sum of the values of Cp_{kj} is. But, as has been mentioned earlier, this has a technological limitation which imposes a minimum value to $\sigma(\Delta t_{ij})$. It can also be observed how an increase in the number of M rows in the array, increases the value of the sum of Cp_{kj}, and therefore of $\sigma(\Delta t_{ij})$. If what we are interested in is the relative standard deviation, $\sigma(\Delta t_{ij})/\Delta t_{ij}$, this is, according to Equations (15) and (16), a function of Ceq_{ij}/C_j and of the probability density function of the resistances. Therefore, a possibility to reduce this, is the increase of C_j, but this is inconsistent with the sensing speed of the array. Consequently, there is a tradeoff between speed and accuracy in the measuring of resistances. This same effect also appears in the case of a sensor with only one component, where an increase of the capacitor to be discharged also implies an increase in accuracy, in this case, related to the reduction of the error term due to quantification [27]. Obviously, this quantification error also appears in the measures of the resistances of the array. Therefore, and in sum, the measuring of the Δt_{ij} time, corresponding to a specific resistor of the array, has as uncertainty resources related to other sensors: the values of the other resistors of the array and the existence of Cp_{kj} capacitors. This causes a first source of uncertainty, crosstalk uncertainty, $\sigma_{crosstalk}(\Delta t_{ij})$. The other sources of uncertainty that a single sensor which is not part of an array would have, must be added to this: $u(z)$ quantification noise and the noise in V_{DD}, VT_j and C_j, which generate what we will refer to as trigger noise uncertainty, $\sigma_{trigger}(\Delta t_{ij})$.

The same procedure can be followed if the aim is to measure the time needed to get a 0-logic for the calibration resistor (Δt_{cj}). This time value will also be a random variable with extreme values determined by:

$$\Delta tmax_{cj} = R_{cj}Ceq_{cj} \cdot \ln\left(\frac{VT_j}{V_{DD}}\right) \tag{17}$$

$$\Delta tmin_{cj} = R_{cj}C_j \cdot \ln\left(\frac{VT_j}{V_{DD}}\right) \tag{18}$$

where:

$$Ceq_{cj} = C_j + \sum_{k=1}^{M} Cp_{kj} \tag{19}$$

and therefore, with its own standard deviation, $\sigma(\Delta t_{cj})$. It must be noted that the differences between the Δt_{ij} and Δt_{cj} values are not only determined by the different R_{ij} and R_{cj} resistance values, but also because, in general, $Ceq_{cj} \neq Ceq_{ij}$ with $i \neq c$, due to the different capacitors associated to the wiring which links each sensor with its corresponding pin in the FPGA (these differences can become very small with a careful design and adequate technology, although they will always exist). This also shows that if the position of the resistors in a column was changed (with the exception of that for calibration) variations in their values would appear. Then, as it has been shown with R_{cj}, there is a third source of uncertainty associated to the difference between the capacitances of the nodes in a column. This will originate a standard deviation in time measurement which will be referred to as $\sigma_{column}(\Delta t_{ij})$.

It could also be argued that there is noise due to the position in the same row but in different columns. However, assuming that the distribution of resistance values is independent of the rows and that the variation in the Cp_{ij} capacitors is identical in all columns, this noise would only be due to the variations in C_j. This is compensated through the calibration circuit of each column, so it is considered as insignificant. Therefore, uncertainty in Δt_{ij} measurement, can be expressed by:

$$\sigma\left(\Delta t_{ij}\right) = \sqrt{\sigma_{crosstalk}^2\left(\Delta t_{ij}\right) + \sigma_{trigger}^2\left(\Delta t_{ij}\right) + \sigma_{column}^2\left(\Delta t_{ij}\right)} \tag{20}$$

as the three terms inside the root are independent of each other.

Noise due to quantification, $u(z) = T_s/\sqrt{12}$, should be added to this deviation [37], where T_s is the period of time of the counter which measures discharge time. Although uncertainty in V_{DD}, VT_j and C_j values can be compensated using Equation (4) (reducing then the values of all terms inside the root), there is still uncertainty in measurement due to RC crosstalk. According to Equations (15) and (16), there are three possibilities to reduce the relative impact of crosstalk: the first one is in the increase of C_j, but, as we have mentioned before, this reduces the scanning speed of the array. The second possibility is the reduction of the Cp_{ij} terms, but this (once a careful design of the system has been carried out) is limited by technology and the number of rows in the array. The third possibility is the limitation in the range of values of the resistors of the array, but obviously, this can only be achieved in some very specific sensors. Therefore, although the RA structure of the sensor array does not show the typical resistive crosstalk, there exists what has been referred to as RC crosstalk and which, therefore, has to be assessed.

Assessment Method of Uncertainty in Measurement

A naive method for uncertainty assessment in Δt_{ij} measurement would be to perform a set of measurements for each of the sensors of the array, varying the resistors of the remaining sensors in each of the measurements, obtaining then the maximum ranges of variation. However, the use of this method of characterization is a difficult task to undertake, as the number of sensors in an array is not usually a small number, even in our case where a "reduced" 7×8 sensor array has been used.

The aim of this paper is to propose the assessment of total uncertainty in the measurement of any resistor of the array based on the independency of uncertainties due to crosstalk, column and trigger by using Equation (20).

For this reason, a circuit simulating by means of resistors the different values which can be read from all positions of an array with RA configuration has been designed. The circuit, obviously, includes an FPGA and the necessary capacitors for time-to-digital conversion.

The assessment method of uncertainty in measurement comprises the following steps: firstly, a set of sufficient resistor values which approximately divide the range of the sensor in uniform sections is selected (8 in our case). Each of these resistors is placed in successive tests in the M rows of a column of the array selected for characterization. The remaining resistors of the column take the mean value in the range. Then, a set of Q time measurements are carried out (Δt_{ij}). The process is repeated M times changing the position of the resistor in a circular manner, in such a way that $M \cdot Q$ time measures are obtained for each resistor. This set of times has a standard deviation which is due both to trigger uncertainty and to position uncertainty in the column, with variance $\sigma^2_{trigger+column}$.

To get the value of each of these terms $\sigma^2_{trigger}$ and σ^2_{column} (for each resistor), the following procedure is proposed. From a set of Q measurements carried out in each row, M $\sigma^2_{trigger}$ values are obtained. These values should be very similar for all rows, and the mean of these values is the final estimation of $\sigma^2_{trigger}$. Because they are independent sources of uncertainty, its variance is $\sigma^2_{trigger+column} = \sigma^2_{trigger} + \sigma^2_{column}$ and therefore σ^2_{column} could be calculated as:

$$\sigma^2_{column} = \sigma^2_{trigger+column} - \sigma^2_{trigger} \tag{21}$$

$\sigma_{crosstalk}$ can be found with the following procedure: a row in the column used for the set of previous measurements is selected. Each of the eight resistors will be placed in this position of the array. Two sets of Q measurements will be carried out for each resistor. In the first set of Q measurements, the minimum value resistor in the sensor is added to the selected resistor in the remaining positions of the column. In the second set of Q measurements, minimum value resistors are replaced by those with maximum values. Then we find the $\Delta tmax$ maximum and $\Delta tmin$ minimum time values for the $2 \cdot Q$ measures, and their difference $\Delta tmax - \Delta tmin$, will be the estimator used to find the range of Δt_{ij} values for each resistor due to RC crosstalk. Finally, assuming that the values of the sensor resistors have a uniform distribution and, therefore, Δt_{ij} as well, for each resistor, the $\sigma_{crosstalk}$ value can be calculated as:

$$\sigma_{trigger+crosstalk} = \frac{\Delta tmax - \Delta tmin}{\sqrt{12}} \tag{22}$$

where again we can solve for $\sigma_{crosstalk}$:

$$\sigma^2_{crosstalk} = \sigma^2_{trigger+crosstalk} - \sigma^2_{trigger} \tag{23}$$

Therefore, the value of $\sigma(\Delta t_{ij})$ can be obtained for each resistor by using Equation (20). Quantification uncertainty, $u(z)$, should be added to this, obtaining then a total uncertainty value, $\sigma_T (\Delta t_{ij})$, determined by:

$$\sigma_T \left(\Delta t_{ij}\right) = \sqrt{\sigma^2 \left(\Delta t_{ij}\right) + u^2(z)} \tag{24}$$

Uncertainty in the whole range will be the maximum value of all uncertainties obtained.

5. Materials and Methods

The circuit in Figure 4 follows the diagram proposed in Figure 1. It has been made using a Xilinx Spartan3AN FPGA (XC3S50AN-4TQG144C) [35] and an oscillator circuit based on crystal quartz with an operating frequency of 50 MHz. Each of the eight capture modules has a 14-bit counter, with a 20 ns time base. The maximum current that an output buffer of the FPGA can sunk for maintaining the digital signal integrity of the outputs is 24 mA. For currents above 24 mA the output voltage of the buffer that drives the row is higher than that identified as a '0' logical value, the output transistor is not in the ohmic region and the resistance of the buffer RB_j changes significantly. However, it is possible to add a resistance in series with the one that is measured to limit the current and this does not alter the procedure described above.

The design rules recommended by the manufactures of the FPGA are applied rigorously so that noise impact affecting VT_j causes the smallest uncertainty possible in measurement. This FPGA works with independent supply voltages for the input/output blocks and the digital processing core where the rest of the circuitry resides. The use of two independent voltage regulators is then necessary in order to reduce the influence of the activity of the core in the input/output buffers. The selected regulators, TPS79633 [38] for the voltage of the input/output banks supplied at 3.3 V and TPS79912 [39] for the voltage of the core at 1.2 V, have extremely low values in their output voltage, both for dropout voltage during maximum charge as well as for output voltage noise (40 µV RMS). Also, they only need a few external components for correct operation, which make them ideal for devices where the area occupied by the circuit is large. For each of the four buffer banks included in the FPGA, a battery of decoupling capacitors of different values in a position very close to the supply inputs are used. These are connected through two supply planes; the first one for the 3.3 V voltage and the other for GND. The printed circuit board where the circuit is mounted on has been manufactured with a FR-4 fiberglass substrate and four layers, leaving internal layers for supply planes and external layers for the remaining signals.

Figure 4. Setup to test the direct interface circuit for resistive sensor array.

The resistor array to be measured is composed of ten rows by eight columns with an RA structure. As has been mentioned, the RA structure makes columns electrically independent between them; therefore the experimental tests are focused on the study of a single column and the ten resistors which can be measured through an only input port, although the design has been made to allow for 80 resistors to be placed.

As indicated earlier, the sensor to be modeled is composed of seven rows and eight columns. Three additional rows have been added in order to asses different calibration methods. Two of them will then be used for a two-point calibration, which will have resistors placed approximately within 15% and 85% of the range where measuring is to be carried out [40]. The third row will use a resistor in approximately 50% of the range which will be used either for a single-point calibration or for the *off-on time* method, using Equation (4).

Figure 5a shows a detail of the implemented electrodes array which is going to be simulated with discrete resistors. Each sensor is connected to a solid and oval inner electrode through which the terminal of the sensor is accessed individually. On the other hand, the outer electrodes enclose the inner electrodes. It can be observed how all outer electrodes in a column (horizontally in the image) are connected between them, forming the common terminal of the column. Electrodes are placed in a PCB with "Rigiflex" technology of four layers and an IPC-6013 reference standard [41].

Figure 5b shows the sheet of discrete piezoresistive material (divided into rectangular sections) which is placed on the electrode matrix. When pressing the sheet, the resistance between the inner and outer electrodes varies. The sheet has been developed by CIDETEC and the range or resistor values can vary approximately between 7400 Ω for pressures of a few kPa to a few dozens of ohms for high pressures.

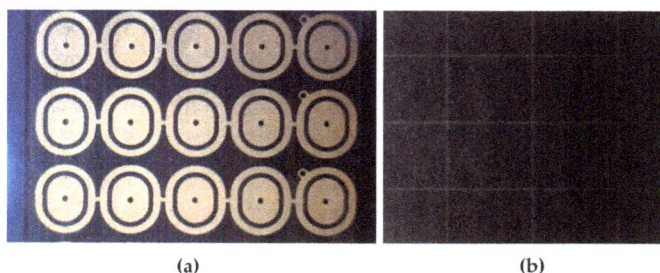

(a)   (b)

Figure 5. Details of the electrode matrix of the finger sensor (**a**) and of the discrete material (**b**).

Experimental tests are performed with 8 resistor values within this range. The resistors used are: 199.96 Ω, 1297.73 Ω, 2401.95 Ω, 3687.49 Ω, 4836.09 Ω, 5810.79 Ω, 6966.87 Ω and 7348.84 Ω (which includes the resistance value of approximately 50% of the range, 3687.49 Ω). As for the capacitor, one with a 47 nF nominal value is selected, which complies with the design rules proposed in [33] for the constant of optimal time which minimizes relative standard uncertainty in measurement. Also, with a capacitor with this value, the sum of time values in the input/output cycles for the ten rows of the matrix allows reaching sampling rates which are higher than 200 tactile frames per second. The power consumptions of the FPGA without connecting the sensor array is 78 mA @ 50 MHz and it increases at a rate of 1 mA per column for a capacitor of 47 nF in our example implementation.

6. Results and Discussion

The aim of this section is to perform uncertainty and accuracy assessment in the measurement of resistance values in the circuit shown in the previous section. For this, the methodology proposed in Section 3 with $Q = 500$ will be followed. Table 1 shows the results for uncertainty obtained with the eight resistors used, where the indications proposed in [36] have been followed to design the capture module.

Table 1. Precision data for test resistors with a 47 nF capacitor and $Q = 500$.

Resistor (Ω)	$\sigma_{trigger+column}$ (ns)	$\sigma_{trigger}$ (ns)	σ_{column} (ns)	$\sigma_{trigger+crosstalk}$ (ns)	$\sigma_{crosstalk}$ (ns)	σ (Δt_{ij}) (ns)	σ_T (Δt_{ij}) (ns)
199.96	13.58	1.62	13.49	2.89	2.39	13.79	14.95
1297.73	18.06	5.24	17.29	11.55	10.29	20.79	21.58
2401.95	18.58	7.93	16.81	25.98	24.74	30.94	31.48
3687.49	29.83	11.49	27.53	30.31	28.05	40.95	41.35
4836.09	32.65	14.29	29.36	43.30	40.87	52.31	52.63
5810.79	40.13	17.77	35.98	56.29	53.41	66.81	67.06
6966.87	38.56	19.98	32.98	59.18	55.70	67.75	67.99
7348.84	38.46	21.42	31.94	70.73	67.41	77.60	77.82

The effective number of bits in time measuring (ENOB) obtained from σ_T is 10.14 bits. To analyze the loss of precision due to the increase in the number of rows of the array, σ_T has been measured for a situation in which there is only one resistor to be measured, as well as those for calibration. Table 2 shows a comparison between the results obtained and those of Table 1 for the resistor showing the worst results in both cases, 7348.84 Ω.

Table 2. Precision degradation due to increasing number of rows.

Rows (M)	$\sigma(\Delta t_{ij})$ (ns)	ENOB (bits)	Resolution (Ω)
1	21.40	12.04	1.70
7	77.82	10.14	6.34

As for accuracy in the estimation of resistance values, the results obtained with the traditional single-point and two-point calibration methods are going to be compared with those of the *off-on time* method indicated in Section 2.

In order to carry out this comparison, as mentioned in Section 5, three additional rows were added for the different calibration methods. In these rows, resistors with 15% (1300 Ω), 85% (5820 Ω) and 50% (3680 Ω) values are placed. Therefore, we can assess all calibration methods in the same test.

For the *off-on time* method, the calibration curve has to be assessed first. This is done by placing the eight resistors shown in Table 1 in the same position. Two sets of tests are performed for each resistor, 500 measurements with the remaining resistors in the column with minimum values and another 500 with the resistors with maximum values. A linear adjustment is carried out from these 8000 measures [36]. Therefore, coefficients A and B of Equation (4) are obtained.

The same tests have then been performed for the three methods. Over a set of 14 resistor values which divide the possible range into approximately equal sections, 500 measurements are carried out with the remaining resistors with minimum values (Rmin = 199.96 Ω), 500 with resistors with medium values (Rmed = 3687.49 Ω) and 500 with resistors with maximum values (Rmax = 7348.84 Ω).

Table 3 shows maximum absolute error values obtained from each of the three groups of 500 measurements. The last two rows show maximum absolute and relative errors in any circumstance.

Table 3. Accuracy data for different calibration techniques with a 47 nF capacitor and $Q = 500$.

Resistor (Ω)	Maximum Absolute Error (Ω)								
	Rmin (199.96 Ω)			Rmed (3687.49 Ω)			Rmax (7348.84 Ω)		
	1 P.	2 P.	Off-On	1 P.	2 P.	Off-On	1 P.	2 P.	Off-On
199.96	13.66	0.88	0.62	13.74	0.90	0.37	13.60	0.77	0.56
763.34	11.70	0.98	0.59	11.59	0.75	0.69	11.44	0.75	0.82
1297.32	10.03	0.65	0.58	10.11	0.76	0.67	9.79	0.65	0.62
1887.55	8.23	1.18	1.22	8.12	1.20	1.11	7.65	1.09	0.63
2401.95	5.69	0.81	0.84	5.42	1.18	1.08	4.98	0.74	1.44
3070.25	4.24	1.93	2.10	3.83	1.81	1.69	3.34	1.81	1.19
3684.25	1.83	2.21	2.21	1.60	1.94	1.98	1.13	2.07	1.51
4083.85	2.04	2.93	3.02	2.31	2.22	1.91	3.13	1.85	1.12
4836.05	5.73	2.19	2.46	5.68	2.32	1.88	6.67	2.02	1.56
5269.05	6.65	3.31	3.49	6.84	3.31	3.12	7.79	3.23	1.89
5813.45	9.32	2.64	2.72	9.88	2.65	2.37	10.63	2.88	1.68
6373.15	12.09	3.07	3.20	12.61	2.95	2.77	13.34	2.80	1.91
6983.15	16.29	2.29	2.35	15.75	2.37	1.81	17.34	2.69	3.41
7349.15	16.52	3.34	3.16	16.55	3.30	2.54	18.86	3.24	3.42
Total Error	124.03	28.39	28.56	124.03	27.67	23.99	129.69	26.59	21.77
Max. Absolute Error	16.52	3.34	3.49	16.55	3.31	3.12	18.86	3.24	3.42
Max. Relative Error (%)	0.225	0.045	0.066	0.225	0.063	0.059	0.257	0.044	0.047

As shown in Table 3, the errors for the two-point calibration and the *on-off time* calibration methods show very similar values. On the other hand, as expected, the errors for the single-point method are of a higher order of magnitude. As the *on-off time* method uses only one calibration resistor, it seems to be a better solution, because it uses simpler hardware and less time is needed to scan the whole matrix. Figure 6 shows the maximum absolute error for any condition using the three calibration methods.

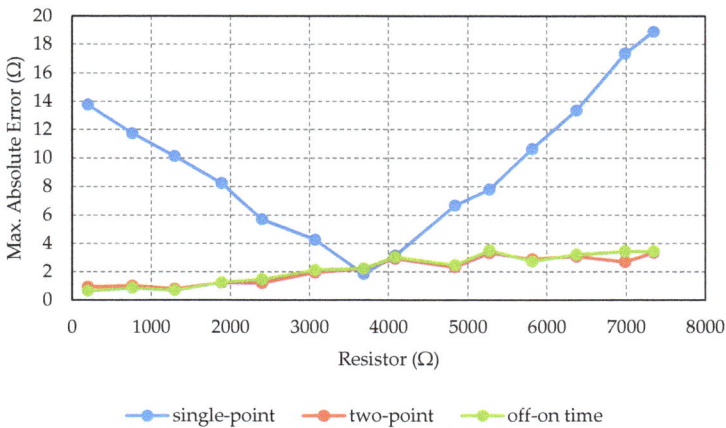

Figure 6. Worst absolute error for each resistor and calibration method.

7. Conclusions

This paper presents a direct connection circuit between an FPGA and a resistive sensor array which allows for parallel measuring of certain resistors of the array (a whole row). The addressing technique used (RA configuration) avoids the typical resistive crosstalk between the resistors of the array. For this, an input/output pin is used in each of the electrodes in the matrix. Although crosstalk is then avoided between resistors measured in parallel (those from one same row), there is another element, RC crosstalk, which affects uncertainty when measuring discharge time in capacitors (and which is used to estimate resistance) causing a reduction in the effective number of bits in time-digital conversion (ENOB).

The implementation of a circuit to perform a set of tests has been carried out. These tests simulate, with discrete resistors of known values, the behavior of a specific tactile sensor. The range of resistor values which has been assessed goes from 200 Ω to 7350 Ω.

In this case, the total uncertainty which includes all studied components ($\sigma_{trigger}$, $\sigma_{crosstalk}$ and σ_{column}) and also quantification noise, results in an ENOB of 10.14 bits. Besides the precision in the estimation of the resistance values, a study of accuracy has been performed by using different calibration techniques avoiding having to know certain values which are difficult to measure (C_j, VT_j and V_{DD}), which could change with time and temperature.

From the calibration methods analyzed, the *off-on time* method, based on an adjustment to a linear function from data obtained in an operation mode previous to the normal mode, provides similar results to those of the two-point calibration technique, but only with the need of a reference resistor. Therefore both the hardware and the sampling time of the whole matrix are reduced. The maximum relative error obtained with this method has been 0.066%

Acknowledgments: This work has been funded by the Spanish Government and by the European ERDF program funds under contract TEC2012-38653.

Author Contributions: This paper is part of the PhD Thesis of Oscar Oballe-Peinado and he is the main contributor. Fernando Vidal-Verdú and José Antonio Hidalgo-López are co-advisors of his PhD Thesis, so both have oriented the work and have helped in the writing and revision of the text. José A. Sánchez-Durán and Julián Castellanos-Ramos have assisted in testing and characterization tasks.

Conflicts of Interest: The authors declare no conflict of interest.

References

1. Tise, B. A compact high resolution piezoresistive digital tactile sensor. In Proceedings of the 1988 IEEE International Conference on Robotics and Automation, Philadelphia, PA, USA, 24–29 April 1988.

2. Castellanos-Ramos, J.; Navas-González, R.; Macicior, H.; Sikora, T.; Ochoteco, E.; Vidal-Verdú, F. Tactile sensors based on conductive polymers. *Microsyst. Technol.* **2010**, *16*, 765–776. [CrossRef]

3. Dahiya, R.S.; Metta, G.; Valle, M.; Adami, A.; Lorenzelli, L. Piezoelectric oxide semiconductor field effect transistor touch sensing devices. *Appl. Phys. Lett.* **2009**. [CrossRef]

4. Drimus, A.; Kootstra, G.; Bilberg, A.; Kragic, D. Design of a flexible tactile sensor for classification of rigid and deformable objects. *Robot. Auton. Syst.* **2014**, *62*, 3–15. [CrossRef]

5. Kane, B.J.; Cutkosky, M.R.; Kovacs, G.T.A. A traction stress sensor array for use in high-resolution robotic tactile imaging. *J. Microelectromechanical Syst.* **2000**, *9*, 425–434. [CrossRef]

6. Kim, M.-S.; Shin, H.-J.; Park, Y.-K. Design concept of high-performance flexible tactile sensors with a robust structure. *Int. J. Precis. Eng. Manuf.* **2012**, *13*, 1941–1947. [CrossRef]

7. Shimojo, M.; Namiki, A.; Ishikawa, M.; Makino, R.; Mabuchi, K. A tactile sensor sheet using pressure conductive rubber with electrical-wires stitched method. *IEEE Sens. J.* **2004**, *4*, 589–596. [CrossRef]

8. Speeter, T.H. A Tactile Sensing System for Robotic Manipulation. *Int. J. Robot. Res.* **1990**, *9*, 25–36. [CrossRef]

9. Zhang, X.; Zhao, Y.; Zhang, X. Design and fabrication of a thin and soft tactile force sensor array based on conductive rubber. *Sens. Rev.* **2012**, *32*, 273–279. [CrossRef]

10. Weiss, K.; Worn, H. Resistive tactile sensor matrices using inter-electrode sampling. In Proceedings of the IECON 2005-31st Annual Conference of IEEE Industrial Electronics Society, Raleigh, NC, USA, 6–10 November 2005.

11. Wu, J.; Wang, L.; Li, J.; Yu, Z. A small size device using temperature sensor array. *Chin. J. Sens. Actuators* **2011**, *24*, 1649–1652.

12. Yang, Y.-J.; Cheng, M.-Y.; Shih, S.-C.; Huang, X.-H.; Tsao, C.-M.; Chang, F.-Y.; Fan, K.-C. A 32 × 32 temperature and tactile sensing array using PI-copper films. *Int. J. Adv. Manuf. Technol.* **2010**, *46*, 945–956. [CrossRef]

13. Fratoddi, I.; Venditti, I.; Cametti, C.; Russo, M.V. Chemiresistive polyaniline-based gas sensors: A mini review. *Sens. Actuators B Chem.* **2015**, *220*, 534–548. [CrossRef]

14. Wang, J.; Chan, S.; Carlson, R.R.; Luo, Y.; Ge, G.; Ries, R.S.; Heath, J.R.; Tseng, H.-R. Electrochemically fabricated polyaniline nanoframework electrode junctions that function as resistive sensors. *Nano Lett.* **2004**, *4*, 1693–1697. [CrossRef]

15. Depari, A.; Falasconi, M.; Flammini, A.; Marioli, D.; Rosa, S.; Sberveglieri, G.; Taroni, A. A new low-cost electronic system to manage resistive sensors for gas detection. *IEEE Sens. J.* **2007**, *7*, 1073–1077. [CrossRef]

16. Lilly, R.V.; Devaki, S.J.; Narayanan, R.K.; Sadanandhan, N.K. Design of a nanostructured electromagnetic polyaniline–keggin iron–clay composite modified electrochemical sensor for the nanomolar detection of ascorbic acid. *J. Appl. Polym. Sci.* **2014**. [CrossRef]

17. Song, E.; Choi, J.-W. Multi-analyte detection of chemical species using a conducting polymer nanowire-based sensor array platform. *Sens. Actuators B Chem.* **2015**, *215*, 99–106. [CrossRef]

18. Wu, J.; Wang, L.; Li, J. VF-NSE method measurement error analysis of networked resistive sensor array. *Sens. Actuators Phys.* **2014**, *211*, 45–50. [CrossRef]

19. Wu, J.; Wang, L.; Li, J. Design and Crosstalk Error Analysis of the Circuit for the 2-D Networked Resistive Sensor Array. *IEEE Sens. J.* **2015**, *15*, 1020–1026. [CrossRef]

20. Saxena, R.S.; Bhan, R.K.; Saini, N.K.; Muralidharan, R. Virtual Ground Technique for Crosstalk Suppression in Networked Resistive Sensors. *IEEE Sens. J.* **2011**, *11*, 432–433. [CrossRef]

21. Saxena, R.S.; Saini, N.K.; Bhan, R.K. Analysis of Crosstalk in Networked Arrays of Resistive Sensors. *IEEE Sens. J.* **2011**, *11*, 920–924. [CrossRef]

22. Liu, H.; Zhang, Y.-F.; Liu, Y.-W.; Jin, M.-H. Measurement errors in the scanning of resistive sensor arrays. *Sens. Actuators Phys.* **2010**, *163*, 198–204. [CrossRef]

23. D'Alessio, T. Measurement errors in the scanning of piezoresistive sensors arrays. *Sens. Actuators Phys.* **1999**, *72*, 71–76. [CrossRef]

24. Kim, J.-S.; Kwon, D.-Y.; Choi, B.-D. High-Accuracy, Compact Scanning Method and Circuit for Resistive Sensor Arrays. *Sensors* **2016**. [CrossRef] [PubMed]

25. Oballe-Peinado, Ó.; Vidal-Verdú, F.; Sánchez-Durán, J.A.; Castellanos-Ramos, J.; Hidalgo-López, J.A. Improved Circuits with Capacitive Feedback for Readout Resistive Sensor Arrays. *Sensors* **2016**. [CrossRef] [PubMed]

26. Ares, L.; Rodriguez-Andina, J.J.; Farina, J. FPGA-based direct resistance and capacitance measurements. In Proceedings of the 35th Annual Conference of IEEE Industrial Electronics (IECON 2009), Porto, Portugal, 3–5 November 2009.

27. Reverter, F.; Jordana, J.; Gasulla, M.; Pallàs-Areny, R. Accuracy and resolution of direct resistive sensor-to-microcontroller interfaces. *Sens. Actuators Phys.* **2005**, *121*, 78–87. [CrossRef]

28. Reverter, F. The Art of Directly Interfacing Sensors to Microcontrollers. *J. Low Power Electron. Appl.* **2012**, *2*, 265–281. [CrossRef]

29. Kokolanski, Z.; Jordana, J.; Gasulla, M.; Dimcev, V.; Reverter, F. Direct inductive sensor-to-microcontroller interface circuit. *Sens. Actuators Phys.* **2015**, *224*, 185–191. [CrossRef]

30. Reverter, F. Direct interface circuits for sensors. In *Smart Sensors and MEMS: Intelligent Devices and Microsystems for Industrial Applications*; Nihtianov, S., Luque, A., Eds.; Woodhead Publishing: Oxford, UK, 2014.

31. Custodio, A.; Pallas-Areny, R.; Bragos, R. Error analysis and reduction for a simple sensor-microcontroller interface. *IEEE Trans. Instrum. Meas.* **2001**, *50*, 1644–1647. [CrossRef]

32. Bengtsson, L.E. Analysis of direct sensor-to-embedded systems interfacing: A comparison of targets' performance. *Int. J. Intell. Mechatron. Robot.* **2012**, *2*, 41–46. [CrossRef]

33. Reverter, F.; Pallàs-Areny, R. Effective number of resolution bits in direct sensor-to-microcontroller interfaces. *Meas. Sci. Technol.* **2004**. [CrossRef]

34. Vidal-Verdú, F.; Oballe-Peinado, Ó.; Sánchez-Durán, J.A.; Castellanos-Ramos, J.; Navas-González, R. Three Realizations and Comparison of Hardware for Piezoresistive Tactile. *Sensors* **2011**, *11*, 3249–3266. [CrossRef] [PubMed]

35. Xilinx, Inc. Spartan-3AN FPGA Family Data Sheet. Available online: http://www.xilinx.com/support/documentation/data_sheets/ds557.pdf (accessed on 6 November 2015).

36. Oballe-Peinado, Ó.; Vidal-Verdú, F.; Sánchez-Durán, J.A.; Castellanos-Ramos, J.; Hidalgo-López, J.A. Smart Capture Modules for Direct Sensor-to-FPGA Interfaces. *Sensors* **2015**, *15*, 31762–31780. [CrossRef] [PubMed]
37. ISO Guide to the Expression of Uncertainty in Measurement (GUM:1995). Availble online: http://www.iso.org/iso/home/store/catalogue_ics/catalogue_detail_ics.htm?csnumber=50461 (accessed on 13 September 2015).
38. Texas Instruments, Inc. TPS795 Ultralow-Noise, High-PSRR, Fast, RF, 500-mA Low-Dropout Linear Regulators Datasheet. Available online: http://www.ti.com/lit/ds/symlink/tps795.pdf (accessed on 6 November 2015).
39. Texas Instruments, Inc. TPS799 200-mA, Low-Quiescent Current, Ultralow Noise, High-PSRR Low-Dropout Linear Regulator Datasheet. Available online: http://www.ti.com/lit/ds/symlink/tps799.pdf (accessed on 6 November 2015).
40. Pallàs-Areny, R.; Jordana, J.; Casas, Ó. Optimal two-point static calibration of measurement systems with quadratic response. *Rev. Sci. Instrum.* **2004**, *75*, 5106–5111. [CrossRef]
41. IPC IPC-6013 Specification for Printed Wiring, Flexible and Rigid-Flex. Available online: http://www.ipc.org/ (accessed on 8 December 2015).

MDPI AG

St. Alban-Anlage 66

4052 Basel, Switzerland

Tel. +41 61 683 77 34

Fax +41 61 302 89 18

http://www.mdpi.com

Sensors Editorial Office

E-mail: sensors@mdpi.com

http://www.mdpi.com/journal/sensors

www.ingramcontent.com/pod-product-compliance
Lightning Source LLC
Chambersburg PA
CBHW051718210326
41597CB00032B/5518